Theory and Practice

Donald P. Warwick, Ph.D.
Professor of Sociology
York University

Charles A. Lininger, Ph.D.
Regional Director for Latin America
The Population Council

With Foreword by
Angus Campbell

McGRAW-HILL BOOK COMPANY
New York   St. Louis   San Francisco   Auckland
Düsseldorf   Kuala Lumpur   London
Mexico   Montreal   New Delhi   Panama
Paris   São Paulo   Singapore   Sydney   Tokyo   Toronto

This book was set in Times Roman by Rocappi, Inc.
The editors were Richard R. Wright and Barry Benjamin;
the production supervisor was Dennis J. Conroy.
The drawings were done by J & R Services, Inc.

The Sample Survey: Theory and Practice

15 DODO 8 9

Library of Congress Cataloging in Publication Data

Warwick, Donald P
    The sample survey.

    1. Social surveys. 2. Social science research.
3. Interviewing. 4. Questionnaires. I. Lininger,
Charles Andrew, date   joint author. II. Title.
HN29.W394   300'.7'23   74-16261
ISBN 0-07-068396-4
ISBN 0-07-068395-6 (pbk.)

Contents

Foreword

The development of the sample survey as a procedure for generating data describing large populations is certainly one of the success stories of social science. When one recalls that the 1930 *Encyclopedia of the Social Sciences* warned that the difficulties of sampling human populations were so great that "complete enumeration is preferred," one can appreciate how far and how rapidly the basic techniques of sample surveys have progressed.

Unlike many methodological inventions, the sample survey cannot be claimed as the product of any one of the numerous academic disciplines that now use it. Neither can its origin be located in any one country. It developed in parts in various places over an extended period of time, usually in response to a pressingly felt need for better information about the condition of society's less-fortunate members.

At the present time "surveys" and "polls" have become household terms in most of the developed countries of the world. Largely through newspaper reporting of survey findings, the concept of taking a small number of people to represent a very large number has become increasingly familiar; national marginals describing political preferences, evaluations of actual or contemplated governmental programs, and human behavior of all kinds appear in the media almost daily. In those countries with a democratic tradition, they are tinged with a certain mystique as representing the "voice of the people."

The basic rationale of sample surveys is so simple and their applicability to various forms of human activity so broad that surveys are carried out at every conceivable level of society by people of every degree of sophistication and experience. Many of these inquiries are very poorly executed, sometimes because their directors do not realize the complications inherent in survey methodology and often because they are attempting to accomplish an enterprise which is far too ambitious for their budget. It is regrettably true that relatively few of all the social surveys that make their way into print have adhered throughout to the highest standards of research methodology.

Of course there is no way to compel prospective survey directors to choose the most effective design, eliminate bias in sampling, develop sensitive and precise questionnaires, and apply the most appropriate analytical procedures to their data. One can hope that a conscientious person will try to meet these standards if he has available an adequate guide to the conduct of these various steps in the surveying process. With this new volume by Donald Warwick and Charles Lininger in hand, he will have little excuse for uninformed errors at the major decision points in the conduct of his survey.

A book which intends to provide an introduction ought to begin at the beginning. Warwick and Lininger have indeed undertaken to describe the successive steps in a survey in terms which assume no prior experience and no knowledge of esoteric vocabulary. They virtually take the beginning surveyer by the hand and lead him through the complexities of planning and carrying through his project. It will be apparent to anyone experienced in these matters that this book was written by people who have themselves been through the survey process many times and are very familiar with the many snares and pitfalls that lie along the way.

This book grew out of a need to provide instructional material to staff people for whom survey research was a totally new experience. It was written originally in Spanish during the 1960s when Warwick and Lininger were located in Lima, Peru, as representatives of the Institute for Social Research. The Institute had agreed to assist the Ministry of Labor of the Peruvian government in developing a sample survey competence. With financial support from the Agency for International Development, the Institute maintained a small staff in Lima for eight years, terminating its relationship after its assignment had been completed. Thanks, in large part, to the talents of the two authors of this book, this enterprise turned out to be an encouraging example of how a relationship between foreign professionals and American consultants can be a successful and rewarding experience for both parties.

The original manuscript written by Warwick and Lininger for their Peruvian colleagues has been greatly revised and expanded in this book, but their continuing interest in the development of social research in the less-industrialized countries of the world is apparent. They have drawn on their years in Peru for specific examples of survey problems and have given their text a generally international orientation.

It is never possible to predict with any confidence that a book on research methods will have a positive influence on the methods which researchers use. It can certainly be said, however, that this book could be used to very good purpose by most people undertaking sample surveys. I hope very much that it is.

Angus Campbell, Director
Institute for Social Research, The University of Michigan

Preface

This book traces its origins to a training program on survey research developed by the authors in 1964-1965 at the Employment and Human Resources Service (SERH) of the Ministry of Labor of Peru. While serving as advisers to a new survey research facility being set up at that institution, we became acutely aware of the shortage of theoretically sound, yet practically anchored, teaching materials on the sample survey. Particularly scarce at that time were books and articles addressing the unique problems of sampling and questionnaire design in the less-developed countries. To fill this gap we prepared a book entitled *Introduction to Survey Research* (Lininger and Warwick, 1965). This was published first in Spanish and then in English by our host organization, and circulated informally by the Survey Research Center of the University of Michigan. Encouraged by the favorable reaction to this effort in both hemispheres, we set out to prepare a more thorough text on the sample survey.

A critical choice in such an endeavor is the balance to be struck between general theory and practical applications. Some colleagues urged us to prepare a kind of "cookbook"—a short, practically oriented manual which would show the neophyte how to do a survey. Others claimed that contemporary students, who were skeptical of academic cookbooks, could profit more from a work *about* survey research—its uses, limitations, and guiding theory. In the end we concluded that neither a cookbook nor a general commentary on this method would be particularly useful. Rather, we decided to prepare an intermediate-level text which would serve as an introduction while treating *both* theory *and* practice and showing their interrelationships in concrete research settings. Hence the title and the constant emphasis on the inseparability of theory and practice in the successful survey.

In writing this text, we have also given attention to three critical issues which are often slighted in general works on survey research. The first is the need for intracultural meaningfulness as well as cross-cultural equivalence.

This emphasis grew out of our experience in translating North American concepts and procedures into appropriate Peruvian equivalents and vice versa, and it was reinforced as we perceived subcultural group identities becoming more pronounced in this and other countries. It is necessary to ensure that the basic notions under study as well as the specific formulation of questions be properly understood by respondents in a single culture and have comparable meanings in subcultures and across cultures.

Second, survey research is located in both a political and an ethical context. We argue, on the one hand, that survey design is subject to bias from numerous sources, including pressures from the sponsor and other outside groups as well as from the investigator's own values and interests. We also emphasize that survey research must respect the freedom, dignity, and privacy of the groups under study. We thus reject the notion that the researcher has a right, in the name of science or some other higher end, to deceive respondents, misrepresent the nature of the study, or prowl unchecked into the private lives of individuals.

Third, the book hopes to show not only the uses but the abuses and limitations of the sample survey. By providing an intimate acquaintance with each stage of the survey, we intend not only to encourage the appropriate use of this method, but to discourage its application when simpler or less-intrusive methods may be adequate, or when there are other techniques better suited to the problem at hand.

This text has been written mainly for two sets of readers: students in the social sciences and allied fields (social work, public administration, education, and marketing, among others) and professionals or informed laymen who wish to acquire a better understanding of phenomena such as public opinion polls. It is more than an introduction and less than an advanced dissection of each part of the survey. To the extent possible, we have eliminated statistical formulas and technical terminology, while still introducing the core concepts and the basic language of the field. Students with no previous work in statistics may find some of the statistical discussions in Chapters 4 and 5 more difficult than the rest of the book. If this problem cannot be surmounted with the assistance of the instructor, it would be advisable to omit these sections and to concentrate on the remainder, which deals with general principles. In particular, we have tried to write Chapter 4 so that it could stand reasonably well without the statistics.

We have incurred more debts in writing this book than we can hope to acknowledge. Our primary indebtedness is to our former colleagues at the Survey Research Center of the University of Michigan. Our years with the Center in Ann Arbor laid the intellectual groundwork for this effort, while our assignment as its first representatives in Peru provided the immediate occasion for writing. In many areas we have codified the cumulative experience and the unwritten laws found at the Center, and in others we have borrowed directly from published material. Particular thanks are due Leslie Kish and Irene Hess, whose advice on sampling is reflected in Chapters 4 and 5; Morris Axelrod, Charles Cannell, and John C. Scott, whose comments on interviewing and field methods have been incorporated into Chapters 7 and 8; Angus A. Campbell and James N. Morgan for their comments and encouragement; and Frank Andrews for a variety of assistance.

We are no less grateful to our colleagues in Peru who stimulated us to write this book and tutored us in the intricacies of research in their country. We wish to thank especially Francisco Codina G., the first Director of the Peruvian Survey Research Center, and Abel Centurión, the present Director. Both were constant companions in the preparation of the first draft. Alberto Insúa, Director General of the Employment and Human Resources Service, and Dr. Benjamin Samamé, then Director of Human Resources, also provided strong encouragement and assistance at each step. We also profited from conversations with Alfredo Balmaceda, Luis Castañeda, and the participants in the training courses in Peru.

In the summer of 1965 the *Introduction* was given extensive pilot testing at the Summer Institute on Survey Research Methods at the University of Michigan. Professor Philip Marcus, then of the Survey Research Center, kindly agreed to use the early version as a text in the course. Each of some 40 students prepared a paper discussing its strengths and weaknesses. These comments, which came from undergraduate and graduate students as well as experienced social scientists, were the single most useful source of criticism received, and had a decisive impact on the present version. We are thus most grateful to Professor Marcus and to each of the participants.

Finally, we wish to thank Graham Kalton for his extensive and constructive comments on the first draft, three anonymous readers for their insightful comments on a recent draft, Betty Jennings for her enthusiastic support, the editors of McGraw-Hill for their patience, and our wives and children for their forbearance.

Donald P. Warwick
Charles A. Lininger

Chapter 1

Introduction

For better or for worse, the sample survey has become a familiar part of the social landscape in dozens of societies. From urban America to rural Australia, teams of interviewers regularly make their way to factories, schools, and households to collect information on a vast array of topics. In the United States, hardly a day goes by without a column summarizing the results of the latest poll by Gallup or Harris. At election times in the United States and Canada, polls are so frequent that the public complains of oversaturation and undue influence on the electoral process. Business organizations, too, use the survey as a standard means of assessing markets and learning about the most and least attractive features of their products. In recent years cadres of technical advisors have been sent to the less developed countries to increase local capabilities for survey research. In short, the sample survey, a method developed only in the last forty years, is here to stay, and is being used increasingly by administrators, planners, and social scientists throughout the world.

What, precisely, is a sample survey, and how does it relate to the census, polls, and market research? A *survey* is a method of collecting information about a human population in which direct contact is made with the units of the study (individuals, organizations, communities, etc.) through such systematic means as questionnaires and interview schedules. A *census* is a survey in which information is gathered from or about all members of a population. A *sample survey* (the subject of this book) is a study in which information is gathered from a fraction of the population chosen to represent the whole. A *poll* is a sample survey dealing mainly with issues of public opinion or elections. The best-known polls are those conducted by commercial survey organizations, such as Gallup or Harris in the United States, and are used mainly for popular articles or confidential reports to sponsors. Sample surveys are also used in *market research* to investigate, among other questions, the nature and scope of markets and the acceptability of products. Usually this research is limited to internal use in a company. In this book the terms *survey research, surveys,* and *sample surveys* will be used interchangeably.

HISTORICAL BACKGROUND

The technique of gathering information through direct contact with individuals has a long history.[1] The ancient empires of Egypt and Rome used periodic censuses as a basis for tax rates, military conscription, and other administrative decisions. It was not until the eighteenth century, however, that the large-scale survey was applied in an organized way to the study of social problems. The British reformer John Howard was a pioneer in this effort with his detailed study of the effects prison conditions (sanitation, ventilation, vermin, rats) had on the health of their inmates. His first study was begun in the 1770s in England, and was later extended to other countries in what may have been the first cross-national application of survey methods. Frederic LePlay, a nineteenth-century French economist, strove even harder to use surveys as instruments of rational social planning. His study of income and expenditures among European families reflected an admirable balance of reformist zeal and scientific objectivity. His efforts to check questionnaire replies against information drawn from independent sources, such as direct observation and the report of other respondents, also showed a rather sophisticated sense of research methodology. But the comprehensive, multipurpose survey that we know today traces its origins most directly to the English statistician Charles Booth. In 1886 he undertook a massive study of poverty which eventually resulted in the 17 volumes of *Life and Labor of the People of London,* the last published in 1897. Like LePlay, Booth was personally aroused by the human misery of the slums and felt that the first step in producing change was to gather complete and accurate data on the problem. The tone of his reports, alternating between controlled objectivity and moral outrage ("an awful place; the worst street in the district") foreshadowed a continuing dilemma for the survey researcher.

[1] The following discussion draws upon the historical material presented in Young (1944, Chap. 1).

The growth of surveys and other types of social research in the twentieth century is closely tied to a heightened emphasis upon the values of knowledge and rationality. The experimental and problem-solving attitude that undergirds scientific and technological progress has carried over to the social sphere as well. Modern man wishes to deal with situations in which he is an actor by developing plans based upon solid information. Although this ideal is often unrealized, the value placed on "getting the facts" is very much a part of contemporary culture. The survey is increasingly seen as a helpful method of collecting information on socially relevant topics.

The combination of techniques that marks present-day survey research can be traced to several related developments in the thirties and forties. One of the most significant was the marriage of probability sampling, perfected earlier in agricultural statistics, with controlled interviewing techniques. Major contributions in this area were made by several groups within the United States government. Rensis Likert and his associates in the Division of Program Surveys of the U.S. Department of Agriculture applied area sampling methods to the study of attitudes and behavior. At the end of World War II they left the Department to form the Survey Research Center at the University of Michigan. The Research Division of the Works Progress Administration experimented with probability sampling in innovative measurement of unemployment during the late thirties, and their work eventually led to the formation of the Current Population Survey in 1947. In the U.S. Census Bureau, Morris H. Hansen and William N. Hurwitz and their numerous colleagues were responsible for major breakthroughs in methodology and related documentation, and for continuing careful attention to problem areas as they were identified (cf. Eckler, 1972). Another critical development, inspired largely by Paul Lazersfeld of Columbia University, was the effort to move surveys from sheer description to causal explanations and the testing of theoretical hypotheses. Lazersfeld was also instrumental in establishing the Bureau of Applied Social Research at Columbia University, a major training center for survey researchers in sociology.

A third quality of the contemporary survey is its strongly interdisciplinary orientation, reflected in the multipurpose questionnaire designed for use by economists, sociologists, political scientists, and other specialists. This tendency can be traced in part to the experience of social scientists during World War II when representatives of various disciplines were called on to work together on common problems in the war effort, such as correlates of civilian morale. This experience convinced many survey researchers that social issues can be approached most effectively by incorporating a variety of disciplinary perspectives into a single study.

Today there are perhaps a dozen major survey organizations around the world, and many smaller research institutes which carry out occasional surveys. The most prominent centers in the United States are the Survey Research Center at the University of Michigan, the National Opinion Research

Center at the University of Chicago, the Bureau of Applied Social Research at Columbia University, and the Survey Research Center at the University of California. The U.S. Bureau of the Census conducts one of the largest survey operations in the world, and is especially well known for its corps of methodological experts, particularly in sampling. The British Social Survey, also a governmental agency, is widely recognized for its substantive and methodological contributions. In Peru a survey research center was begun in 1965 within the government, and both authors served as resident advisors to the center for several years. National and urban samples were developed, and full-scale surveys are now being carried out on a variety of topics. In Canada, the Survey Research Centre of York University has recently developed a similar survey research facility. Excellent survey operations have also been developed in Germany, Norway, Sweden, India, and other countries.

THE USES AND ABUSES OF SURVEY RESEARCH

The sample survey has many uses, and also generates various abuses. The uses generally involve one or more of the following objectives: the *description* of populations, hypothesis-testing and other forms of *causal explanation*, the *prediction* of future conditions, the *evaluation* of social programs, and the development of *social indicators* (see Chapter 3). The unique contribution of the survey lies in its ability to provide systematic answers to questions such as:

Questions	Examples
Who does *what?*	What are the social characteristics of those who voted for Richard Nixon and George McGovern in the 1972 United States presidential elections? Which parts of the population are most likely to be involved in the various forms of drug usage?
Why—what are the *reasons* for certain kinds of behavior?	Why do some couples prefer to have large and others small families? Why do some women accept and others refuse to accept contraception? Why do students drop out of high school?
How?	What is the process by which change was adopted in an organization? How would families cope with gasoline rationing?
How well?	How well did the New Jersey–Pennsylvania Income Maintenance Project meet its original objectives (see Chapter 3)? How well did an organized program of contraceptive counseling and services succeed in reducing unwanted births?
With what effect?	What were the direct and indirect effects of a literacy instruction program? What were the effects of a change in supervisory styles on the productivity of an organization?

Although the last two questions are closely related, the second is broader than the first. The question of "how well" usually means that an outcome is related to the intended effects of an intervention. "With what effect," by contrast, often implies a more extensive consideration of consequences.

The specific applications of survey research range from highly pragmatic public opinion polls and market research studies to highly theoretical analyses of social influence. Planners and administrators in many countries have used surveys as a rapid and effective means of gathering base-line information for policy decisions. For centuries the primary sources of this information, when it was used at all, were the census and administrative statistics such as those growing out of customs and taxation records. Today governments often supplement these data with surveys covering population matters, unemployment, income, housing, migration, education, public health, and other policy-related areas.

Social scientists have applied the sample survey to an even broader range of theoretical and practical problems. Among these are voter behavior; psychological influences on the spending and saving behavior of consumers; attitudes, values, and beliefs related to economic growth; the concentration of political power among elites; the relationship between religious beliefs and economic or political behavior; the effects of different styles of supervision on productivity and morale in large organizations; the correlates of mental health and illness; and the effects of different patterns of child rearing on personality. Economists rely on regular consumer surveys for information on family financial conditions, and surveys of business establishments to measure recent investment outlays and short-term plans. Demographers and other population researchers have developed a special application known as the KAP (knowledge, attitude, practice) survey. This attempts to collect information on reproductive behavior and contraception.

As the uses of the survey have expanded, so has controversy over its misuses and confusion over its purposes and effects. In the United States, the most surveyed nation in the world, citizens complain that too-frequent surveys are a public nuisance, that some questions invade privacy, that polls are "rigged" and try to influence the very events they are studying, and that increasing numbers of salesmen gain entry into private homes under the guise of "a few questions that will take only a minute of your time." In Chile a survey sponsored by the U.S. Army (Project Camelot) precipitated an international incident when it was accused of infringing on Chilean autonomy. More recently sample surveys and even the 1970 census in the United States have been plagued by refusals and omissions because of suspicions, fear, and other forms of resistance, particularly in central city areas.

These reactions underscore one of the central assumptions of this book: survey research must be viewed not only as a fruitful method of scientific inquiry, but also as a form of social intervention touching sensitive personal and political nerves. Its application involves ethical and political considerations as well as the mastery of basic techniques. Beyond technical proficiency, the survey researcher must understand the sociopolitical and sociopsychological situation of the population being studied, and be fully aware of the limitations of this methodology. Moreover, professional ethics requires that deception and falsification be avoided, and that promises made be honored. It is particularly important that guarantees of anonymity or confidentiality be scrupulously observed. Deception, manipulation, and abuses of trust not only raise serious moral questions in themselves, but may provoke a public backlash against all surveys (cf. Shils, 1959; Carlson, 1967; Kelman, 1968; and Warwick, 1973).

Researchers who hope to collect survey data from cultures or countries other than their own should be alert to another possible abuse or source of criticism: the failure to provide some form of compensation or quid pro quo to respondents. American Indians, for example, have complained that their collaboration with anthropologists and other social scientists has produced little of tangible benefit to their societies. Social scientists from the less developed countries have issued similar complaints about foreign researchers who enter their countries, use local "hired hands" to do research, and then leave with the data to be heard of again only when the results are published. While there are no simple solutions to these questions, researchers entering such delicate situations would do well to consider arrangements resulting in a more equitable distribution of benefits. Cash payments to respondents may sometimes be necessary, particularly when the individuals involved are put to considerable inconvenience, but may also create problems. The ideal solution is to carry out research which is genuinely collaborative, in the sense that both members of the local culture and the outsider have an equal voice in the design of the research and the use of the findings. When this is impossible the outsiders can at least try to provide valuable experience for local professionals, and to arrange for sharing the results with local institutions. As a further contribution to equitability, arrangements could be made for local participants to receive training and financial assistance for subsequent analyses of the data (Lininger, 1973).

THE SURVEY AND OTHER RESEARCH METHODS

Every method of data collection, including the survey, is only an approximation to knowledge. Each provides a different glimpse of reality, and all have limitations when used alone. Before undertaking a survey the researcher would do well to ask if this is the most appropriate and fruitful method for the problem at hand. The survey is highly valuable for studying some problems, such as public opinion, and almost worthless for others. Decisions about research methods involve many considerations, including costs, time, the researcher's own experience and qualifications, and the availability of trained staff and facilities. Nevertheless, in coming to such a decision it would be helpful to consider the following six criteria.

  1 *Appropriateness to the objectives of the research.* The essential issue here is whether the method will produce the kinds of data needed to answer the questions posed by the study. Is the purpose of the research to generate hypotheses, to test hypotheses, to generate projections, to evaluate an action program, or what? Does the study call for extensive historical data which may not be accessible by survey methods, or which may be collected more easily through archival research? Are measures of change required, and can they be obtained? What level of detail is needed, for example, for geographic or social subgroups in the society? Will the research budget allow for enough cases to provide this detail? Does the research call for an intensive examination of ongoing social interactions, as in a tribal ritual? If so, the researcher might better consider participant observation or some other qualitative method. The basic point is that the study directors should be clear about what they hope to accomplish and what can be accomplished by survey methods.

  2 *Accuracy of measurement.* Broadly speaking, accuracy is attained through an objective portrayal of the true situation under study. While objectivity is an elusive notion in survey research as elsewhere, it remains an ideal toward which most researchers strive. More specifically, we may define accuracy as the generalizability of the measurements taken to all of the relevant measurements that might have been taken of the phenomena in question, such as voting behavior or social class. The essential question is one of sampling: Are the measures or observations used as the source of data representative of the broader universe of measurements and observations which they are supposed to portray (cf. Cronbach, Rajaratnam, and Gleser, 1963; Campbell and Fiske, 1959)?

  Several factors contribute to accuracy. The first factor is *quantification,* or the availability of reliable and valid empirical indicators. Statistical measures such as income scores, prestige ratings, and attitude scales allow for the objective comparison of individuals, communities, and even total societies. The second condition is *replicability.* This is achieved when the measuring instruments and conditions of research are arranged so that they can be repeated either in the same place at a later date or in a different setting. Replicability contributes to accuracy by allowing other researchers to determine whether a given measuring instrument (including the participant observer) taps the same behaviors in other relevant situations, and whether the relationships found in one set of circumstances hold up in others. A third contributor to accuracy is *qualitative depth* in measurement. Decisions about the validity of even quantitative indicators are often based on qualitative assessments of the congruence between this and other information. Thus the greater the range of data available on a given subject, the better the chances for accuracy in measurement and interpretation. Findings do not interpret themselves—they must be set in an analytic context, and this context is often suggested by qualitative data such as clinical impressions, ethnographic observations, or historical records. Finally, accuracy is enhanced by *control over observer effects.* The researcher should always ask if the presence of an observer or interviewer changes the reality observed to such an extent that the sample of behavior obtained is not representative of normal conditions. In some studies, especially those using observational methods, reactions to the outside observer can be an important source of information about community norms and attitudes. More often, however, because the precise nature of these reactions in unknown, they become a source of measurement error and thus reduce accuracy.

   3 *Generalizability of the results.* In addition to the generalizability of the measures taken to other measures that might have been taken (accuracy), most social research involves the generalization of results based on one group to a broader universe or population. Even the anthropologist who observes a remote village usually wishes to generalize not only about that village, but also about the tribe, the region, the country, the social structure, or even human nature. Thus it is essential to know if the sample group itself is representative of the population about which conclusions are sought.

   4 *Explanatory power.* The goal of many, if not most, social scientific studies is to move beyond description to the analysis of causation. The question then becomes not only "who" or "what," but also "why." Some methods of social research are much more suitable than others for analyzing causes in depth. The sample survey, as we shall see, may hold the edge in this regard when compared with aggregate statistics.

   5 *Administrative convenience.* Decisions about research methods often hinge on three administrative considerations: cost, speed, and organizational complexity. Most researchers will want methods which provide high accuracy, generalizability, and explanatory power with low cost, rapid speed, and a minimum of management demands. The trick, of course, is to arrive at an adequate balance between the first three factors and administrative convenience.

   6 *Avoidance of ethical and political problems.* In most studies it is possible to distinguish between two sets of ethical and political difficulties: those which could have been avoided with proper judgment and planning, and those which are inescapable and thus must be confronted as an inherent part of the research. Here we are concerned only with the former, and how they differ from one method to another. In Chapter 2 and elsewhere we will comment on the latter, particularly those arising from the sponsorship of research and relationships with respondents.

   The most common ethical problems in social research are deception in gaining the cooperation of participants, invasions of privacy, violations of confidentiality or other promises made to respondents, and injury or harm to participants, whether physical, psychological, economic, or political. Methods involving essentially administrative statistics or other unobstrusive sources of data clearly present fewer problems in this regard than surveys, participant observation, or other methods involving live contact (cf. Walton and Warwick, 1973). Nevertheless, even with archival material of fairly recent vintage there may be ethical problems posed by damaging revelations about individuals still on the scene.

Political difficulties are most likely to stem from charges that the research serves foreign or outside sponsors, or that the results will be biased toward a certain set of interests. In many cases the researcher will want to pursue a certain line of investigation despite the risks involved. But even here he will want to choose methods which provide a high level of accuracy and yet keep the level of controversy about the research process at a minimum. There is no virtue per se in avoiding controversy, particularly at the expense of the truth, but there is considerable wisdom in choosing methods which yield the necessary data without leaving a wake of conflicts and resentment.

The relative advantages and disadvantages of survey research can be highlighted by using these six dimensions to compare it with three alternative methods: participant observation, the census, and aggregate data.

SURVEY RESEARCH AND PARTICIPANT OBSERVATION

When the objectives of the research require that people be studied in their normal surroundings, the choice of methods often comes down to survey research, participant observation, or some combination of the two. The first makes use of structured questions to a carefully controlled sample, while the second elicits information through the investigator's intensive, but less structured, interactions with a group. This comparison is particularly helpful in showing the uses and limitations of the sample survey, for its strengths are the prime weaknesses of participant observation, and vice versa.

The sample survey is an appropriate and useful means of gathering information under three conditions: when the goals of the research call for quantitative data, when the information sought is reasonably specific and familiar to the respondents, and when the researcher himself has considerable prior knowledge of particular problems and the range of responses likely to emerge. All of these conditions are met in the areas of research that have been the traditional strongholds of the survey—public opinion, voting, attitudes and beliefs, and economic behavior.

Participant observation is usually more appropriate when the study requires an examination of complex social relationships or intricate patterns of interaction, such as kinship obligations or gift exchange in tribal villages; when the investigator desires first-hand *behavioral* information on certain social processes, such as leadership and influence in a small group; when a major goal of the study is to construct a qualitative contextual picture of a certain situation or flow of events; and when it is necessary to infer latent value patterns or belief systems from such behavior as ceremonial postures, gestures, dances, facial expressions, or subtle inflections of the voice.

Under such complex conditions the sample survey may be of limited value. It would depend, first of all, on whether the researcher knows enough about the situation to ask the right questions, and whether the respondents are willing and able to provide appropriate answers. In many cases, people are simply not conscious of what is happening around them, or may not be able to translate their experience into words. Second, the typical interview is arranged so that the interviewer arrives, spends anywhere from fifteen minutes to two hours asking questions, and then departs. A short, structured interview conducted by a relative stranger may not generate enough confidence and motivation for the complex tasks involved, particularly when the study probes into highly sensitive matters. Also, in some parts of the world this type of interaction may be so strange that it arouses hostility and suspicion among respondents.

Third, because the survey obtains reports of individual respondents at a single moment in time, it is often much better for providing a picture of reality at that moment than for describing the flow of events or assessing causal connections. It is true that a careful analysis of survey data can suggest important leads to causality, and that the respondents themselves can be useful sources of information on social processes. Also, the analytical potential of surveys is greatly enhanced when there is a time series of cross-sectional surveys or when the same respondents are reinterviewed later (see Chapter 3). Nevertheless, this method is frequently less useful than participant observation for studying ongoing processes which may not be evident to the participants themselves.

Finally, while one of the prime advantages of the survey lies in the structure which it provides for classifying information, this structure is necessarily attained at the expense of intricacy and complexity in the data. Various compromises can be worked out between open-ended and closed questions, but these fall far short of the freedom enjoyed by the participant observer to develop organizing schemes on the spot.

In the case of accuracy, the sample survey normally holds the edge over participant observation on quantification. There is no inherent reason, however, why participant observers cannot count, classify, and report at least some of their findings in tabular form (cf. Becker and Greer, 1960). This advantage of surveys should not be overstated, for the problems studied through participant observation often do not lend themselves to quantitative analysis, for reasons noted earlier. The sample survey also offers greater possibilities of replication. Because of the highly personalized and often intuitive nature of his work, the participant observer may find it difficult to lay out the exact conditions under which he collected his data and the chain of inference followed in reaching a certain conclusion. The tables are turned, on the other hand, when it comes to qualitative depth. While a well-designed survey can obtain ample qualitative information through open-ended questions, probes, and the summary impressions of interviewers, observational techniques offer still greater opportunities on this score by allowing great flexibility in adapting the approach to the problem at hand. The observer often will have greater exposure to qualitative cues—verbal and nonverbal—that may elude the net of the survey researcher. Finally, both the survey interview and participant observation are vulnerable to observer effects, and it is difficult to state in general terms which is more so.

The greatest single advantage of a well-designed sample survey is that its results can be generalized to a larger population within known limits of error. The greatest weakness of most observational studies is that the limits of generality are unknown. An intensive study of a single community may be beautifully designed and masterful in its portrayal of social life (a good example is W. F. Whyte's *Street Corner Society*), and yet we can never be sure of how well it represents other communities.

Generalizability is limited in most observational studies for three reasons. First, participant observers rarely follow the survey researcher's strategy of defining in advance the universe about which conclusions are to be drawn and then using random methods to pick the specific group to be studied. In anthropological studies it is common practice to choose a village which is not too close to a city, mission, or other contaminating influence, and which is judged to be reasonably similar to other villages in the area. Similar processes of ad hoc "expert" selection are followed in sociological surveys carried out in a local factory or school that happens to be convenient. In all these cases, generalizability is severely limited by the nonrandom character of the first stage of sampling. Second, even if probability methods were used at this first stage, generalizability would be further limited if the sample consists of a single case. There is ample evidence that suburbs, villages, and other social groups assumed to be alike, such as Indian communities in highland Peru, actually show great differences in social structure and values, conflict, mutual trust, and the capacity for community action. The researcher can never assume without further evidence that one suburb, village, or commune really represents some broader universe. Finally, generality is reduced by the sampling of observations within the unit studied. Where the data are gathered by a single observer and perhaps a local assistant, unknown bias may be introduced by their selective perceptions and their distinctive impact upon the group. At this point the accuracy of measurement, especially the problem of replicability, is closely related to the generalizability of the findings. The conscientious observer can help us estimate the extent of selectivity by honest and open reporting of the details of field work, but this does not solve the problem completely. Selective perception and personal impact are also found in the survey interview, although they are reduced to some extent by the use of structured questions and randomized by the use of several interviewers.

In general, the sample survey is more costly and requires more elaborate administrative arrangements than observational studies. Costs are high in the survey because of the relatively large number of people required, the overhead expenses involved if it is administered through a research institute, and the need for high-priced data processing services. Moreover, in the large-household survey the efforts of many people must be planned and coordinated. These administrative problems are greatly reduced for the typical participant observer, who often works alone and requires relatively little institutional backstopping.

The ethical and political difficulties noted earlier can arise in both methods; their gravity will depend on the specific circumstances involved. However, the sample survey seems more vulnerable to political difficulties because of the more impersonal nature of the data collection, its greater visibility over a large area, and the complexity of its administrative arrangements. Large surveys often focus on sensitive or controversial topics within the public domain, and are thus more likely to raise questions about the origins of the study and the uses of the data. However, observational studies may present their own brand of political problems. Anthropologists have been banned from some countries on the grounds that they are serving "imperialist" interests, or because their emphasis on "primitive" peoples is considered an insult to local populations.

In short, there is no magic to either of these methods, or to any other. Each is useful for some purposes and useless for others. The strength of the sample survey lies in its potential for quantification, replication, and generalizability to a broader population. Participant observation normally has the edge on qualitative depth and flexibility for the observer. In many studies the ideal solution is to develop a methodological mixture which will capitalize on the strengths of each approach. A design which combines participant observation or other qualitative methods with a sample survey provides opportunities for cross-checking and for a much more complete picture of the situation being studied.

THE SURVEY AND THE CENSUS

A census, as noted earlier, is a survey covering all members of a given population, such as a nation, a community, or a large organization. The greatest appeal of the census is that it eliminates questions about the representativeness of the information obtained. Because of its cost and the other resources required, however, it is not a realistic alternative for most social researchers. Moreover, it is becoming increasingly clear that a complete census is often unnecessary, wasteful, and a burden on the public, and also less effective than a survey for gathering certain kinds of information.

Specifically, the sample survey enjoys five advantages which make it an attractive alternative to the census, even in population and economic studies commonly regarded as the primary domain of the latter. First, and most obvious, the sample survey is less expensive. An adequate sample typically covers only a small fraction of the cases included in a census, and thus requires a smaller staff and administrative apparatus. The Survey Research Center at the University of Michigan successfully uses samples of about 2000 households to obtain a picture of voting behavior and economic attitudes in the entire United States, a country of well over 200 million people. Second, the survey permits greater speed in collecting and analyzing data. Because of its scope, a nationwide census requires a year or more of preliminary work to set the stage for the actual field operations. Enumerators must be recruited and trained in all parts of the country, maps must be prepared, and administrative centers set up. These disadvantages would be greatly reduced, of course, in a census of a single organization or a small community. Once the field work is completed, it usually requires months (and in some countries years) before census data are processed and published.

Third, since the number of interviewers used in the survey is much smaller, it usually is possible to provide more intensive training and careful supervision to ensure high quality in the data. Population censuses in many countries suffer from the poor qualifications of the enumerators themselves, and from the lack of adequate supervision. Fourth, the higher qualifications and greater training of the survey field staff together with the smaller size of the study allow greater flexibility in the topics covered by the survey. Well-trained survey interviewers can use more complicated questionnaires, spend more time with each respondent, and explore topics in detail through intensive probing. Finally, the sample survey is less visible than the census and not as great a burden on public good will. A national census of population not only involves contact with each household in the country, but typically requires a nationwide publicity campaign to stimulate interest and collaboration. Because a survey involves only a minimum of publicity (see Chapter 8) and contact with a fraction of the population, it is less likely to arouse overt opposition and feelings of saturation with social research.

For all these reasons many countries have chosen to supplement basic censuses and censuslike operations with smaller-scale but better-controlled sample surveys. In Peru, for example, the periodic surveys of business establishments conducted by the Technical Office of Manpower Studies (OTE-

MO)—studies which formerly covered all but the very small establishments—are now done on a sampling basis. The success of these operations,
of course, presumes the availability of competent specialists, a condition
lacking in some countries for both surveys and the census.

Perhaps the overriding advantage of the census is that it can provide
information on small geographic or political subdivisions, an important feature usually not present in the survey. Such information is commonly used
for assigning political representation (e.g., legislature), for distributing resources from the national government to political subdivisions, and in making other kinds of decisions. The emerging pattern in much of the world is to
invest in periodic censuses of population, housing, agriculture, and business
every five or ten years to obtain bench-mark data, and then to supplement
this information with more frequent surveys of greater complexity.

THE SURVEY AND AGGREGATE DATA

One of the least expensive forms of social research is that involving aggregate data—information summarizing characteristics of a total group or geographic region. The information is called "aggregate" because it includes
figures only for total groups, such as an entire country, rather than for the
individual units comprising the group. Since this information is often a by-
product of administrative operations such as customs and income tax collection, central bank operations, and the registration of births or automobiles,
it is typically less expensive than a census or a large survey. The following
are some examples of aggregate data: the total number of automobiles produced in the United States by year; birth and death rates by countries,
states, or counties; gross national product; and the total value of imports
and exports for a given year, again by country.

What are the relative merits of aggregate data as compared with the
sample survey? Because aggregate data are usually published, they are less
expensive and more readily accessible than information from surveys. This
advantage is gradually being reduced, however, as survey data are being
stored in specialized "data banks" and made available to scholars.[2]

_____

[2] Increasingly, "data banks" are bringing together aggregate, survey, and often other kinds of data.
Some of the major centers are the Inter-University Consortium for Political Behavior and the Economic
Behavior Program at the Institute for Social Research of the University of Michigan (cf. Economic Behavior
Program, 1960); the Roper Center at Williams College, which includes data over many years from the Roper
polls; the Survey Research Center of the University of California at Berkeley; the Institute for Behavioral
Research at York University in Toronto; and the United Nations Demographic Center (CELADE) in San-
tiago, Chile. Aggregate statistics are also regularly published by such international organizations as the
Population Division of the United Nations, the Inter-American Statistical Institute, and the Organization for
Economic Cooperation and Development (OECD) in Paris.

At the level of the actual information provided, the greatest asset of aggregate data lies in plotting rates and trends for total units, such as corporations, countries, and cities. Such information is especially helpful to the administrator or analyst concerned primarily with *what* is happening rather than with detailed explanations of *where, why,* and *to whom.* The vice president for sales of a corporation may be perfectly content with figures showing that the sales of a product increased between this year and last. The head of the marketing division will probably be less satisfied with this information, for it provides no clues about the types of people who account for the rise in sales. In the automobile industry it would be important to know if the increase stemmed from high school graduates purchasing their first car, middle-aged parents acquiring a second car, or more frequent turnover of cars among essentially the same buyers as before. A compilation of different types of aggregate data may provide some leads, but still cannot answer this basic question (cf. Morgan, 1967).

The strength of the sample survey in this case is that it does show the relationship between rates of behavior (such as the purchase of automobiles) and the characteristics of the individuals involved. It thus normally holds greater explanatory power. A good survey can answer questions such as: What is the average income of the Chevrolet owner? Are owners of the Ford Mustang more likely to be men or women? At what age is the driver in this country most likely to purchase his or her first new car? What percentage of the families purchasing new cars already own at least one car? Moreover, panel surveys (studies involving reinterviews of the same individuals) can help to explain changes in behavior over time, such as shifts in party preference from one election to the next, occupational mobility, or decisions to buy cars, houses, or take expensive vacations. Aggregate statistics on voting will tell us that more people in a certain precinct voted Republican in 1974 than in 1972, but they cannot identify the type of person who switched and why. A sample survey conducted in that precinct only in 1974 would be better than aggregate data for the latter purposes, but cross-sectional samples at both times would be even better.

Finally, in some cases aggregate data are more accurate than survey data. Such is often the case, for example, with information on the ownership of financial assets, life insurance, or other assets which tend to be concentrated in the highest income levels. The typical national sample of 2000 cases would normally draw few individuals with extremely high incomes, so that the information on such ownership would be highly unreliable. Reports compiled on these individuals by financial institutions or government agencies are likely to be superior to surveys. Even so, the survey would still be helpful and normally better than aggregate reports in providing information on the *distribution* of such assets.

THE SURVEY IN COMPARATIVE INTERNATIONAL RESEARCH

In the past decade the survey has been used increasingly as a vehicle for collecting data and testing hypotheses in a variety of cultural settings. One of the best-known cross-national surveys is the five-country study of political culture, socialization, citizenship, and political participation directed by Almond and Verba (1961). The research by Daniel Lerner (1958) on communication and modernization in the Middle East has also received wide attention. More recently Alex Inkeles and his colleagues at Harvard University (cf. Smith and Inkeles, 1966) have carried out a six-nation study of social and cultural factors in modernization. Similarly, surveys on fertility, reproductive behavior, abortion, and related topics have been launched in many countries, often with explicit coordination and guidelines for ensuring equivalence (Baum, Dopkowski, Duncan, and Gardiner, 1974 a, b, c, d; Mauldin, 1965). Each of these studies has met with formidable difficulties in data collection and analysis, and yet each has helped to broaden the horizons of social research—even by its mistakes.

Systematic comparison on a cross-societal basis, whether through the survey or other methods, offers several advantages to the student of man and society. First, cross-cultural research places pressure on the investigator to clarify the meaning of his guiding concepts, even for use in his own society. Even seemingly obvious concepts such as "marriage" and "divorce" may present enormous definitional problems as one moves across societies and cultures. Similarly, the widely used concept of "unemployment" may prove inadequate to capture social reality in many cultures. The commonly accepted dichotomy between the employed and the unemployed person in industrialized western countries has little meaning in societies in which large parts of the population are underemployed or in nonmarket economies. Second, cross-cultural research extends the range of variation on the phenomena an investigator wishes to explain (dependent variables) and on the conditions used to explain them (independent variables). If a social scientist wishes to examine the correlates of political unrest, he will find much more variation in this dimension if he uses the full range of nation-states than if he confines his attention to the North Atlantic region. In the same vein, a study of the relationship between religion and economic growth will obviously cover a broader spectrum of explanatory conditions if it includes India, China, Europe, North Africa, and Israel than if it is limited to Catholic-Protestant differences in the West.

Third, comparative analysis provides the only means for testing the generality of *relationships* between variables, such as social class and voting behavior. Bendix and Lipset show, for instance, that "upward mobile people in the United States (the middle-class sons of workers) tend to be more conservative than those who have inherited high social status. However, in a number of European countries such as Sweden, Finland, and Germany, the upward-mobile are more radical than stable individuals. . . . Teachers and physicians tend to be on the right in Germany, while as compared with other professionals, they are rather on the left in France and Britain" (1957, p. 90). This example illustrates the many areas in which cross-national survey research can shed light on the generality of findings drawn from single-culture studies. Fourth, comparative analysis is often a source of new concepts, theoretical insights, and hypotheses about the phenomena under study, even in one's own society. The search for explanatory factors to make sense of findings emerging from several nations, for example, may reveal a causal variable, such as some property of entire societies, that is not evident in single-culture studies. Typically, properties of entire countries become clear only when these units are compared with others unlike them. Similarly, exposure to other societies and cultures may have a kind of reactive effect through which the investigator becomes aware of certain critical influences at work in his own culture. Research on nationalism in the developing countries, for example, has provided numerous insights into the growth of "black nationalism" in the United States.

Many problems arise in carrying out explicitly coordinated cross-cultural surveys, and we shall comment on these at appropriate points. One of the most difficult challenges lies in developing concepts and operational definitions that are equivalent not only in translation but in real *meaning* from culture to culture. Beyond this there are numerous practical problems in ensuring comparability in all stages of the research, including adequate sampling and equivalent interviewer training, coding, and quality control of the information (cf. Warwick and Osherson, 1973). Nevertheless, significant comparative surveys have been carried out and point the way to improved design. It is our hope that social researchers in the future will be even more imaginative, intrepid, and yet careful in applying survey methodology to the intriguing task of comparative cross-cultural analysis.

TYPICAL STAGES IN THE SAMPLE SURVEY

The one-time sample survey usually involves seven related stages. It should be emphasized that these stages are interdependent—each affects and is affected by the others. An awareness of the whole process allows the researcher to anticipate the implications of the decisions taken at any one stage. A brief review of these stages follows.

Planning involves setting the goals for the survey and devising a general strategy to obtain and analyze the data. It should begin with an explicit attention to the concepts and hypotheses guiding the research, and a careful review of the literature. It should also include the setting of specific objectives for the study and other tasks reviewed in Chapter 2.

The elaboration of a **research design** (Chapter 3) is closely related to planning, and usually occurs simultaneously. This stage should also include attention to the kinds of data needed to satisfy the basic objectives of the study.

Sampling (Chapters 4 and 5) is the process of choosing certain elements in the population to represent the whole. At this stage the researcher must carefully define the population to be studied and the generalizations from sample data which it will permit. The sample design consists of procedures for selecting the population elements and for converting the sample data to estimates about the total population. A major challenge in sample design is to tailor the procedures to local conditions and available resources while maintaining the advantages of probability sampling.

Questionnaire design (Chapter 6) is a process of translating the broad objectives of the study into questions that will obtain the necessary information. This is typically a process of trial and error involving long hours of discussion and several pretests. Usually the basic procedure, such as a personal interview or mailed questionnaire, is laid out in the general plan for the study. The main effort then centers on the number and type of questions, their sequence, and the means of motivating the respondent and maintaining his interest.

Field work (Chapters 7 and 8) includes the recruitment and training of interviewers as well as the actual interviewing and supervision of interviewing. It may also include various tasks related to the development of a sample, locating sample members, and pretesting draft questionnaires. Field work will be reduced, of course, in studies relying primarily on telephone interviews or on mailed or other types of self-administered questionnaires.

Editing and coding (Chapter 9) are processes for converting the responses recorded in the questionnaire into categories (usually expressed numerically) which can be counted or tabulated. Editing is sometimes introduced prior to coding in order to keep the latter stage as simple as possible. Editors may consider and classify several related responses taken from different parts of the questionnaire, or undertake other complicated tasks requiring greater training and judgment. Questionnaires which permit answers to be recorded in predetermined categories combine interviewing and coding into a single process.

Preparation for analysis (Chapter 10) covers a variety of related tasks: key punching the coded responses, carrying out machine consistency checks to determine the compatibility of related answers, dealing with unresolved problems of missing data, assigning different weights to the interviews when this step is called for by the design, the creation of scales and composite variables, and the actual statistical tabulations required for the analysis.

Analysis and reporting (Chapter 11) may consist only of the presentation and interpretation of simple distributions and cross tabulations of information collected in the survey, or they may involve complex statistical treatment and elaborate theoretical interpretation of the findings. The nature of both the analysis and the report of the findings will depend on the precise objectives of the study.

FURTHER READINGS

Abrams, M., *Social Surveys and Social Action* (London: Heinemann, 1951).

Campbell, A. Angus and Katona, George, "The Sample Survey: A Technique for Social Science Research," in L. Festinger and D. Katz, *Research Methods in the Behavioral Sciences* (New York: Dryden Press, 1953).

Glock, C. Y. (ed.), *Survey Research in the Social Sciences* (New York: Russell Sage Foundation, 1967).

Young, P., *Scientific Social Surveys and Research* (New York: Prentice-Hall, 1944).

Chapter 2

Planning the Sample Survey

No step is more vital to the success of a survey then careful planning. In some types of research, such as participant observation, the investigator is relatively free to improvise and change course as the study moves along. In a survey, once the sample is drawn, the questionnaires printed, and the interviewers on their way, changes are both difficult and expensive. Put simply, a survey that cannot be planned should probably not be done.

Albert Hirschman (1958) uses the terms *forward* and *backward linkage* to show the interdependence of various elements in the process of economic growth. The same notions can be applied to the stages of survey research discussed in Chapter 1. Figure 2-1 shows both the chronological overlap and the functional interrelatedness among these stages. The notion of forward linkages suggests simply that the earlier stages of research affect the latter. Thus the sample design and decisions about eligible respondents will (or should) influence the type and wording of questionnaire items as well as the selection of interviewers; the coverage of the questionnaire, such as whether it deals only with the respondent or asks about other household members,

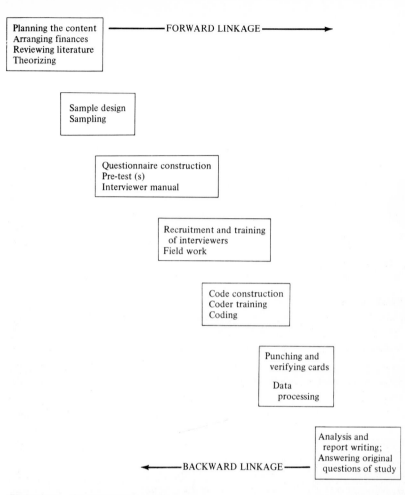

Figure 2-1 Stages in the sample survey.

will affect the content of the questions, the organization of the codes, the populations available for analysis (e.g., individuals versus family units), and so on.

But an even greater challenge in survey planning lies in seeking out the backward linkages. In a very real sense the planning process should begin at the end, with the final needs for data, and work back to the beginning. With a clear picture of the information required and its probable uses, the planning process can move back to the analysis and tabulations necessary to satisfy these needs, the variables and sample groups that will yield the tables, the codes that will produce the variables, the sample that will cover the appropriate population, the interviewers and field supervision needed to handle the questionnaire within the designated sample, and, in general, the direction required to coordinate a large study.

BASIC QUESTIONS IN PLANNING THE SURVEY

The sample survey provides ample opportunities for both innovation and error. The highest challenge in survey research is to translate a vaguely formulated concern with a problem into an original, well-conceptualized, and methodologically sound study. The remainder of this book will try to show how the transition from problem to survey can be made with as little wasted effort and as few errors as possible. Some of the more common mistakes include the omission of essential questionnaire items through a failure to define the exact information needed from the study; failure to anticipate the full costs of interviewing, coding, data processing and analysis; carelessness in designing or implementing the sampling plan; and inadequate specification of the population and subgroups about which generalizations will be made. To avoid these errors and to draw maximum benefit from the survey, both study directors and sponsors should give careful attention to the following basic questions before finishing a questionnaire or drawing a sample.

What Are the Objectives of the Study?

The first question to be raised in planning a survey is, "What do we hope to learn from this study—What do we expect it to do for us?" Most surveys are motivated by the search for more information on a subject, the desire to test some theoretical hypothesis, or a practical problem in need of a solution. One of the most practical exercises in planning is to translate these broad needs, hypotheses, and concerns into topics amenable to study through the survey. When the sponsor or client is not familiar with survey research it may be necessary to spend a great deal of time helping him to decide exactly what he wants and can reasonably expect from the study. Dexter (1966) argues that a large part of the work of the social science consultant is to help clients decide what to want from social science research. This process is especially necessary when the client is primarily concerned about the policy decisions rather than information gathering. Administrators who commit their institutions to sponsor a survey usually hope that it will provide some basis for deciding whether one program or option is better than another. Under these conditions it is both practically and ethically imperative for the researchers to point out the likely uses and limitations of survey data in answering policy questions. A well-designed survey could identify the populations that would be affected by different poverty programs, and may show how poverty program X has different effects from poverty program Y, but it cannot, in itself, say whether one program is *better* than another in some absolute sense. Decisions about good and bad or better and worse are value judgments rather than empirical observations. The most that a survey can do is to say: "Given the following yardsticks of performance (outcome variables), the effects of this program seem to be . . ." Joint discussions between the sponsors and the researchers are helpful in avoiding misunderstandings about the uses of survey data, and also in establishing agreement on the areas of information required.

The most direct and helpful means of clarifying the objectives of a study is to prepare a list of the practical and/or theoretical questions which the interested parties would like to have answered. The aim at this stage is to specify the major *categories* of data needed. The process can be illustrated by a hypothetical national urban survey which will serve as a continuing example in this book. Let us assume that we have the responsibility of designing a multipurpose survey which will provide information of interest to urban planners, administrators, and other officials in the country of Pacifica. A crucial step in planning the study is to meet with the potential users of the data to discover and clarify their needs for information. In the case of the Pacifica survey the discussions, held over a period of several weeks, produced the following questions, among others:

Do residents of urban areas feel that their economic and social situation is improving or deteriorating?

What is the overall pattern of housing in the major cities, and how does it differ from city to city? Specifically, do most heads of households own their own homes, rent homes or apartments, or work out other arrangements? Is there crowding?

On the average, how much do families spend per month to rent a home? for mortgage payments?

What is the rate of unemployment among household heads in urban areas? Does this differ by educational levels? age?

What is the pattern of employment? In what occupations do employed residents work? What are the differences in employment patterns across the major cities?

What are the major social characteristics of the residents, including age, sex, education, and family composition?

How long have residents lived in their homes? How many are likely to be moving within the next year?

What is the actual family size of urban residents? the ideal family size?

How well informed are household heads about family planning methods?

What are the attitudes of the population toward social change? Do urban residents incline toward traditional or modern attitudes?

At the end of Chapter 6 we will translate these broad concerns into specific questionnaire items for the urban survey. In Chapter 9 these items will be used as the basis for building code categories. Chapter 5 will deal with the problems of sample design to meet the needs of this national urban survey.

Apart from meeting the objectives of the sponsors, a large-scale survey can increase its usefulness to the scholarly community and other interested parties by including certain standard demographic questions or broadening its coverage in other ways. The standard questions might cover age, sex, race, education, family size, occupation, income, and perhaps religion. The economics of survey research show that the major expenses in most studies are for sampling, interviewing, and administration. Thus *a few* items can be added to most questionnaires with little effect on total cost. We are decidedly not proposing that "interesting" items be added indiscriminately, but rather that the needs of other potential users be taken into account. Some additions could have the beneficial side effect of attracting joint financing for the study, and also of allowing more than one major report to be prepared on the findings.

Is This Survey Necessary?

The sample survey is a relatively conspicuous and obtrusive research method (cf. Webb, Campbell, Schwartz, and Sechrest, 1966). Poorly designed or repeated surveys run the great risk of exhausting the limited reserve of public tolerance for social research, contaminating the atmosphere for future studies, and even generating pressure for legislation restricting household (door-to-door) surveys. Thus a survey should be launched only if it is clearly needed to satisfy a study's objectives, and if its potential benefits outweigh its likely nuisance value. The burden of proof in this case should be on the researcher. The following questions should be considered in deciding whether a given survey is necessary.

Will It Provide the Information Needed? The survey method, as we have seen, is not appropriate to all kinds of problems. But, unfortunately, it has become so popular in some quarters that it is the first alternative considered even when it may be unsuitable. The uses and limitations of the survey can be illustrated in the area of public policy.

The growing field of "evaluation research" provides numerous examples of ways in which a well-designed survey, often used in conjunction with other methods, can be helpful to policy makers. In the evaluation of the New Jersey–Pennsylvania Income Maintenance Project, for instance, survey methods were incorporated into an unusual field experiment. The project itself

> . . . involved experimental treatments which were combinations of guaranteed income levels and "tax rates" (rates by which payments are reduced if incomes are augmented above the guarantee levels). The population was households in five locations in New Jersey and Pennsylvania. . . . Eligible households were allocated at random to treatment and control groups, with treatment groups given checks bringing their incomes up to guarantee levels (or above, less "tax rate") upon submission of monthly income reports. Both treatment (experimental) and control group households are visited quarterly for extensive interviews (for which both are paid). [Rossi, Boeckmann, and Berk, 1973, p. 3.]

The interviews were designed to answer the main question of concern in the Income Maintenance Experiment: whether workers receiving income maintenance would drop out of the labor force to enjoy greater leisure. The evaluation study also explored a number of other potential effects of the experiment. By all accounts the study was admirably designed to provide information of direct relevance to policy makers.

Governmental administrators, by contrast, often complain that research sponsored by their agencies does not help them to reach policy decisions. The most common reasons are that the sponsor expected too much, or did not have a clear understanding of the research process, or that the researcher created false expectations about the practical relevance of the study. Survey data often cannot be translated into policy statements, at least not without a generous dose of value judgments. Also, as Carter points out, ". . . while survey results can be useful, policy decisions usually hinge on matters of public philosophy more than on indications from survey results" (1963, p. 549). Similarly, if the United States government must choose between continuing or closing a military base or a poverty program, the most critical consideration may well be constituency interests rather than "objective" information about other effects of the decision. The astute politician often has enough of a sense of constituency reactions to guide his actions without taking a survey.

Is the Survey the Best Way of Obtaining This Information? Sometimes the survey will provide the information required by the objectives of a study, but the same information could be obtained equally well and, perhaps, more simply by other methods. For example, if a city wishes to develop a classification scheme covering its major types of housing problems, but is not interested in the *frequency* with which each problem appears, it may be better to interview a few civic leaders and urban experts and then carry out some nonrandom observations in deliberately chosen, diverse areas of the city.

The data provided by the sample survey in this case would exceed the needs of the client. Survey data, on the other hand, might be *insufficient* for zoning reform, for they would not provide adequate detail on the small geographic areas in need of reform. A zoning study might well require a complete enumeration of the city, or all the areas being considered for reform.

Can This Survey Be Merged with Some Other? Even when it seems clear that new survey data are necessary, the investigator should still ask whether a separate survey is called for. One option increasingly attractive in studies requiring modest amounts of data consists of "buying into" an "omnibus" survey conducted by a large research organization. Often by adding a few questions to a larger survey, one's requirements can be satisfied at a lower cost and with higher quality information over a broader population than in a privately sponsored single-purpose study. This alternative also helps to reduce the survey fatigue brought on by the proliferation of studies, especially those conducted by amateurs. These, we realize, can sometimes be justified on the grounds that they provide firsthand experience and excitement for the coming generation of social scientists. One wonders, however, if they could not receive equally valuable training by helping to construct and pretest the questionnaires, and then later joining in the analysis of the data gathered by the research organization. In the United States, two university centers which have used the collaborative omnibus survey with great success are the Survey Research Center at the University of Michigan and the National Opinion Research Center at the University of Chicago. Other commercial and university-based survey organizations are also open to this possibility.

How Much Is Already Known about the Subject?

The early stages of planning should include a careful review of the literature on the topics to be covered, including books, articles, and reports of similar studies. Occasionally this search may have the happy outcome of making a new study unnecessary by uncovering data which satisfy the investigator's objectives. More often it will suggest fruitful ways of conceptualizing the problems, approaches which have worked particularly well or failed in other studies, tantalizing but untested hypotheses, and questionnaire items particularly well suited to the present study. Moreover, while no survey should be completely beholden to the past, there is great value in attempting to build on existing knowledge rather than to replace it. This cumulative quality of social research is enhanced by using comparable questions and procedures whenever possible. A thorough understanding of the literature is helpful both in increasing comparability and in reaching decisions about new directions of inquiry.

What Is to Be Measured?

After defining his research objectives, the investigator is usually left with a number of concepts or terms the general meaning of which are understood but which are not clear as to details. These serve as his road map to the aspects of reality on which the study will focus, and also reflect an implicit theory of how reality is constructed. When an economist says that he wishes to study unemployment in a given society, he is assuming not only that unemployment can be defined, but also that the society is organized in such a way that the concept is meaningful. Thus every survey, whether it is billed as descriptive or theoretical, begins with a set of concerns as well as theoretical preconceptions. As Myrdal points out in the prologue to *Asian Drama*, "A 'purely factual' study—observation of a segment of social reality with no preconceptions—is not possible; it could only lead to a chaotic accumulation of meaningless impressions" (1968, Vol.I, p.24). The next challenge in planning is to surface one's preconceptions and to clarify one's central concepts. This task requires both conceptual and operational definitions of the phenomena to be measured.

Conceptual Definitions The concepts used to guide survey research vary greatly in their generality. Some, such as *age, sex,* and *level of education,* refer to relatively concrete phenomena and present few problems in definition. Others, such as *power, authority, deviance,* and *aggression,* operate at a very high level of generality and are extremely difficult to define for the purpose of a sample survey. Between these poles there are many terms, such as *unemployment* or *the labor force,* which on the surface seem to require little inference from observable facts, but which in fact require very careful definition.

Consider the seemingly straightforward concept of *the labor force.* Few would disagree that such a concept is needed to cover those who are either employed or unemployed, and that the residual category of "not in the labor force" is useful in describing those who are not covered by the other two. But in designing a survey, we must also decide if the labor force should include persons temporarily ill, on vacation or leave, in the armed forces, or in mental hospitals. Moreover, how should we classify those who are seasonally unemployed, handicapped, or not available for full-time work, such as housewives? What is the minimum age for including a person in the labor force? Why? If this is a cross-national study, should the age be changed from country to country to reflect local conditions?

Or take the familiar notions of *employment* and *unemployment.* A clear conceptual definition of these terms requires answers to these questions: How many hours per week must a man work before he is considered employed? Is a salesman who sells on commission employed or unemployed if he earns no income during the period of reference? Does this rule apply to real estate salesmen, streetcorner peddlers of pencils, and door-to-door salesmen of brushes? Should wives and children who work in a family store be considered *employees* of that store? What about family farm workers? The seasonally unemployed?

Fortunately, in the field of manpower, much of the work on definitions has already been carried out by official bodies such as the International Labor Organization, the Inter-American Statistical Institute, and, in the United States, by the Bureau of Labor Statistics and the Bureau of the Census. But sometimes these "official" definitions convey a false sense of security, leading one to apply concepts where they may be inappropriate, or to ignore others that may be more appropriate. For example, in the United States official statistics have concentrated on employment and unemployment while neglecting the concept of underemployment. In the less developed countries this approach would miss a major socioeconomic problem, as thousands of people work long hours for little or no income, or put in less than a full workweek when they would like to work more. Underemployment is especially common among peddlers and itinerant artisans who cannot find stable jobs and thus turn to a largely unremunerative self-employment. A study in Lima, Peru, showed that these independent workers accounted for almost 20 percent of the labor force, a high percentage of whom were classified as underemployed (Centro de Investigaciones Sociales por Muestreo, 1967).

In other comparative or international studies, the conceptual baggage brought in by the Western investigator may be inappropriate to the local situation. Gunnar Myrdal, himself an economist, points up this problem in his monumental study of South Asia:

> There is a conservatism in methodology in the social sciences, especially in economics, that undoubtedly has contributed to the adherence to familiar Western theories in the intensive study of underdeveloped countries. Economists operate to a great extent within a framework that developed early in close relationship with Western philosophies of natural law and utilitarianism and the rationalistic psychology of hedonism. . . . That economists work within a methodologically conservative tradition is usually not so apparent to the economists themselves, especially as the tradition affords them opportunity to display acumen and learning and, within limits, to be inventive, original, and controversial. Even the heretics remain bound by traditional thought in formulating their heresies [1968, Vol. I, p.17].

Myrdal points specifically to the biases implicit in the concepts of employment and unemployment which, he argues, "rest on assumptions about attitudes and institutions that, though fairly realistic in the developed countries, are unrealistic in the underdeveloped countries" (1968, Vol. I, p.19). In short, an essential task in arriving at conceptual definitions lies in determining whether the concept itself fits the reality to be investigated.

A further challenge in comparative cross-national studies is that of arriving at conceptual definitions that are *equivalent* from culture to culture. In fact, one of the prime advantages of comparative studies is that they often make us aware of how loosely we have used a concept in our own society. The term *divorce* provides a good example. While working only in their home countries, social scientists rarely stop to work out a conceptual definition of divorce, for this seems obvious. In comparative research, a serious problem is raised by this term: does it refer to the breakup of a marital union in general, or only to marital dissolutions based on court decisions? In commenting on this problem of international comparisons with demographic data, Kingsley Davis writes:

> It is possible to gather an infinite quantity of seemingly comparable information, but unless it is comparable with respect to analytically significant features of the material, it will be useless and, in some cases, misleading. For instance, it is relatively easy to gather statistics on the number of divorces granted by a nation's courts. . . . In one sense, the data are comparable—that is, they do concern marital dissolutions granted by courts in the countries concerned. However, this basis of comparability is so meager that the uncritical use of the data for demographic and sociological purposes would, and often does, lead to absurd results. The tendency is to use the figures as an index of the breakup of marital unions, because this is what is socially significant; but reproductive unions between men and women break up regardless of whether or not a court has authorized and recorded it. They break up by informal mutual consent, desertion, religious sanction, or prolonged separation [undated, pp. 4-5].

Similar definitional problems arise in cross-national surveys on crime, delinquency, suicide, and other problems that are defined administratively rather than theoretically.

Even more explosive definitional problems arise with highly charged concepts such as *intelligence* or *authoritarianism*. In such cases, the problem of conceptual equivalence is relevant whether one is comparing entire societies or culturally distinctive subgroups within a single society. In the United States, for example, a debate has raged for several decades over whether standard measures of intelligence are fair to blacks and other cultural minorities. In view of the political uses of findings showing group differences in mental ability, this is more than an academic question. Similarly, it is not difficult to define authoritarianism so that certain groups, such as Catholics or fundamentalist Protestants, almost automatically emerge with high scores on this trait.

Operational Definitions   Clear conceptual definitions greatly facilitate
the development of operational indicators—behaviors that can be observed
and measured with the tools available, such as a survey. The ultimate success of an operational definition in a survey depends upon the extent to
which the behaviors sampled by the survey interview represent the total
universe of behaviors embraced by the concept. With concepts such as age,
sex, and nationality, the fit between the behaviors and the concept is usually
quite good, though some ambiguities may remain. In the case of highly
abstract concepts, such as *power, social mobility,* or *traditionalism,* the survey
interview may tap only a fraction of the relevant behaviors, and in a way
that requires a high degree of inference from the observed behaviors to the
original concept. To cite an extreme example, one would be on very precarious grounds to assume that a person who *says* he is powerful during an
interview actually holds power in some objective sense. The inferential leap
is much smaller with questions about education, employment, and marital
status.

This process of translating a concept into an operational definition can
be illustrated with the problem of unemployment. According to the conventional international definition, a person above a certain age (e.g., 14) is
considered unemployed if he is not working, is able to work, and actively
looking for work. In the United States this individual could be identified
operationally as one who did not work at all during a defined reference week
(e.g., the week before he is interviewed), and who was actually taking steps
to obtain employment, for example, by registering at an agency or writing
letters. The following groups are also considered unemployed: those waiting
to be called back to a job from which they were laid off; those waiting to
report to a new job beginning in 30 days or less; and those who would have
been looking for work, except that they were temporarily ill, or believed that
no work was available. But even this operational definition leaves some
questions unanswered. Is a person considered to be "looking for work" if he
wants to work only a few hours a week, or if he turns down a job because he
wants higher pay than was offered? How should a person be classified who
is out of work and looking for employment, but who cannot accept many
kinds of work, perhaps because of a physical handicap or certain kinds of
mental illness? These problems should be resolved before the questionnaire
is printed.

The question of what is to be measured also raises the issue of exploratory items in the survey interview. Here the problem comes down to balancing three, sometimes incompatible, objectives: (1) keeping the interview as
short as possible, (2) including enough items to satisfy the data requirements
of the study and to make it useful to a broader audience of scholars, and (3)
using the questionnaire as a means of testing new ideas which may be in-

completely formulated but still of interest to the investigator. In most surveys there is no need to have an exact, watertight rationale for every item unless the entire study is likely to come under the type of political attack where such a rationale may be essential for self-defense. If the investigator has a feeling that a certain question could prove valuable in testing a hypothesis or in shedding light on a problem, he may want to include it even though its credentials may not be in perfect order. It is often better to be vaguely right than exactly wrong about a question. At the same time, many items tried "in the hope that they might be useful" are often nothing more than ill-conceived shots in the dark. When used in quantity, they make the survey impossibly long, and lead to an unjustifiable waste of everyone's time. At the very least, an item considered for inclusion should have some likelihood of being analyzed and some explicit, if incomplete, theoretical or practical rationale.

What Is the Target Population?

The findings of a sample survey will relate only to the population from which the sample is selected. It is imperative, therefore, to arrive at a clear definition of the target population—that is, the body about which one wishes to draw conclusions. If the study calls for generalizations about the United States, Canada, or Chile, it is not enough to study Skokie, Saskatoon, or Santiago. Though ideally the target population for sampling should be identical with the population about which conclusions will be drawn, compromises must often be made for practical reasons. For example, in most national studies it is necessary to exclude transients from the target group for sampling, despite the fact that they may be of interest in generalizing about a country. Chapter 4 explains how a part of the population can be chosen to represent the whole; here the emphasis is on deciding which "whole" is of interest.

The first step in defining a target population is to specify the geographic area or physical limits (e.g., a factory or school) covered by the study. The next task is to define the population or populations to be studied within these limits. This does not mean specifying the persons to be interviewed, but rather the group or groups about whom one wishes to generalize. Many household surveys investigate several target populations simultaneously, using the same set of respondents to provide information. The first might consist of the *dwelling units* within the area, usually excluding group quarters such as dormitories, prisons, hospitals, and hotels. The study may call for tables on the number of rooms, bedrooms, bathrooms, and people per room in the privately occupied houses and apartments. A second target population might be *persons living in the dwelling units*, including children, boarders, servants, and other habitual residents. In most surveys, the respondent is asked to provide information *about* these individuals since the cost of tracking down each person and the problems of talking with children make interviewing each one impractical. Information about this second population might be used in estimating the total population in the area, household and family composition, and various other characteristics. A third population might involve the adult residents of the dwelling unit, but here one must be very careful, for there are several subpopulations within this category. Is the desired target group *all adults*, or a subpopulation of *heads of families, heads of households, males twenty-one years of age or over, women of child bearing age, persons in the labor force*, or some other group? There is overlap between these categories, but each is different for purposes of generalization. If the survey obtains information about a more inclusive group, it will still be possible to analyze and present data on selected subgroups; however, if the survey obtains information only about heads of households, this will not permit generalizations about all adults, and so forth. These issues should be confronted squarely in the planning stage to avoid many headaches later.

Who Will Be Interviewed?

This decision depends upon the target population, the nature of the information desired, and considerations of convenience. If the target population is all adults and if the information sought includes attitudes and opinions that cannot validly be provided by other household members, the sampling procedures should either cover all adults in the selected dwellings or include a method of random selection of adults within the household. The reason is that probability sampling methods provide a solid basis for generalization about a target population only when applied at all levels. If adults are selected within the household according to availability, serious biases will be introduced into the results. Female adults may be more available than males, less educated adults may defer to those in the household with more education, and so on. One might estimate the population of an area to be 80 percent female based on the sex of those easiest to interview; one would obtain a similarly biased measure of attitudes in the population by using the same respondents.

When accurate information about an individual in the target population can be obtained only from the individual himself, as in most attitude surveys, it is essential to have a predetermined plan for the random selection of persons within the household. If, on the other hand, the survey deals mainly with information *about* the household and its residents, and if valid information can be provided by any of the adults, random selection within the household is less important. Decisions on this matter should be made explicit during the planning stage, rather than left to the interviewer on the doorstep.

What Is the Overall Design of the Study?

The precise study design should flow from the objectives of the study, or some compromise between them and such constraints as time and costs. The major issues of design, including the relationship between research objectives and data-collection procedures, will be taken up in Chapter 3. Here we might note that a crucial question for planning is whether the survey will require data collection at one or more than one point in time, and if more than one, with the same sample (a panel study), separate samples, or both. Unfortunately, present research technology does not permit us to conduct a survey now and repeat it five years later. Though it is tempting to use the technique of retrospective introspection as a means of gathering instant data on change, methodological studies have repeatedly shown that for most topics the past is recalled with considerable bias. Respondents will freely tell what they think they thought several years ago, but the answers are usually poor reflections of what they actually thought at the time. The point to be emphasized is that the planning process should pay close attention to the relationship between the ultimate needs for data and the research design.

What Will Be the Personnel Needs and Total Cost of the Study?

Very early in the planning process it is important to consider such practical questions as the likely cost of the projected survey as well as the time and personnel needed to carry it out. In view of the fact that personnel and travel costs are usually the largest single part of the budget for a sample survey, we will combine our discussion of cost and personnel. The following are the principal expenditures likely to be made in the survey:

1 Salaries and other personnel costs
   a *Administrative staff:* project director, for general administration and coordination, and others as needed for supervising the work in the study. Estimates should include time for planning and reviewing the literature, sampling, questionnaire design and pretest(s), interviewing, code preparation, editing and coding, consistency checks and "deck cleaning," tabulations, analysis and report writing, and publication.
   b *Clerical staff:* secretaries, sampling clerks, accountants and record keepers, etc.
   c *Field staff:* field supervisors, interviewers, drivers, others needed to collect the data. Budget estimates should allow for training and prac-

tice interviews, as well as any field work that may be needed for the sample.

   d *Consultants:* general consultation and specialists such as sampling experts and computer programmers needed only in certain stages of the study.

2 Travel costs and living expenses in the field

The travel costs and maintenance of study directors, supervisors, samplers, and interviewers during sampling; pilot tests of the questionnaire; interviewers' training; and actual field work. Estimates should include transportation to group meetings and the costs of maintaining the staff while away from home (per diem).

3 Services

   a Printing of questionnaire and instructions.
   b Vehicle operation and maintenance; insurance.
   c Coding of the data: personnel.
   d Machine consistency checking and corrections.
   e Data processing: personnel, computer time or other equipment expenses.
   f Publication costs: editing, typing, printing.

4 Equipment and supplies

Vehicles (including mileage charges by the staff for the use of their own cars as well as rentals in the field); office equipment; paper; printing the questionnaire and miscellaneous printing and reproduction costs; telephones and other communications expenses.

5 Other costs

   a *Overhead:* if the survey is being carried out by a research organization, from 5 to 50 percent of the total personnel costs may be added to the budget to cover indirect costs of maintaining the buildings, administrative staff, library, auditing, etc. Many foundations provide up to 10 percent for indirect costs of the research they support.
   b Publicity for the study; conferences during the planning stages or later to discuss the results.
   c Transportation of materials and equipment to and from field sites.
   d Rent for temporary office space during field work.

The sad fact is that most researchers preparing their first budget grossly underestimate the costs of the sample survey mainly because they have not planned it out step by step, or because they unrealistically expect considerable help free of charge. At the other extreme are occasional budgets grossly inflated because of staff inexperience and its resulting inefficiencies and learning costs. The first error creates special problems when the study is being supported by an outside agency which approves one budget, only to be met with a request for supplementary funds a few months later. A foundation executive in the United States is quoted as saying, "I've learned to apply a factor of 160 per cent to the estimated cost of a social science research project, and a factor of 180 per cent to the proponent's estimate of the time it will take to bring the project to the stage where it is already for publication" (Brunner, 1962, p. 97); however, others take a hard line on supplementing grants and expect original budgets to be realistic and firm. As Brunner points out, "Too many social scientists appear to lack budgeting sense in searching for research dollars" (1962, p. 97).

Given that interviewer costs are usually one of the largest single items in the budget, it is helpful in planning to know how interviewers allocate their time. A study carried out at the National Opinion Research Center showed that only 25 percent of interviewer costs result directly from interviewing, 40 percent are due to travel time and related expenses, and the remaining 35 percent cover study, editing of questionnaires, and other routine clerical or miscellaneous tasks (Sudman, 1963, pp. 633-634). This cost breakdown also revealed that "interviewers in metropolitan areas spend a greater proportion of their time in traveling than do interviewers in nonmetropolitan areas, since they have more difficulty in finding respondents at home, although nonmetropolitan interviewers travel longer distances" (ibid., p. 633). These figures have obvious implications for the preparation of a survey budget. They especially point up the fact that interviewers often will not be able to complete an interview with an urban respondent on their first call, and may have to return many times.

How Long Will the Study Take?

The answer to this question was suggested above: longer than anyone expects. The inflation factor of 180 percent proposed by the dour foundation representative may be extreme, but in most studies conducted by a staff with little prior experience a figure of 125 to 150 percent would not be unreasonable. So many things can go wrong in a survey that it is impossible to state fixed rules for making time estimates. Moreover, the time required for analysis and report writing varies widely from investigator to investigator. It is possible, however, to list the normal steps which should be included in a time budget. Appropriate allowances can then be made for obstacles and delays. These steps include:

1 Formulating the plans for the study, preparing research proposals, and working out financial arrangements.
2 Preliminary discussions and negotiations with officials or other persons whose cooperation will be needed.
3 Sample design, gathering materials (maps, census data, etc.) for the sample, drawing and recording the sample, and writing instructions for its field use.
4 Quality control measures in sampling (see Chapter 4).

5 Developing an interview schedule or questionnaire.

6 Pretesting the interview schedule or questionnaire. Studies contemplating complex measures of attitudes, opinions, self-esteem, and the like, should allow time for several pretests.

7 Recruiting the interviewers, supervisors, and other field staff.

8 Training the field staff, writing instructions for the interviewers on the use of the questionnaire and on administrative matters.

9 Collecting the data in the field and getting the last interviews in.

10 Quality control of interviewing, e.g., through a 10 percent reinterview program.

11 Checking the interviews returned against the addresses selected.

12 Editing the questionnaires.

13 Coding the data from the interviews and about the noninterviews.

14 Key punching and verification.

15 Consistency checking and corrections.

16 Processing and tabulating the data, including the construction of composite variables and scales.

17 Interpreting the results.

18 Preparing and publishing one or more research reports.

19 Conferences with interested persons (especially the sponsors) about the findings.

20 Completion of administrative and accounting records, storage of the questionnaires, and other final "housekeeping" details.

Some of the more common obstacles slowing the pace of a large sample survey include sickness, fatigue, and adverse weather. Snowstorms or a rainy season may set the field work back from a day to a month or so, or cause it to be rescheduled. Also, most interviewers will lose their initial enthusiasm for field work and correspondingly decrease their output, especially if they have experienced several refusals and reach the point where most of their remaining assignments are fourth and fifth call-backs.

Seasonal factors may also be important in planning the timing of a survey. In university communities, the residents available during the summer months may represent a rather different population than those present during the rest of the year. A survey which cuts across both periods may mix the normal population of students with an amalgam of visiting teachers, engineers, physicians, and others returning for a summer of graduate work or refresher courses. The summer is also an undesirable time to conduct surveys in many areas because of the large number of families absent on vacations. These absences present special problems in cross-section surveys because of the fact that upper-income groups are more likely to be travelling or on vacation than those in lower-income brackets. The spring and fall are often inconvenient times to interview farmers since these are peak times in their work schedules, while in the winter they may have more time available. Holidays may create difficulties in surveys of employment and unemployment which use "last week" as a fixed period of reference in determining the number of hours worked. One way to cope with this problem is to randomize the interviews over time. Thus, in addition to all of its other demands, survey planning requires careful attention to the calendar.

Can This Survey Be Completed?

As a final question the investigator should ask explicitly if he and his staff have the necessary resources to bring the study to a successful conclusion. The answer will depend upon the response to several other questions: Are there sufficient funds to collect *and* analyze the data? Will it be possible to recruit enough interviewers to finish within the time available? Will the respondents cooperate? Is there danger of violent political opposition and will that threaten its completion or affect the results? Is there a reasonable margin for delays or unforeseen costs so that if they occur it would still be possible to satisfy the original requirements for data? If on balance the responses are not favorable, the research team owes it to themselves, their sponsors, the respondents, and the survey profession at large not to undertake the study at all, or to seriously restrict its scope so that it can be successfully carried out.

BIAS IN RESEARCH DESIGN

Survey research is never an aseptic process executed in a sociopolitical vacuum and guided solely by the canons of the scientific method. It is very much a human endeavor influenced by the researcher's own values and expectations as well as external obligations, conventions, and pressures (cf. Sjoberg, 1967; Horowitz, 1971; Friedrichs, 1970; Phillips, 1971). The question of bias is a very broad topic which will be taken up at appropriate points throughout this book, particularly in the chapters on sampling and estimation, questionnaire design, and interviewing. In the following pages we will focus on possible bias from the influence on research design of the investigator's own values as well as actual or perceived pressures from the community and research sponsors. In planning the sample survey and in analyzing and presenting its findings, it is extremely important to be aware of and, to the extent possible, correct for the biasing tendencies emanating from these sources.

One potential source of bias is the sponsorship or research support of a study. Many small-scale surveys, of course, are conducted by university classes or independent researchers on a shoestring budget, often from teaching or research funds at the university. The problems arising from sponsorship are minimal in such cases, although bias may still arise from the investigator's own values and social pressures. Large-scale household surveys, on the other hand, are very expensive and usually require outside support. In the United States and Canada budgets of $50,000 to $250,000 or more are typical. Such amounts are beyond the reach of most individual scholars and private institutes.

Though large-scale funding may be obtained in a variety of ways, two of the most common are *grants* and *research contracts*. In the first, the researcher typically chooses a problem growing out of his or her own interests and then seeks support from a governmental granting agency, a foundation, or the like. With a research contract, the initiative is more likely to be with the sponsor or funding agency which sets the problem of interest and often specifies the kinds of relevant information. The patterns of initiative and sponsor involvement differ markedly from one study to the next, and there are also other patterns. The basic question here, however, is the degree to which the researcher's own values and interests as well as the interaction between these and the supporting agency may influence a sample survey. The influence may extend from the choice of topics to the reporting and uses of the survey data.

The Choice of Topics

The patterns of sponsorship in social research and the interests of the researchers themselves affect both the subjects which are chosen for study and those which are not. Harold Orlans writes: "Every active researcher has an extensive agenda of work he would like to do sometime, but how often are his priorities the same as those of the granting agencies, and how far should he accept theirs?" (1967, p. 11). Obviously, the availability of funds draws survey researchers who find it difficult to operate as "independents" toward those areas of inquiry receiving support, and indirectly diverts them from others. Social scientists, for their part, often have a more activist and interventionist sociopolitical orientation than the population as a whole (Friedrichs, 1970). These perspectives clearly lead them to focus on areas fitting their definitions of "problems," and to steer away from less interesting terrain.

A more subtle problem related to the choice of topics is the restriction of the scope of the questions studied. This is a common problem with evaluation research on politically sensitive subjects. In the New Jersey-Pennsylvania Income Maintenance Experiment, for example, there were many possible outcome variables which could have been evaluated such as political alienation, marital satisfaction, anxiety, participation in voluntary organizations, reading habits, the time spent by parents with their children, and so forth. Yet, largely for political reasons, the evaluation conducted by Mathematica focused very heavily on the work-leisure tradeoff: ". . . whether workers would drop out of the labor force under income maintenance, trading leisure at slightly reduced income levels for work at slightly higher income levels" (Rossi, Boeckmann, and Berk, 1973, pp. 8-9). The reason for this emphasis seemed to be that the planners of the experiment expected that the work-leisure issue (whether individuals raised above the poverty level would cease to work) would be the foremost policy question raised by Congress and the American public. Such a limited focus arising from perceived pressures from sponsors would be attacked as bias by some and as the legitimate exercise of administrative judgment by others. But in either case, the scope of the research design is more narrow than it might have been under less restrictive conditions. Such a restriction of focus could also occur for career or personal reasons (promotions, gaining an academic reputation, etc.) unrelated to sponsorship. This example serves to illustrate the difficulty and complexity of conducting fair and impartial social research.

Another pitfall for social researchers is seen when, in their eagerness to demonstrate a hypothesis rather than test it, they omit from the research design variables which might provide disconfirming evidence. This could happen in a study of shift work if the research did not include questions about the advantages of night shifts (e.g., easier shopping during the day, more opportunities for hunting), and focused instead only on the liabilities. Researchers who see their studies primarily as an instrument for social action or alleviating social problems are particularly prone to the bias of one-sided design.

Conflicts of Goals

A frequent problem in applied or policy-oriented studies arises from the disparity between the goals of the sponsor and those of the researcher. Clients often want information that is "practical," with clear implications for action. Researchers, on the other hand, may share this concern for practicality, but also want to test theory or advance basic knowledge in the field. During our stay at the Survey Research Center of the University of Michigan, this tension between the theoretical and the practical was regularly satirized in the annual Christmas skits presented by the staff. One song went: "We don't test no theory with someone else's dough." In a concrete project this tension may be experienced by both parties. As Herbert Hyman suggests, "The analyst will no doubt feel some pressure as a result of this potential conflict with sponsor. In addition, insofar as he is also action-research oriented, he will incorporate some of this conflict within himself. Consequently, he must balance the gains from findings that have more immediate application but less theoretical value against the gains of findings with less immediate applicability to action or policy but with more relevance to the growth of fundamental knowledge" (1955, p. 46).

Bias In Carrying Out the Study

The relationships among the investigator, his or her research institute, and a sponsor, may influence many aspects of the survey process, including the conceptualization of the problem, the choice of hypotheses, the selection of the sample, the framing of the questions, the analysis of the data, and the interpretation and uses of the findings. In one sense there can be no unbiased research with survey methods or any others, for a study will always reflect the predilections of the investigator(s), the research technology and guiding concerns of the relevant disciplines at the time of the study, the constraints imposed by sponsorship, and various other influences. If we define "bias" to include such factors, the quest for unbiased research will be unending. At the same time, it is possible to provide a clear and open definition of a research problem, to draw up a sample meeting rigid standards of selection, and to collect information on the problem which satisfies interested observers on all sides. In other words, while every survey may be biased to the degree that it studies some questions and not others, and conceptualizes them in particular ways, it is still possible to meet reasonable standards of objectivity in sampling, measurement, analysis, and other aspects of the research. In practice this means taking concrete steps to avoid or counteract bias or distortion at critical stages in the research process.

The controversial Project Camelot represents a clear case of bias in the earliest stages of research, the conceptualization of the problem. The study was sponsored by the U.S. Army and carried out by the Special Operations Research Office (SORO) of the American University (see Horowitz, 1967). It was promoted by the Army as a means of collecting policy-related information on internal revolutions or "insurgency" in the less developed countries. Camelot ran aground, however, when it touched off an international furor in Chile, where critics denounced it as an affront to national autonomy.

The design of Camelot can justifiably be criticized on two grounds. First, the investigators introduced obvious bias by conceptualizing the research problem in a highly politicized manner, focusing on *insurgency* rather than on *social change* or some other more neutral problem. This bias was congenial to the sponsoring agency, which was then greatly concerned about revolutionary movements in Latin America. The study was also attacked for working with assumptions and hypotheses containing an implicit endorsement of stable and "moderate" Latin American regimes, such as those in

Venezuela and Chile in 1965. The lesson of Camelot is that when a researcher allows a study to take sides in an ideological conflict, he or she loses personal credibility and jeopardizes the research, educational, and technical assistance efforts of other social scientists.

Actual or perceived pressure from sponsors, as well as the researcher's own orientations, may also contribute to bias in the sample design of a survey. An example is seen in the study known as "A Profile of the Aging: USA" (Cain, 1967). This project received funds from the Foundation for Voluntary Welfare, a group generally regarded as opposed to public welfare programs. The study was billed as a nationwide survey to determine the needs and resources of the aged. Yet upper-class respondents were overrepresented in the sample, while the design excluded nonwhite persons over 65, individuals in nursing homes, and those receiving old age assistance—all groups in which the problems of aging are relatively severe. The results, published in 1960, suggested that older citizens were more healthy and more satisfied with existing programs than had been indicated in earlier studies. The research received national attention when the findings were widely broadcast by the American Medical Association in its campaign against federal programs for medical care of the aged (Medicare). By contrast, no charges of bias were raised against a study of medical expenses, insurance, and debts carried out by the Survey Research Center at the University of Michigan with direct support from the American Medical Association (Katona, Lininger, and Kosobud, 1963). The absence of charges of bias does not, of course, certify objectivity, but the example does suggest that surveys can be carried out on sensitive topics to the satisfaction of parties with widely varying ideological viewpoints.

Analysis and Publication

Once the basic data have been collected, the researcher must face a series of decisions about how they will be analyzed and interpreted, and where the results will be published. At each choice point, pressures from the sponsor or other external agents as well as the investigator's own values can influence the outcome of the decision. For example, in attempting to reduce a mass of survey data to manageable proportions, the analyst may adopt a global index that arbitrarily places one group in a more favorable light than another. Some composite measures of "authoritarianism" seem to portray members of the working class, Catholics, and other groups as relatively authoritarian when different indicators of the same concept would produce different results. Similarly, to maintain a cordial relationship with a sponsor, the study directors may soft-pedal those portions of the results which might jeopardize the sponsor's interests or tarnish his public image.

Finally, decisions about whether, when, and how to publish the results of a survey may be affected by similar pressures. Should the findings be published at all, or should they be confined to confidential reports to the sponsor? Surveys resulting only in confidential reports are likely to raise questions about the scientific integrity of the data. Some investigators and research organizations solve this dilemma by restricting access to the data for a period, such as a year, and then publishing the results and/or opening the files to interested researchers. Then again, should brief, popular summaries be issued at the risk of oversimplification or misinterpretation (both real possibilities with highly controversial studies), or should the findings be confined to scholarly journals, at the potential risk of oblivion? Should the data be released while they are fresh and vibrant, or should they be allowed to "cool" for a period? These are common dilemmas in studies bearing on public policy or group interests; they become even more complicated when there is a close relationship between the investigator and one or more of the interested parties.

The problems of bias in the analysis and reporting of survey data are particularly serious when the sponsor is an agency directly responsible for policy decisions in the area covered by the research. The greatest danger is that the researchers will unwittingly allow themselves to be used to further the mission or distinctive viewpoint of the agency. The temptations in this regard are greatest when the investigators write to *persuade* rather than to *report*. In persuasion, the data are treated as a vehicle for moving reluctant officials to action. To achieve this end the writer often condenses, omits technical details, and relies on prose appealing to moral sentiments and political sensibilities. In the case of writing to report, the approach is one of balanced presentation, with due attention to the complexities of the data. The result is usually more accurate and more dull. Simplification and rhetorical appeals clearly have their merits in the advocacy of policy, but they also invite charges of nonobjectivity and attacks by parties who feel that their image was misrepresented or their interests compromised in the report.[1]

Preventing or Counteracting Bias

At a very general level one can readily prescribe intellectual honesty and openness as the first line of defense against bias in survey research. However, since many of the personal and external pressures toward bias are often

---

[1] The advantages and disadvantages of persuasive reporting are illustrated by the controversial "Moynihan Report" in the United States. Its actual title was *The Negro Family: The Case for National Action*. The author, then an Assistant Secretary of Labor, chose to use existing data to make a persuasive "case for national action." The ensuing debate is reviewed in detail in Rainwater and Yancey (1967).

unrecognized by the researcher, and since most of the distortions are unintentional, this advice is of little practical value. More concretely, several operational steps can be taken to counteract bias.

First, at each relevant stage, including the initial formulation of the questions to be studied, the research plans should be subjected to the scrutiny and comments of a broad spectrum of interested parties. These commentators can be asked, for example, if the way in which the problem is posed, or the concepts to be translated into questionnaire items, or the preliminary interpretations of the findings carry ideological overtones. As a rule of thumb, if groups that are ideologically opposed to each other, or otherwise in dispute on the questions at hand, such as Medicare for the aged or income maintenance, agree that the research is fair and will produce objective information, the study is not likely to suffer from significant ideological bias.

One of the authors was engaged in such a procedure in a study of the disincentive effects of unemployment compensation benefits in which interviewing was carried out by the Michigan Survey Research Center (Lininger, 1963). At this time (1955) the question of unemployment compensation was politically sensitive in Michigan, and of immediate and direct interest to the principal parties in a major group under study, the auto industry. Before the research could proceed it was necessary to obtain concurrence on the questionnaire from both the auto manufacturers and the major contending power, the auto workers union (UAW/CIO). Following careful review, both groups approved the questionnaire as well as the interviewing, and the study moved ahead. Significantly, however, the union decided to attack the research as a preventive measure in case the results came out unfavorable to its interests. In the end, neither side quarreled with the findings (Yntema, 1957; Lininger, 1963).

Second, expert opinion should be sought on the adequacy of the sample design (see Chapter 4). The best assurance of a sample not biased in favor of one or another group is strict adherence to the canons of probability sampling. Consultation with a sampling specialist, which is advisable in any event, will quickly uncover potential sources of bias in the design.

Third, the draft questionnaire should also be circulated to a group of technical experts for comments on the objectivity of the items. Colleagues might be encouraged, for example, to take the role of devil's advocates of various sorts in criticizing the items. The interest group representatives mentioned earlier could likewise be brought in at this stage, perhaps with the technical experts. In short, the most effective means of promoting objectivity and fairness in survey research, apart from personal and professional integrity, is to build in a process of consultation with experts as well as interested (and preferably opposing) parties at each stage. If this is left to the end it may be too late.

EQUIVALENCE IN CROSS-CULTURAL SURVEYS

This book is written with the hope of contributing not only to conventional single-country surveys, but also to comparative cross-cultural research. One of the most difficult problems faced in planning a comparative survey is that of ensuring equivalence in concepts and research procedures across the various cultural units studied. Some of these problems will be discussed more thoroughly in later chapters. Here we will suggest the range of issues to which the investigator should be sensitive in the early stages of planning. To oversimplify somewhat, we may classify these issues under the headings of conceptual and methodological equivalence.

We have already alluded to some of the difficulties in attaining comparability in the concepts used in multicountry studies. Sometimes a concept such as "God" exists in all of the cultures in question, but the problem for research purposes is that it means different things in different places or is more salient in one than the other. Moreover, in certain cases, concepts may lack exact equivalents in all of the units to be investigated. Alex Inkeles (1971) points up this problem in describing his six-nation study of modernization:

> Consider, for example, the idea of "influencing" someone. A number of our questions about politics and the realm of opinion required our interviewees to say whether or not they had ever tried to "influence" a politician, or whether people like them could "influence" the course of government action. It proves almost impossible to render the concept of influence, as we meant it, in Yoruba. The culture seems to be quite familiar with the idea of commanding someone, or forcing him, or begging him, or pleading with him, but not of influencing in the sense in which we commonly use the word in English . . . The idea could be expressed only by a substantial circumlocution, and then only very inexactly. Similar problems existed in other societies in rendering concepts such as skill, equality, dignity, human nature, responsibility, and numerous others.

This example illustrates the blurred line between concepts and methodology in cross-cultural surveys, for the nonequivalence of concepts becomes most apparent during the technical stage of translation. Nevertheless, we might note the principal ways in which the methods used in each country should be equivalent (rather than formally identical).

First, explicit attention should be given to the choice of countries or other cultural units. For example, the inclusion of Nigeria as the only African country in a five-nation survey would not permit us to generalize about sub-Sahara Africa. Similarly, if a study hopes to draw conclusions about "developing nations," it is essential that the countries included be chosen in such a way that they represent this population. Five countries can be chosen by experts to represent a range of differences in geography and culture, and still not permit generalizations to all "developing countries." Second, the sampling carried out within countries must be comparable for all of the units under study. One of the major drawbacks in the well-known study, *The Civic Culture* (Almond and Verba, 1961), is that the sampling frames and selection procedures varied considerably across the five countries involved. For instance, though the target population was defined as adults eighteen years of age and over, the Mexican sample excluded persons living outside of cities with a population of 10,000 or more. Third, the questions should be equivalent in meaning from culture to culture, rather than formally identical in translation. Fourth, the criterion of equivalence should also be applied to the characteristics of the interviewers, the training and supervision which they receive, and the actual field procedures used. For example, serious bias may be introduced if the interviewers work in close and visible cooperation with local governmental officials in one country and not in the others. Fifth, the study plans should devote explicit attention to the question of minimal rates of completion and maximal rates of refusals. The analysis of the data will obviously be complicated if the response rate in one country is 50 percent and in the others 85 percent. Sixth, plans should be developed to ensure comparability in the coding, keypunching, and statistical analysis of the data if these tasks are to be carried out at the local research sites. For example, the coding of "open-ended" questions inevitably requires on-the-spot decisions about the classification of ambiguous responses. In studies where the coding is carried out in five or ten different locations with no constant supervision it is more than likely that these ambiguities will be resolved differently in each country to the detriment of equivalence.

After reading this discussion the reader may feel that survey planning requires an omniscience far beyond the reach of most mortals. It is true that careful planning is an extremely challenging task, and that there is a seemingly infinite number of factors to be taken into consideration, ranging from the vicissitudes of culture to local weather conditions. But it is equally true that as one gains experience with surveys many of these questions will arise automatically, and others are solved before they arise. Nevertheless, at the beginning of a study it is important to be as comprehensive and panoramic as possible in anticipating ultimate needs and possible difficulties. We hope that this discussion contributes toward that end.

FURTHER READINGS

Hyman, H. H., *Survey Design and Analysis: Principles, Cases, and Procedures* (Glencoe, Ill.: Free Press, 1955).

Sjoberg, G. (ed.), *Ethics, Politics, and Social Research* (Cambridge, Mass.: Schenkman, 1967).

Sudman, S., *Reducing the Cost of Surveys* (Chicago: Aldine, 1967).

Chapter 3

Survey Design

The design of a survey is a prearranged program for collecting and analyzing the information needed to satisfy the study objectives at the lowest possible cost. There is no ideal survey design in the abstract, divorced from the particular goals of the research. Some investigators planning their first survey may regard the cross section at a single point in time as the only practical option, given their inexperience, or the most appropriate because it is the most frequent. In fact, many possibilities are open, and even the onetime cross section admits of several variations, though some options may be out of the question for political, ethical, or financial reasons. The design finally chosen should be consciously tailored to the overall objectives of the study and the exact types of information needed, and should also take account of the various methods to be used in gathering this information, such as a personal interview, a mail questionnaire, or a telephone interview.

RESEARCH OBJECTIVES

Most sample surveys pursue one or more of the following objectives: exploration or clarification of the dimensions of a problem; description of groups, individuals, events, or phenomena; causal explanation; hypothesis testing; evaluation; predicting or forecasting future events, and the development of social indicators. The designs required to attain each of these objectives may be similar or different, depending on the exact kinds of information that they require from the survey. A design appropriate for description may be too sophisticated and costly for the clarification of a problem through exploration and too simple for causal analysis. Let us briefly consider each of these objectives.

Exploration

Humility is a virtue in survey research as elsewhere. There are times when we may be vitally interested in a problem, but simply not know enough about it to design even a rudimentary descriptive study. In these cases it is better to clarify the problem itself before taking to the field. Selltiz et al. suggest a number of functions that may be served by an exploratory study:

> . . . increasing the investigator's familiarity with the phenomenon he wishes to investigate in a subsequent, more highly structured study, or with the setting in which he wishes to carry out such a study; clarifying concepts; establishing priorities for further research; gathering information about practical possibilities for carrying out research in real-life settings; providing a census of problems regarded as urgent by people working in a given field of social relations [1959, p. 51].

The sample survey is usually not the most appropriate vehicle for pursuing essentially exploratory research. The informational payoff is too low given the high costs and imposition upon the public. One can, of course, embark on a nationwide or even cross-national study based upon ten or twelve questions pitched at this level: "What sorts of things upset you most these days?" or "What do you think is the main thing wrong with this country?" or "How do you think children should be raised?" However, such "dragnet" questions are deceptively simple and relatively wasteful unless the range of responses is known in advance. One reason is that the interviewers must be instructed on the level of information that constitutes an *adequate* response. Individual respondents will cover the full range from the laconic to the garrulous, so that a policy must be set on when to probe and when to stop (see Chapter 6). If interviewers in California reach one decision and those in New York another, or worse, if it is left to the discretion of each person, the resulting information will be extremely difficult to code. Usually it is more productive to conduct a few informal, unstructured interviews with knowledgeable persons before attempting a large, cross-section sample.

This procedure was followed in a four-year study at the Michigan Survey Research Center on the effects of shift work (Mott, Mann, McLoughlin, and Warwick, 1965). During the first year the investigators worked at defining and clarifying the various ways in which an afternoon shift, night shift, or rotating shift might affect the worker and his family. Interviews were held with managers, workers, union leaders, and others who helped to articulate the dimensions of the problem. Subsequent pretests combined structured questions with further qualitative probing. After the staff was satisfied that they were familiar with the range of physical, psychological, and social consequences of shift work, the results of their exploratory probing were translated into a series of structured questionnaire items and the study moved to several large field sites.

Description

The aim of descriptive surveys is to arrive at a precise measurement of certain phenomena, such as political party preference, neurosis, racial prejudice, or divorce. This type of survey is often the subject of an invidious comparison with explanatory studies, with one caricatured as "fact-grubbing" and the other exalted as "development of theory." Such a dichotomy is misleading. Descriptive and explanatory surveys are often identical in their operations, distinguished more by intent and use of the data than content. One man's description is grist for the next man's theory, as happens when an essentially descriptive study of attitudes in one country is replicated in another to test the effects of culture or social structure.

An essential task shared by descriptive and explanatory studies is conceptualizing the phenomenon in question. If a psychologist wishes to determine the frequency of neurosis in the United States population, he must first clarify what he means by the term. This task will rapidly take him into the realm of theory, for he will have to decide on *theoretical* grounds if neurosis consists of a set of presently observable behaviors, or only those presently observable behaviors produced by early experiences. If he decides on the second alternative, he may have to abandon the survey and seek out a method that is better equipped for assessing childhood experiences and their effects. The problems of definition in this example are the same whether one wishes to describe or explain neurosis. The relationship between description and explanation in survey research has been aptly summarized by Herbert Hyman:

> The descriptive survey is thus a training ground for the development of skill in conceptualization of the phenomenon and in the treatment of the findings in relation to error factors, both essential to effective analysis of explanatory surveys. But there is still another feature of the descriptive survey which is especially valuable for ultimate work in the design and analysis of explanatory

surveys. Out of the findings of such surveys often comes the basis for the formulation of fruitful hypotheses about phenomena, or at least for some reduction in
confusion about phenomena [1955, pp. 77–78].

Description, in other words, can lay the groundwork for the pursuit of other
objectives, including explanation and hypothesis testing, evaluation, prediction, and the development of indicators.

Causal Explanation

A sample survey may take as its main goal the causal explanation of a
phenomenon or situation. Both the terms "explanation" and "cause" have
had a stormy history in philosophy as well as the social sciences. At the
simplest level, to *explain* something means to make sense out of it, to make
it comprehensible. But there the trouble begins, for there are no fixed criteria
as to what constitutes comprehensibility. Some sociologists and historians,
such as Dilthey and Weber, place strong emphasis on the role of understanding or *verstehen* in providing adequate explanation. Others, including
many adherents of logical positivism, hold that the true test of explanation
lies in prediction. In their view a phenomenon, such as suicide or murder
rates, is not explained until it is set within the context of a predictive theory,
including a logically related set of hypotheses. Hence, it often happens that
what makes sense to one social scientist, and thus satisfies his or her standards of explanation, may remain quite incomprehensible to another. For
present purposes, we would only note these different views of explanation,
and urge the potential user of survey research to be clear about his or her
own criteria (cf. Sjoberg and Nett, pp. 288–298).

The word "cause" has been no less troublesome in philosophy and the
social sciences, particularly since David Hume's demolishing attack on the
very notion of causation. Nevertheless, the concept has persisted, and is
often linked to that of explanation. For many social scientists the most
convincing evidence of explanation is the establishment of a causal relationship between two or more variables. For example, the study of shift work
mentioned earlier (Mott et al., 1965) sought to explain how an employee's
work schedule affects, i.e., is causally related to, several aspects of his attitudes and behavior, including difficulty in carrying out his obligations as a
husband and father, marital happiness, job satisfaction, and physical health.
The resulting survey of industrial workers in the United States was designed
to obtain a variety of evidence, both quantitative and qualitative, which
would allow the establishment of causal links.

To establish a causal connection three conditions must be satisfied.
First, the assumed cause and effect must be associated with each other—

when A is present B must also be present, at least under specified conditions. Second, the cause must precede the effect, or at least the effect should not precede the cause. Third, all other possible explanations of the effect must be ruled out.[1] Thus, to conclude that shift work causes marital dissatisfaction, the researchers had to show that work schedules and marital satisfaction are correlated; that dissatisfaction increased after a man moved to an afternoon, night, or rotating shift; and that this heightened dissatisfaction was not produced by factors related to shift assignments, such as age, seniority, education, or physical health. Close analysis did suggest, in fact, that some workers had chosen a night or afternoon shift as a refuge from preexisting marital problems. When they were removed from the analysis, there was little evidence of negative effects of shift work on marital satisfaction. This finding underscores the importance of establishing the time order of the events under consideration. If marital dissatisfaction exists before a worker moves to a nonday shift, and is not aggravated after the change, it is difficult to make a case for the causal impact of job hours on this condition.

In survey research there are various approaches to causal explanation, some depending on the practical opportunities open to the researchers, others on philosophical definitions of explanation. One, to be discussed shortly, is formal hypothesis testing. Another, which is almost always combined with hypothesis testing, is the use of experimentation involving survey methods. For some investigators, on the other hand, the essence of explanation lies in developing a rich and nuanced qualitative picture of the events under study, rather than in testing a preconceived hypothesis. Data from survey interviews, particularly those including a number of open-end questions, might thus be treated as a source of qualitative as well as quantitative insight into the problems under study. Indeed, there are many survey researchers who feel that the most compelling explanations are those which effectively draw together numerical data and qualitative materials.

Hypothesis Testing

The testing of hypotheses is one of the most frequently used approaches to causal explanation. It is particularly favored by those whose prime criterion of explanation is a logically interrelated set of hypotheses leading to accurate prediction. An hypothesis is, in essence, an empirically testable statement, that is, one which can be refuted or supported by empirical data. Most often it is expressed in the form of a proposition stating a causal relation-

---

[1] A thorough discussion of causality, which is beyond the scope of this book, would require numerous qualifications and extensions of these three statements. For a relatively brief and useful discussion of causality and social research see Selltiz et al., pp. 80–88. For more comprehensive treatments see Blalock (1967, 1971).

ship, such as: the movement of a worker from a day to a nonday shift will cause an increase in marital dissatisfaction. For those who place a high value on the development of axiomatic theory, an hypothesis should be derived from a set of postulates and assumptions about the sphere of behavior in question, such as working hours and marital satisfaction.

Whatever one's predilections for theory building, the statement of hypotheses has great practical value in planning and designing the sample survey. The main advantage is that it forces the investigator to be clear about what he expects to find through the study, and raises crucial questions about the data that will be required in the analysis stage. As a practical exercise in planning, therefore, it is often helpful to spell out the predicted or expected lines of influence among the major variables in the study including their reference periods or the time order among them. If the study directors of an explanatory survey cannot set forth at least a few hypotheses of this type, the research may require further conceptualization before it is ready for concrete discussions of design.

Evaluation

The sample survey is increasingly being used, often in conjunction with other methods, as a tool for evaluation research. This research can be defined as the assessment of the process and/or consequences of deliberate and planned interventions. The most common application involving surveys is with social programs such as Head Start in the United States, organized efforts to help couples plan their families, and organization development programs in many countries.

Edward Suchman points out three elements that must be present for an evaluation to take place:

> (1) an objective or goal which is considered desirable or has some positive value; (2) a planned program of deliberate intervention which one hypothesizes is capable of achieving the desired goal; and (3) a method for determining the degree to which the desired objective is attained *as a result of* the planned program [1972, pp. 53-54].

We would add that in some evaluations, particularly those launched by the auditing agencies of governments, the emphasis is as much on the *process* as the *product* of planned programs. In assessing a program of foreign assistance, for example, the General Accounting Office of the U.S. Government (an agency which rarely uses sample surveys) may be as much concerned with the degree to which the program under scrutiny adheres to proper procedures in spending money as in the final outcome of the expenditures.

From the standpoint of survey design, a great deal will depend on the precise objectives of a given evaluation study. The following distinctions suggest different objectives with crucial implications for design.

1 *Formative versus summative evaluation.* Evaluation research is *formative* to the extent that it attempts to provide a diagnostic basis for revising, improving, or guiding an ongoing program. *Summative* evaluation, by contrast, aims to provide an overall appraisal of a program, indicating the major outcomes, the relationship between the outcomes and the intended goals, with perhaps some comment on unintended consequences, and the process by which the outcomes were achieved. While the former places great emphasis on feedback to an action program, with the possibility of shifting course on the basis of the results, the latter focuses mainly on assessment at a given point in time. Recommendations oriented to the future (e.g., continuing or discontinuing the effort, expanding or contracting, etc.) may follow from both, although the summative evaluation more likely will have this as a primary purpose. Given these differing emphases, the design of a formative evaluation survey would call for data collection at several points in time, and would make it difficult to conduct a tightly controlled experiment aiming to measure the precise effect of an intervention.

2 *Evaluation of effects versus process.* The most common type of evaluation study is one which emphasizes effects—that is, the extent to which the intervention achieved its original objectives. One of the greatest challenges in designing surveys to measure effects is to create the conditions which will allow for the attribution of causality to the program. For example, if a government introduces a massive family planning program, and the country's fertility subsequently declines, it is not necessarily true that the fertility decline was an *effect* of the interventions. The same decline may have occurred if the government had done nothing, or more likely, may have occurred more slowly or at greater personal cost. Other studies are more concerned with the process or means by which the intervention was accomplished. An organization development program might be evaluated, for instance, by showing how many executives participated in problem sessions or sensitivity training or by reporting the major topics discussed and interactions that developed, rather than by documenting changes in productivity or some other measure of performance. For purposes of evaluation, the process would be treated as the product.

3 *Short-term versus long-term evaluation.* Evaluation studies also differ in their temporal focus. Some are concerned only with the impact of an intervention on immediate targets, while others relate it to long-range goals. The time horizons of the evaluation have obvious implications for design. In many respects the short-term assessment presents fewer problems, but runs the risk of excessive narrowness. An evaluation of a national family planning program after one year, for example, may provide optimistic results by giving undue weight to "early acceptors"—those with a strong inclination to use family planning who had lacked only the means to do so. If the same study were extended over a period of five or ten years the results might be far less positive. On the other hand, studies which attempt to project program impacts far into the future come closer to the goal of prediction than evaluation.

Evaluation research involves many more intricacies than can even be alluded to in this short discussion. The interested student is invited to consult the growing literature in this field, which now amounts to a subdiscipline in the social sciences.[2] One question, however, is particularly salient for the design of an evaluation survey. This concerns the precise variables that are to be examined in assessing the outcome of a program—*what* is to be evaluated. This, in turn, requires a clear conception of the goals or values the program is attempting to promote. Suchman writes:

> This problem of defining the values underlying a program is perhaps the most subjective and difficult aspect of evaluative research. For one thing, values differ for different people. The values of program personnel may differ from those of their clients or of the public. Furthermore, different program personnel and different segments of the public may have conflicting values among themselves, as witness the current controversy over "middle class" professional values as opposed to the "lower class" values of the poverty class [1972, pp. 67-68].

The question of what is to be evaluated raises fundamental problems of design, as well as of politics and ethics. Concretely, should the evaluation researcher accept the criteria of success provided by the program administrators, or should he or she apply independent criteria as well? If the intervention was aimed at a variety of social groups, such as blacks, Chicanos, and native Americans in the United States, should their criteria of success be incorporated as well? These are questions which should be faced squarely in the planning stage, but they will also have critical implications for design. For example, an educational program designed for poor children might be evaluated on the basis of standardized scores obtained from tests given before and after the intervention. This design would typically reflect the criteria of success held by school administrators and perhaps other professionals in the field of education. But it would also be possible to broaden the scope of the research by obtaining the qualitative reactions of the children's parents to the program, covering such questions as the child's apparent attitudes toward learning, relations with other children in the family, self-respect, and other more elusive criteria. The second design, which might be termed multiperspective evaluation, would obviously be much more com-

---

[2] See, for example, Rossi and Williams (1972); Williams (1971); and Zurcher and Bonjean (1970). The collection edited by Weiss (1972) provides a helpful overview of the most significant issues.

plex than the first, but might also provide a more sound basis for evaluating the educational program.

It should also be emphasized that the choice of criteria involves basic political decisions. Almost any action program will appear better on some criteria than on others. As suggested in Chapter 2, there may be very strong pressures from the sponsors to choose those criteria that will cast the intervention in the most favorable light possible, particularly if the program is politically controversial. Similar pressures may arise from the investigator's own values, particularly if he or she is partisan to the program or the executing agency (cf. Weiss, 1972; Warwick and Kelman, 1973). In designing an evaluation survey, therefore, it is essential for the independent researcher not to become an unwitting victim of cooptation. The best defense against a design biased toward the interests of one or another partisan group is extensive consultation with all interested parties, particularly the program's opponents.

Prediction

Another common objective of survey research is to provide the basis for predictions about some future event or situation. The most familiar example is the public opinion poll used to forecast election results. In this case, a survey of attitudes and expected behaviors on the part of a sample of respondents forms the basis for statements about the probable voting patterns of an entire electorate. Other predictions simply involve the extrapolation of descriptive data about the present to a specified point in the future. In demographic forecasting, for example, information about the present size, age distribution, and age-specific fertility and death rates of a country's population are used to estimate the total population at some future date, such as the year 2000. There are also various other modes of prediction (Bell, 1964).

One issue that should be of particular concern in designing a predictive survey is the impact of the survey itself on the behavior to be predicted. There is no doubt, for example, that preelection polls in the United States and other democratic nations are often used as a vehicle for influencing the outcomes of elections. In *Silent Politics* (1972), Leo Bogart catalogues the many ways in which polls can not only inform but shape public opinion and related electoral decisions. For instance:

    1   They provide candidates for office with intelligence about where they are strong and where they are weak.

    2   They offer guidelines as to what candidates should and shouldn't say to win voter approval.

3 More questionable use of preelection research is the carefully timed, last-minute release of polls to influence convention delegates with demonstrations that Candidate X couldn't possibly win, whereas candidate Y might.

4 A less guileful but more important effect of polls as an influence on elections comes about when they create the impression that the outcome is a sure thing three or four months before the election itself. What this does to demoralize party workers and to drive away potential contributors to party funds represents a grim example of self-fulfilling prophecy [1972, p. 27].

The use of polls takes on particular political and ethical significance in closely contested elections where the survey results may sway the independent or undecided voters. Needless to say, the sample design as well as the way in which the questions are written will have a great deal to do with the outcome of the poll. During the 1968 United States presidential elections, George Wallace became so frustrated with preelection surveys that he cried: "They lie when they poll. They are trying to rig an election."

There are many other problems in using survey or census data to predict the future. One is that the factors accounting for a trend uncovered in the survey may themselves change over time. This problem is particularly evident in the field of population projections, where unexpected shifts in attitudes and socioeconomic conditions may produce marked changes in fertility. The recent history of population projections in the United States is but one example of the precarious nature of demographic forecasting. Another difficulty is the self-defeating prophecy. This occurs when the survey results lead to a change in public or official awareness which, in turn, reduces the likelihood of the predicted behavior. For example, after being told by a survey that drug use is increasing in high schools, a community may take positive steps, through legislation, law enforcement, education, or other means, to reverse the trend. Another example comes from short-term economic forecasting. A survey may show that families are becoming increasingly concerned about the future in ways that make them cautious of making major new purchases. These findings might convince policy makers to take offsetting action to revive consumers' willingness to maintain their previous spending rates. Nevertheless, with proper design, awareness of reactive effects on the environment, and due recognition of the limits of forecasting, the sample survey can be a helpful tool in making predictions.

Developing Social Indicators

Beginning in the 1960s a group of social scientists, legislators, and others launched a movement for the development of "social indicators." The aim, in general, was to provide the social and political equivalent of such familiar economic indicators as unemployment rates or the wholesale price index (cf. Bauer, ed., 1966). While there has been considerable ambiguity from the beginning as to just what differentiates a social indicator from the more ordinary statistic, the consensus seems to be that an indicator must be a "socially significant" measure repeated at regular intervals. "In other words, social indicators are time series that allow comparisons over an extended period and which permit one to grasp long-term trends as well as unusually sharp fluctuations in rates" (Sheldon and Freeman, 1972, p. 166).

In these discussions the sample survey is often proposed as one of the most promising sources of indicators. Through repeated surveys, it might be possible to obtain a continuing series on such measures as the health and educational attainments of different groups in the society, job satisfaction, leisure-time activities, crime, satisfaction with governmental services, and the adequacy of public transportation. The survey, as noted in Chapter 1, is indeed an appropriate method for gathering information on many of these subjects.

As in the case of evaluation research, however, the indicator movement has strong political overtones. *What* sphere of activity is chosen as an indicator and *how* it is defined and measured will reflect sociopolitical values (Henriot, 1970). The design of the samples and the timing of the repeated surveys are also not likely to be matters of indifference to politicians and interest groups.

> In short, when used for purposes of setting goals and priorities, indicators must be regarded as inputs into a complex political mosaic. That they are potentially powerful tools in the development of social policy is not to be denied. But they do not make social policy development any more objective. . . . In a situation where all sides have equity of resources to gather, interpret, and communicate indicator information, it could be argued that social indicators can serve to develop a more rational decision-making process in social policy development. But this is unlikely to be the case very often, and in instances of unfair competition indicators are essentially a lobbying device [Sheldon and Freeman, 1972, p. 169].

Here and elsewhere both the designer and the consumer of survey research should be alert to the political ramifications of the content as well as the methodology of a given study. Precisely what constitutes a social indicator is not self-evident and should be subject to careful scrutiny.

STUDY OBJECTIVES AND SURVEY DESIGN

Few surveys have but a single objective. Many try to gather data that can be used one way for description, another for causal explanation, and perhaps a third for evaluation or prediction. Similarly, the design of research on social indicators must include not only the definition and construction of the indicator in the first survey, but also the design of subsequent surveys in a time series. The challenge in survey design is to devise a program of data collection which fulfills the greatest number of central objectives without exceeding the resources available.

As should now be apparent, there is no one-to-one fit between the study objectives reviewed thus far and the various types of survey design. Ideally, for example, an evaluation should be carried out by a carefully controlled experiment in order to obtain solid evidence on causality. However, for political and ethical reasons such experimentation with human subjects is often impossible. Hence, the best design available, under the circumstances, may be the quasi experiment (Campbell and Stanley, 1963), or even an ex post facto analysis. The following discussion considers two major categories of survey design which can be applied to a variety of objectives: the single cross section and designs for assessing change.

The Single Cross Section

This design involves the collection of information at a single point in time from a fraction of the population selected to represent the total. Within this design the researcher must determine whether to use mail questionnaires, telephone interviews, or personal interviews at each selected household, or to tailor the study to his objectives in other ways. The results can be used, with varying degrees of confidence, for describing the population under study, attempting to trace causal connections, evaluating an action program, predicting a future state of events, or for the development of one or more social indicators.

The shift work study cited earlier was a cross section at a single point in time designed primarily for explanatory analysis. The aim was to trace the impact of nonday working hours on the worker's health, attitudes, leisure-time pursuits, and family relationships. The difficulty with this design, and all other cross sections at only one point in time, was that it did not provide direct, time-ordered evidence on causality as would be available from an experiment. Instead, in seeking out causal connections the study directors had to rely on respondents' recollections of key conditions and other indirect information, such as comparisons of the responses of men on shift work with those on daylight schedules. Thus, while there were no before and after measures showing that men *changed* when they moved from a day to an afternoon or night shift, the findings did show significant differences between day and shift workers on health and social relationships. These findings strongly supported the attribution of causal influence to working hours.

The most critical limitation of the single cross section, as suggested by the example, is the absence of independent measures from different points in time. Nevertheless, social science methodologists have been able to devise helpful and sometimes ingenious means for inferring causality from single cross-section data. The most obvious means is to divide the sample into meaningful subgroups and then to seek out similarities and differences. For example, it is possible to obtain a rough indication of the effects of education on income by comparing the levels of income of persons with different levels of education. Higher average incomes at successively higher educational levels would, other things being equal, provide evidence of the economic consequences of education. In Chapter 11 we will review some statistical techniques which will help to account for those things that we can expect *not to be equal* in order to obtain a more refined test of the impact of education on income.

Several variations may be introduced to enhance the usefulness of the cross-section design at a single period in time. The first consists of *parallel samples* covering several different "target populations" in the same study. This approach is well illustrated in Frey's study of Turkey:

> The fundamental sampling unit was to be the individual villager, or peasant, and not the family or the household head. Even so, the study was constructed so that we would emerge with three separate samples rather than merely the one sample of the peasantry. Our teams traveled to 458 villages, completing in each case a separate schedule of information about the village as a whole, thus giving us, after some statistical adjustments, comprehensive data on a sample of Turkish *villages*. Moreover, in addition to the designated set of interviews with the sample of villagers in each of these villages, our teams also were instructed to obtain a series of *elite* interviews in each sampled village. These additional interviews were four: with the village head man (*muhtar*), with the village priest (*hoca* or *imam*), with the legal wife of each, regardless of whether such individuals turned up in the regular sample. Thus, the investigation was constructed so as to yield (1) a regular sample of Turkish peasants, (2) an elite sample of certain formal village leaders and their spouses, and (3) a sample of the village communities of the country. The findings on the two added samples were obtained at very low marginal cost and markedly increased analytic opportunities [1963, p. 339].

Another variation is *oversampling* certain subgroups in the population in order to obtain enough cases to carry out a separate analysis. For example, if an investigator wishes to carry out a cross-sectional study of Toronto, Ontario, and at the same time compare the Portuguese minority with other citizens, it would be necessary to use a higher sampling fraction in known Portuguese districts than elsewhere. An equal probability sample would produce so few cases that detailed analysis of this minority would be impossible unless the total sample size was very large, e.g., 10,000–15,000. This procedure is discussed in Chapter 4.

A third option lies in the use of *contrasting samples* in addition to or in place of sampling from the entire population. This is basically an extension of the principle of oversampling but is applied to those groups which are likely to show the greatest differences on the central variable of the study. A contrasting design was used by the Survey Research Center in a study of public attitudes toward the proximity of atomic energy installations. The sample consisted of communities located very close to these installations and those at a considerable distance. The advantage of this approach is that it will often sharpen the differences seen in the results. The drawback is that such sharpening may be achieved at the expense of generalization to the entire population (unless a regular cross section is included as well).

In another approach, both descriptive and explanatory analysis as well as evaluation can be improved by combining the usual cross section with an intensive study of *natural clusters* such as schools, factories, hospitals, or army units. For example, a national survey of the attitudes and values of college students might include two interpenetrating parts: (1) a probability sample of 2000 individual students drawn from a master list of persons registered at all colleges and universities in the country, and (2) a probability sample of 20 colleges and universities for use in an intensive analysis of institutional or contextual effects on attitudes and values. The interviews in the second sample might collect several types of information not available through the first, including sociometric data on the network of interpersonal relations in the school, and quasi-anthropological data on the norms, social pressures, and general academic climate in the student community. This latter type of information was most helpful in establishing apparent causal connections in a study of Bennington College (Newcomb et al., 1967), though this study focused on only one school. The "Coleman report" on the equality of educational opportunity in the United States approximated this design by collecting data on the overall characteristics of schools as well as their students (Coleman et al., 1966).

Designs for Assessing Change

The sample survey may be used to describe, explain, evaluate, or predict change. Explanatory and evaluative studies of change fall broadly into two categories: *experimental* analyses attempting to appraise the effects of an event which occurs between the first and later measurements, and *nonexperimental* causal analyses aimed at showing the dynamics of change when the variables in the study have not been systematically manipulated. Experiments may be *natural*, as when an economist studies the effects of an economic recession through repeated measurements, or *arranged*, as in a before-after study of the effects of a public information campaign. Nonexperimental studies are by far the more frequent. Examples include studies of changes in spending and saving habits among consumers, shifts in voting, and changes in attitudes among students during the college years. The major design options for studies of change include successive independent samples, panel studies, and other combined designs.

Successive Samples   Successive samples involve the selection and interviewing of two or more different samples drawn at different times from the same population. For example, a study might use different samples of United States citizens in 1965, 1967, 1969, 1971, and 1973. When data from these have been collected at several points in time, rather than just once, they are said to form a *time series* and may be used for *trend analysis*. This design may be summarized schematically as follows:

Successive Samples from the Same Population

Time 1	Time 2	Time 3	Time 4
Sample A	Sample B	Sample C	Sample D, etc.

The successive sample design is most appropriate when: (1) the aim of a study is primarily to describe changes in the attitudes and behavior of a population; (2) when the surveys are used to carry out a largely descriptive analysis or evaluation of the effects of a natural or arranged intervention, such as a war, a change in business conditions, an election campaign, or a governmental change program; and (3) when a population is to be monitored repeatedly over a long period of time. This design was used by the Survey Research Center in two long and well-known series: (1) the national studies of consumer attitudes and behavior begun in 1946, and (2) studies of voting behavior, with the first in 1948. Other examples include a study by the Survey Research Center on changes in public attitudes toward atomic energy and atomic bombs following the Bikini tests in 1946, and an evaluation of the impact of a publicity campaign about the United Nations on the residents of Cincinnati, Ohio (Star and Hughes, 1950). In evaluation studies such as these, it is possible to use two equivalent samples from the same population, with the "event" whose impact is under study occurring in the intervening period.

Another advantage of successive but independent samples is the avoidance of the conditioning or contaminating effects of a prior interview. This can be a drawback in surveys that interview the same sample of individuals at different points in time (panel studies). There is evidence that after a first interview people are more aware of the subject matter covered in the study, such as atomic energy, and may develop a different set of attitudes than others in the population. This is seldom a problem when interviewing successive samples at specified time intervals.

Successive samples also may suffer from several limitations, particularly in explanatory analysis. One that is shared by all repeated surveys, including the panel study, stems from possible changes in the influence of interviewers across the several periods of data collection. Both experimental and practical evidence suggest that the attitudes and expectations of interviewers can influence the results of a study (cf. Rosenthal, 1966). As Campbell and Stanley point out:

> If the same interviewers are employed in the pretest and posttest, it usually happens that many were doing their first interviewing on the pretest and are more experienced, or perhaps more cynical, on the posttest. If the interviewers differ on each wave and are few, differences in interviewer idiosyncrasies are confounded with the experimental variable. If the interviewers are aware of the hypothesis, and whether or not X (the experimental intervention) has been delivered, then interviewer expectations may create differences [1963, pp. 53-54].

An optimal solution that is rarely feasible involves the use of different but equivalent (same sex, age, training, race, etc.) interviewers in each wave of the study. Whether or not this policy is adopted it is usually advisable to keep the interviewers ignorant of the major hypothesis in the study. Sometimes, of course, this is impossible.

A second and major limitation of the successive independent sample design is its inability to account for gross changes (as contrasted with net changes). For example, if the Cincinnati study showed that 30 percent of the citizens favored the United Nations before the publicity campaign and 40 percent after (these are not the actual figures), we would know only that the net change was 10 percent. Without further information it is impossible to determine the gross changes behind this figure. It could be that the final 40 percent consisted of the original 30 plus an added 10 percent, with the rest showing no change, or the campaign might have had the bizarre effect of producing a 40 percent conversion rate, while alienating the original 30 percent, in which case the gross change would have been 70 percent rather than 10 percent. With data from independent samples, the investigator can only speculate about these questions.

Third, with surveys based on successive, independent samples it is difficult to account for the dynamics of change. Let us assume that we use this design to measure shifts in voting intentions prior to a national election, with one survey six months and the other one week before the election. If there is a change in the proportion intending to vote Democratic between surveys one and two, we can compare the differences between the two surveys in the proportions expressing Democratic intentions by age, race, education, region, and other traits. In this sense, the successive sample design can shed light on the characteristics of the changers. However, without panel data, individual changers and nonchangers cannot be described directly, nor can they be related to knowledge and perceptions of major events that occurred between the two times or other potential explanatory conditions, such as watching television and attendance at party rallies.

Finally, when successive samples are used in measuring experimental changes, as in the Cincinnati study, it is difficult to control the influence of factors in addition to the experimental variable. A shift of attitudes in favor of the United Nations may result from the publicity campaign, from other uncontrolled conditions, such as a "secular effect" (a general change in population attitudes which has nothing to do with the experiment), or from various incidents on the international front, such as a rise or decline in cold war tensions. If factors such as these can be predicted before the study, questions and probes can be built into the questionnaire in order to assess the respondent's awareness of their existence to provide some basis for estimating their impact. Without appropriate controls, the investigator cannot rule out or make adjustments for the effects of extraneous variables.

Panel Studies   A panel survey is one which collects information on overlapping sets of questions from the same sample on two or more occasions. The sample in question may consist of individual respondents, families, dwelling units, or other units, though the term "panel" usually refers to individuals. This design is contrasted with successive samples in the diagram below:

	Time 1	Time 2	Time n
Successive	Sample A	Sample B	Sample n
Panel	Sample A	Sample A	Sample A

Panel studies enjoy several advantages over successive samples and the cross section at a single point in time:

1   The panel technique permits the investigator to estimate the extent of gross change in the population. This is an important advantage even in descriptive studies which try to show how many individuals enter and leave a given category within a certain period. Students of religion, for example, may wish to estimate not only the net change in membership for a denomination over the past five years, but also the turnover—the number of former members who have left and the number of new members who have joined. The panel provides this information.

2   Panel data are particularly useful in answering questions about the dynamics of change, such as: Under what conditions do people shift their party affiliation? Are there peculiar factors at work in this election, such as the issue of religion in the 1960 Kennedy-Nixon campaign? What is the role of interaction with friends in changing political attitudes? Information on both the characteristics of changers and the factors likely to account for change provide important insights into the dynamics of change. As noted earlier, some of this information can be obtained from successive, independent samples, but it is far less complete than with panel studies.

3   By adding a few new questions to each interview, it is possible to accumulate more information about each respondent than in the single cross section or in successive samples. "Many such items change very little or not at all with the passage of time, so that a comprehensive case history of each panel member can be built up. These data can be cross-tabulated with the more limited material secured in a single interview for a better understanding of the phenomena being studied" (Parten, 1950, p. 98).

4   As in the case of successive samples, the panel can be used to measure the degree and direction of change following the introduction of an experimental variable. In addition to the advantages already noted regarding change, the panel technique permits a more intensive understanding of the process of change, especially when the respondents are followed over a relatively long period. Studies of attitude change can determine whether initial shifts hold up in succeeding years, whether there are peaks and troughs in the variables studied, and whether individuals who did *not* change at first begin to do so over time. This advantage is illustrated in a restudy of Bennington students who graduated from college around 1938. In the first study, Theodore Newcomb (1943) measured attitude change in the student body annually from the freshman through the senior year. In the follow-up research carried out between 1959 and 1962, Newcomb and his coworkers (Newcomb et al., 1967) contacted over 90 percent of the graduates of the earlier period and repeated many of the same or equivalent measures. In this way, they were able to determine whether changes during the college years generally persisted into adulthood (as they seemed to), and under what conditions persistence or change was most likely in the intervening years.

5   Repeated contacts with the respondent may help to reduce initial suspicions and reservations about the survey so that in later interviews he or she provides more information than previously. The reverse side of this benefit, however, is the "conditioning effect" to be discussed shortly.

Though the panel survey represents a great improvement over the cross section and successive samples in explaining change, it has four drawbacks:

1   It is often difficult to develop a sample of respondents who agree to be reinterviewed. Individuals who are normally quite cooperative in responding to a single cross section may resist being contacted two, three, or more times. Losses at the beginning reduce representativeness, for it is likely that those who refuse to participate are different from those who agree. The

former may be less literate, busier, with larger families, and show less interest in surveys (Moser, 1958, p. 112).

2 A further problem is *sample mortality*. Panel members die, move, become ill, or lose interest in projects which once excited them. Kish (1965) estimates that in a sample of United States families, about 20 percent will have changed addresses in a year's time. If these respondents cannot be followed, the sample base will drop to 80 percent in the second year and still lower in the subsequent years. Combined with whatever biases may exist in the sample from refusals or not-at-homes, these losses can present extremely serious problems of representativeness and validity in the data, especially since the movers are likely to be different from the rest of the sample on many characteristics. Some survey organizations have tried to reduce sample mortality through a concerted effort to follow movers to their new addresses. Others have attempted to develop a "team sense" among panel members through such devices as an occasional newsletter, prizes or other rewards for persistence. These latter techniques may succeed in cutting losses, but they are rarely feasible in studies with a limited budget, and can produce even further conditioning effects.

3 The problem of conditioning or contamination through measurement has already been noted in connection with successive samples. The essential problem is that the more individuals are subjected to repeated measurement, the more likely they are to develop a "guinea pig" reaction (see Campbell and Stanley, 1963). This conditioning, contaminating, or educational effect may take at least two forms. First, as noted earlier, participation in the panel may increase the respondent's awareness of the issue or problem area covered by the survey. For individuals in many countries, the United Nations is both a vaguely perceived ideal of international communication and a spatially remote organization. An hour or more of questioning about its functions and effects may not only produce baseline information about the respondent's initial attitudes, but also heighten his consciousness of the UN and make him more attentive to news about its work. One experiment on the effects of panel membership suggests that this influence may not be great (Lazersfeld, 1941). However, this study would have to be repeated on a broad range of issues before the problem of conditioning could be discounted. Second, the respondent's initial statement of attitudes, opinions, or beliefs may lead him to be more consistent than he usually is. An individual who says, initially, that the UN is useless may hold to this view even in the face of information that might otherwise have produced an attitude change. The greater the pressures toward consistency produced by the initial interview, the less representative the panel member may be of his more vacillating counterparts in the larger population, and the lower the generalizability of the findings on change.

4 As the panel moves into its later stages, with three or four repeated interviews, some members may find the process increasingly bothersome and lose interest in the substance of the study. The same can happen with the interviewers. The result may be a reduction in the quality and depth of the information. Respondents may give perfunctory or stereotyped responses, unthinking repetitions of earlier answers, and shorter interviews. These respondents may be, in effect, latent dropouts who do not wish to offend or displease the interviewer by stating their disaffection openly.

Special problems arise when the panel technique is used to evaluate experimental changes, such as a movie, an informational campaign, or some other stimulus. These problems are less serious when the investigators randomly assign those who will be exposed to the experimental condition (such as the movie) and those who will be assigned to the control group. But in many cases, especially in natural experiments, such assignment is impossible so that the investigators must use a later interview to determine who has and has not been exposed to a stimulus such as a movie. This approach is very appealing because of its methodological simplicity and convenience, but it can also be very misleading. One simply cannot assume that those who have exposed themselves to a movie or some other self-selected stimulus are in all ways similar to those who have not had the exposure. It is quite likely that those respondents who choose to listen to a television program on the United Nations are, on the average, more interested in the UN than those who do not, and perhaps more ready for a change in attitudes. For this reason, great care must be exercised in using the panel design as a quasi experiment.[3]

Combined Designs   Some of the difficulties in sample design discussed thus far can be resolved by combining the principle of successive samples with that of the panel technique, or using other variations tailored to the needs of the study. Two possibilities that deserve serious consideration when conditioning effects are expected to be important are the panel with one control group and the panel with two control groups.

Panel plus One Control Group   We have already seen that one of the most serious drawbacks of the panel method is the effect of the initial interview on later measurements of change. One way to determine indirectly whether or not there has been a conditioning effect is to carry out the second (or "after") interview with the panel and with an equivalent sample not previously interviewed. This design is summarized below:

Panel plus One Control Group

Time 1	Time 2
Sample A	Sample A
	Sample B

[3] The problem of self-selection in exposure to an experimental influence is discussed in detail in Campbell and Stanley (1963).

By comparing the average responses of Samples A and B at Time 2, the investigator can infer the extent of the conditioning effects produced by the interview at Time 1 and make appropriate adjustments in the interpretations.

Panel plus Two Control Groups   In surveys used to evaluate the impact of an experimental intervention, a further problem arises in separating the effects of the experimental variable from those of the initial measurement. A change in the panel following exposure to a movie may be the result of the movie, the conditioning produced by the first interview, or some interaction between the two. One way of determining the relative impact of the experimental variable versus conditioning is to use a control group which, like the panel, is measured at Time 1 and Time 2, but which is not exposed to the experimental stimulus, in this case the movie. This design is an improvement on the panel used alone, but it still does not permit the investigator to assess the effects of conditioning. A design which allows for an assessment of both the experimental variable and conditioning involves the use of the panel, a control group which is not premeasured but which is exposed to the experimental variable, and another control group which is neither premeasured nor exposed to the experimental variable. This arrangement can be represented as follows:

Panel plus Two Control Groups

Time 1	Exposure to experimental stimulus	Time 2
Sample A	Yes	Sample A
	Yes	Sample B
	No	Sample C

This design permits the investigator to answer three basic questions. First, by comparing the results for Sample A before and after the introduction of the experimental stimulus, he can judge whether or not it had the expected effect. Second, by comparing the results of Sample A and Sample B at Time 2, he can estimate the extent to which the reaction of panel members to the experimental stimulus was influenced by the measurement at Time 1. If Sample B is significantly different than Sample A, it is likely that the initial measurement of Sample A did affect the outcome of the experiment. Third, a comparison of the difference between Sample A at Time 1 and Sample C at Time 2 with the difference between Sample A at Times 1 and 2 provides some information on general shifts in population attitudes that have nothing to do with the experimental stimulus. If the resources of the survey are large enough, even more complex designs can be introduced to assess the relative effects of the experimental stimulus, preconditioning, and general changes in the population (cf. Campbell and Stanley, 1963).

Other Combinations Another combined design was used successfully with the national consumer surveys conducted by the Michigan Survey Research Center from 1960 to 1962. Experience with earlier studies of consumers' economic behavior suggested that the conditioning effects of participation in a panel were limited to greater consciousness of economic activities and better recall on the part of respondents. In the new design the same sample of addresses was maintained for each of the three annual surveys, except that the sample was updated each year with addresses representing new construction, new occupancy, and the conversion of buildings to dwelling use. The procedures called for interviews with those who had recently moved into sample addresses, and reinterviews with former respondents who had not moved. The result was a complete cross section each year which shared the advantages of successive samples as well as a panel design. The complete cross section covering recent arrivals as well as nonmovers provided trend data on attitudes, income, and other financial indicators. The panel component, on the other hand, provided data on savings through a comparison of the holdings reported by the same respondents in interviews held two years apart. Since the questionnaire for the movers-in was shorter because of the absence of the savings questions, it was possible to ask them a special sequence of questions on migration and family formation.

The technique of having different question sequences for different sets of respondents was also used to good advantage in recent surveys in Peru. The overall design called for general surveys in several metropolitan areas not previously studied. The samples covered adults, and the main questions had to do with labor force participation and underemployment. Because many adult females had little work experience, only a few of these questions were applicable. Instead, since most had borne one or more children, the questionnaire included a sequence of items on the fertility history and related attitudes of female respondents. Given that the women who were actively employed were younger and had no or few children, the two question sequences typically offset each other in their demands on respondent time. This general approach can often be applied with great advantage in other types of survey design.

Finally, we might mention the possibility of splitting a large survey into successive samples covering six months or a year. Rather than gather data on a sample of 2400 in a one-month period the researcher can divide the sample into 12 independent monthly samples of 200 households. As Kish states:

> The sum of repeated surveys over the entire period can lead to better statistical inference than a single, concentrated *one-shot* survey. *Probability selection of time segments from an entire interval permits statistical inference from the sample to an average condition over the interval* [1965, pp. 475–476].

A prime advantage of repeated samples is that they provide explicit information on phenomena known to vary by season, such as employment and unemployment, and also on the effects of unusual events such as international incidents, storms, or disasters. Nevertheless, the one-time survey concentrated in a period of one or two months will probably remain the most popular because it allows for a more rapid analysis and presentation of the data, and does not require the employment and supervision of interviewers over a long period.

In sum, a survey can be designed to satisfy one or more of the following objectives: exploration of a problem, description, causal analysis, hypothesis testing (a form of causal analysis), the evaluation of an intervention, and the development of social indicators. Normally, however, a more simple and less expensive method is suitable for exploratory analysis. The chapter argues that the distinction among some of these objectives may be more apparent than real, and that most surveys can be designed to pursue more than one objective. Particular attention is devoted to the costs and benefits of the cross-sectional design at a single point in time versus designs geared to assessing change, including successive, independent samples, panel studies, and combined designs.

FURTHER READINGS

Bauer, R. (ed.), *Social Indicators* (Cambridge, Mass.: M.I.T. Press, 1966).

Campbell, D. T., and J. C. Stanley, *Experimental and Quasi-Experimental Designs for Research* (Chicago: Rand McNally, 1963).

Hyman, H. H., *Survey Design and Analysis: Principles, Cases, and Procedures* (Glencoe, Ill.: Free Press, 1955).

Weiss, C. H. (ed.), *Evaluating Action Programs: Readings in Social Action and Education* (Boston: Allyn and Bacon, 1972).

Sampling and Estimation

The foundation of a good sample survey is the sample. A *sample* is some part of a larger body specially selected to represent the whole. *Sampling* is the process by which this part is chosen. While not every survey that uses proper sampling methods will provide adequate data, a study that does not will be seriously impaired from the outset. For a sample to be useful, it should reflect the similarities and differences found in the total group. A person who is studying a factory with 15 work groups cannot assume that one of these groups will be representative of all the others. In most factories work groups differ in age, attitudes, productivity, the ethnic background of their members, and other qualities. A representative sample of workers or work groups should capture the diversity of the whole.

In designing a survey sample, it is best to assume that people are different from one another in attitudes and behavior, even if the extent of differences is unknown and an object of study. Thus, in the example cited, the safest course is to assume that the 15 work groups in the factory show significant differences on any given characteristic, and to sample accordingly. This chapter argues that the most reliable way to assure a representative sample is to use chance procedures for choosing the units to be studied. These procedures and the logic behind them will be laid out in some detail. Nonchance procedures are also used in survey research, but they may introduce unknown or unanticipated bias into the findings. As a result, the data based on the sample may differ systematically from the information that would have been obtained from the entire group. Worse yet, unless there are independent data to provide a check on the findings, the biases may never be detected or even suspected. In many cases, the survey results themselves are the only source of statistical information about the subject under study, so that there are no other validation checks.

An example will illustrate the dangers of using nonchance sampling procedures, or of mixing them with chance procedures. A study of women began with a chance selection of dwellings in the study area. However, bias arose when chance procedures were abandoned in selecting the respondents. If the women in the selected dwellings were not at home when the interviewer called, next-door neighbors were substituted. At first sight, this may seem like a perfectly reasonable procedure, especially in a study with a limited budget and a tight deadline. The results, however, revealed a decided bias in the respondents actually chosen for interviews. Working and socially active women were seriously underrepresented by a sampling method favoring those who stayed at home during the day. Hence a seemingly innocent departure from chance procedures was highly damaging to the results of the study.

Another important point is that a sample is never adequate or inadequate in itself. It must always be judged by its usefulness for a given purpose. Sometimes it makes little difference which part of the whole is selected for the sample. The host at a dinner party tastes the first glass of wine to determine if the entire bottle is suitable for his guests. A medical technician takes a blood specimen from any convenient area of the body. These "samplers" assume that the wine is *sufficiently* similar throughout the bottle and the blood *sufficiently* similar throughout the body to yield samples that are useful for their purposes. In other words, they assume a high degree of *homogeneity* or sameness in the whole from which the sample is drawn. In point of fact, the wine at the top of the bottle may be slightly different in taste from that at the bottom, and the blood drawn from an infected area may not be exactly the same as specimens from other parts of the body. But for the purposes for which the host sips the wine and the medical technician draws the blood, the samples chosen are representative enough to be useful. Under conditions of high homogeneity, almost any sample is as good as any other. Unfortunately, in samples involving people such homogeneity can never be taken for granted. The procedures involved in sampling individuals and social groups are thus considerably more complicated than those for wine and blood.

BASIC TERMS AND CONCEPTS

In survey research the sample is selected to be counted, measured, or questioned in order to obtain data from which to make inferences about the total group under study. Three interlocking processes are involved: *sampling,* the process of selecting a part from the whole; *measurement,* the intermediate step of counting and asking questions; and *estimation,* the process of making inferences about the total group from sample data. Sampling and estimation together make up the *sample design,* the main theme of this chapter. Measurement is treated elsewhere in the book, especially in Chapters 6 and 9. In the following pages we will define and illustrate the basic terms and concepts involved in sample design, introduce some basic symbols and equations, and later move to problems of practical application.

Population, Elements, and Universe

A *population* is any complete group, whether of people, houses, farms, or pigs. *Elements* are the individual units making up the population. While a population typically refers to a finite group, the term *universe* covers events or things that are without numerical limit, such as all possible rolls on a pair of dice. Following this convention, all dwelling units in the Tokyo metropolitan area, as of a certain date, and all workers at Factory Z during the study week would be populations. This discussion focuses primarily on the sampling of populations.

Probability

Probability is the proportion of times that a particular outcome may be expected to occur out of many repetitions of the event. It is usually expressed as a decimal fraction from 0 to 1. Zero indicates that an event will not occur at all, 1 that it will occur with certainty. In flipping a fair coin, the probability of a head on any flip is the number of heads that might come up (one possible outcome, head) over the total number of possible outcomes (two, head or tail), or $1:2 = 0.50$. For a two-headed coin, the probability of a head is $2:2 = 1.00$, or certainty, and of a tail $0:2 = 0.00$, or impossible. When there are several possible outcomes, the probability of an event occurring in a particular way is the quotient of the number of ways the event can occur in the designated way divided by the total possible ways for the event to occur. For example, a *seven* has a higher probability of occurrence than a *six* on a roll of a pair of dice. The reason: there are six ways for a seven to occur (1-6, 2-5, 3-4, 4-3, 5-2, 6-1) but only five possibilities for a six to appear (1-5, 2-4, 3-3, 4-2, 5-1). With 36 possible outcomes from the roll of a pair of dice, the respective probabilities of a seven and six are 6:36 and 5:36. These elementary notions of probability form the basis for much of the theory and practice of sampling.

Probability and Nonprobability Sampling

Probability sampling is a process of sample selection in which elements are chosen by chance methods such as flipping coins, drawing numbered balls from an urn or through tables of random numbers. There are several variations in probability sampling, but all share a common trait: the selection of the units for the sample is carried out by chance procedures and with known probabilities of selection.

Nonprobability sampling includes all methods in which units are not selected by chance procedures or with known probabilities of selection. Sometimes they are also called *nonrandom* sampling methods. In this context, however, the word *random* may be confusing. In its technical sense, random refers to chance occurrences. But in popular usage, it has come to embrace everything from a true probability sample to a haphazard or diverse collection of people. Our preference is to use *random* only in its technical meaning, as in the later discussion of simple random sampling. The following are the most common types of nonprobability samples:

1 *Haphazard collections.* These are samples made up of individuals casually met or conveniently available, such as students enrolled in a class or people passing by on a street corner. Such samples generally do not permit generalization beyond the collections themselves and are seldom of scientific interest.

2 *Judgment sampling.* In this case, sample elements are chosen from the population by interviewers or other field workers using their own discretion about which informants are "typical" or "representative." The instructions, for example, may call for interviews with a representative mixture of successful businessmen, campus radicals, nonelective politicians, or any other target group. A critical weakness of this method is that different interviewers may have a different notion of what constitutes a "successful businessman," or even a "businessman." And even if the definition of the target group is reasonably clear, the procedures used in drawing a sample may vary so much from one judge to the next that there would be no comparability in the samples chosen. Quota sampling basically grew out of attempts to refine and eliminate the more gross errors in judgment sampling.

3 *Quota sampling.* This is a process of selection in which the elements are chosen in the field by interviewers or other field workers using prearranged categories of sample elements to obtain a predetermined num-

ber of cases in each category. The quotas are established on the basis of known characteristics of the population to be studied. For example, a recent census may show that of the total population in a certain city, 30 percent lives in District X. The simplest kind of quota sampling would operate by specifying that 30 percent of the sample would come from that district with the actual selections made by the interviewers in the field. In fact, quotas may be established with considerable sophistication on the basis of census and other information. But there are practical limits imposed by the types of information available. In practice, quotas are usually based on a few characteristics such as neighborhood type, age, sex, educational level, employment status, and rent level or house value. A typical assignment to an interviewer might be to interview 16 employed males in a certain suburb, choosing half from the areas containing better homes and apartments and half from the poorer areas. The instructions might further specify that half of the 16 should be under thirty-five years of age, and the other half thirty-five years or more. One immediate difficulty is that the interviewer must decide on *which* homes to choose in the better and poorer areas. The instructions may give general suggestions about how to obtain representativeness, but unless strict probability methods are followed there is ample room for arbitrariness. Next, the interviewer must determine the age, sex, and employment status of the respondents, and eliminate those who do not meet the criteria. Here, too, the process of selection and elimination is likely to vary from one interviewer to the next.

The weakest link in the quota sampling chain lies in the selection of individuals to fill the quotas. We simply do not know if the people chosen fully represent the similarities and differences existing among the elements in a given category. The interviewers' selections are likely to be biased toward the more accessible or attractive elements of the population. They may choose disproportionate numbers of those who are close to the research center, those who are at home during the day, individuals who would seem to be cooperative, or households without dogs. Interviewers may feel that if one element in the category is as good as another, why not pick those that cause the fewest headaches? Payment "by the interview" may place a premium on such convenience factors.

Proponents of quota sampling will argue that this method produces reliable generalizations about the population represented even if it lacks a solid theoretical basis for them. They will also argue that because interviews typically cannot be completed for all sample cases selected by probability methods for reasons such as illness and refusals, the resulting cases do not qualify as a probability sample. The advantage of probability sampling, however, is that even with a 10-20 percent noncompletion rate it is still possible to arrive at reasonably good estimates of sampling error. With quota samples, even with the full quota of interviews completed, there is no theoretical basis for calculating the range of error to be expected in the estimates produced. The completion rate also is of dubious value in evaluating the results of interviewing for a quota sample since the interviewers' reports of the number of homes or potential respondents contacted in order to obtain the quota of completed interviews cannot be checked or controlled effectively by the study directors.

4 *Expert sampling.* This is a process in which elements are chosen on the basis of informed opinion that they are representative of the population in question. For example, a specialist on secondary education in the United States may decide that four schools across the country adequately represent the range of variation seen in teaching methods. Other experts may decide that, on the basis of previous elections, the voting returns of a single district are a sufficient basis for making predictions about election results for the entire nation. While these selections are often shrewd and canny, they provide a very risky basis for generalization. Judgments based on past performance, for instance, may be invalidated by shifts in critical conditions between past and present. Expert samples often provide fascinating case studies and may generate important hypotheses, but they are not reliable bases for statistical estimation.

5 *Purposive samples.* These are selections from certain subgroups in the population chosen for their importance in testing hypotheses. In a study of the adaptation of rural migrants to urban conditions, for example, it may be important to test the reactions of individuals living in older, more established communities and those in very new areas. The sample could be designed to yield enough households in both the old and the new areas to allow such comparisons to be made with statistical reliability. The validity of this approach depends on the methods used to draw the purposive samples. It is often possible to apply strict probability methods at each stage in the sample design. Haphazard selections, on the other hand, suffer from the limitations indicated for other nonprobability methods.

Values, Estimates, Error, and Bias

The next few concepts refer to measurements taken on either a population or a sample. Let us begin with a simple example. A social scientist wishes to study age distributions in the population of her country. Because there has been no census in the country for eight years, and because of some doubts about the validity of information in the last census, she decides to analyze data from a recent national probability sample. In the case of age the *population value* is the value that would result from the measurement of the item (age) for every element in the population. However, even in a complete census of the population there would be measurement error. Some individuals may not know their age, others may lie about it, and still others may be omitted from the census. The population value must thus be distinguished from the *true value*—that which is being measured and is therefore without error. The *population estimate* is an inference about the population value based on sample data. For the country in question, the population value and the true value for age are unknown. The best source of information may be the population estimate, but it must be treated with care, for it is subject to both bias and sampling error.

Bias refers to systematic error leading to a difference between the population estimate (or the population value) and the true value. Bias may result from nonrandom errors in measurement, estimation, or other sources. Its critical quality is that it is systematic error—it pushes the estimate in one or another direction. An age estimate may be biased downward, for example, if there is a tendency for respondents to portray themselves as being younger than they are.

Sampling Error and Standard Error

Even without bias and other measurement error, an estimate derived from a sample is not accurate in the same way as a complete enumeration or census. In our example, a probability sample of individuals might have an average age exactly the same as the population from which it was selected. Assuming that the population value is available, both might come out to 24.3 years. This correspondence between the population estimate and the population value is likely to be a coincidence. It would be similar to an exact 50:50 split between heads and tails on 100 tosses of a coin. If the sampling process could be repeated to select many samples of the same size from the same population, the age distributions in these samples would vary. Each sample would produce an estimate of the mean (average) age for the population, but some would be higher than 24.3, some lower, and some the same value, with the differences arising from the chance composition of each sample. These differences between the population estimates from different samples and the population value are called *sampling error,* usually expressed as the *standard error.* The standard error is a measure of the variability, around the population value, of the population estimates from repeated samples. In simple terms, it gives us a clear notion of how far and with what probability an estimate based on a sample departs from the value that would have been obtained with a complete census. In practice, of course, the researcher usually has only one sample to work from, and therefore only one population estimate available for a given variable. However, the expected relationships between population estimates and population values are well established in probability theory. The techniques for working out these relationships will be reviewed shortly.

Lists, Frames, and Sampling Units

The remaining definitions deal with the materials actually used in drawing samples. A *list* is an inventory of the units in a population or a subpopulation with a direct, one-to-one correspondence between each item listed and the unit it represents. The simplest example is a list of all members in a population, such as a trade union, in which each member is identified by name. If the listing units are population elements, it is an *element list*. Sometimes the listing units may be groups of elements found in convenient, identifiable, and unambiguous natural groups, such as apartments in an apartment building or homerooms in an elementary school population. In such cases, the inventory would consist of a *cluster list*.

Sampling units are those elements or groups of elements which form the basis of sample selection. They may or may not be identical with the listing units. When a complete list of population elements is available, it is often most convenient to draw a sample directly from this list, in which case the sampling and listing units would be the same. For example, the sampler may choose every fifth name on a complete list of union members; in other cases, several listing units, such as houses, may be combined to form a single sampling unit. For instance, where houses are quite far apart, the study directors may decide that it is necessary, for identification purposes, to list each one separately, but that to save on travel costs for interviewing, houses will be selected into the sample in contiguous clusters of four.

A *sampling frame* consists of the materials and procedures used fully to account for the population when complete element lists are not available. A sampling frame often will consist of maps, sketches, lists, aerial photographs, and instructions on the manner in which they are to be used. It is basically the operational procedure and materials used to account for the population in drawing the sample.

SIMPLE RANDOM SAMPLING

The special form of probability sampling called *simple random sampling* (srs) is an excellent starting point for more detailed discussion. Because srs has the least complicated formulas for sampling and estimation, it provides a useful base for comparing and explaining other forms of probability sampling. Hansen, Hurwitz, and Madow (1953) provide the following rigorous definition:

> A procedure of sampling will be called simple random sampling if, in a sample of size *n*, all possible combinations of *n* elementary units that may be formed from the population of *N* elementary units have the same probability of being included. [Vol. I, p. 110.]

It may also be said that simple random sampling is *a process of sample selection in which the units are chosen individually and directly through a random process in which each unselected unit has the same chance of being selected as every other unit on each draw*. An important theoretical issue is whether an element is to be eligible for selection into a sample only once or is to be replaced in the pool for possible reselection into the sample. These options are called sampling *without replacement* and *with replacement*, respectively.

The usual practice in sampling from finite populations is to sample without replacement, that is, with an element eligible to be selected only once.

SRS Selection

The main practical requirement for the application of srs is that each element in the population be clearly and unambiguously identified. This is necessary to permit the independent and direct selection of the individual elements, often through a list that uniquely identifies each element.

If the elements are people, there must be sufficient information about each person to allow him to be identified individually. This condition is generally met only when lists are available which inventory the population, such as registration lists of university students. When this information is available and complete, all of the units in the list can be numbered sequentially and the sample can be chosen by applying a chance selection process to the range of numbers corresponding to the list. The selection might be made by a mechanical process such as numbered balls in an urn or, more commonly, through a table of random numbers. In sampling without replacement it is usual to obtain as many different random numbers from a random number table as cases are desired for the sample, excluding numbers which are outside the range of numbers corresponding to the population. Those elements are selected that correspond to the random numbers that were obtained.

This procedure is illustrated in Table 4-1, which shows a hypothetical population of 12 families and their monthly incomes. We will estimate the mean level of monthly income of this population by sampling with different samples sizes, and show how the estimates will vary according to size.

Table 4-1 Monthly Income of 12 Families

Family number	Monthly income
1	$ 500
2	450
3	375
4	475
5	1350
6	900
7	675
8	325
9	550
10	1200
11	225
12	775
Total monthly income	$7800
Mean monthly income	$ 650

In order to estimate from a sample of two families, we could number each family from 1 to 12, put balls numbered from 1 to 12 in an urn and then choose two in a blindfolded selection. Alternatively we might use a table of random numbers, a portion of which is reproduced as Table 4-2.

First we need agreement on how to read the numbers in the table. Since our highest number is 12 we will read them in pairs, with 01, 02, . . ., 12 representing the 12 families. We will agree to read them as a book, that is from left to right and then to the left on the next lowest line, although another system might also be used equally well. Then we need a random start, which is the circled pair, 32. Since it is outside of our range of interest we continue with other numbers discarding those outside the relevant range. The first selected number to fall within our range, 10, enters *family 10* in the sample.

We would then continue to read the random number table until we arrive at a second previously unselected number within the range, 02. This selects *family 02* as the second family for the sample of two elements from the population of 12.

With families 10 and 02 as the sample, we can proceed to calculate the mean monthly income of this sample, the sample mean, which is $825. By comparing this figure with the complete list of the population and their income in Table 4-1, we can see that this sample mean differs from the population value ($650) by $175.

This, however, is not the only sample of size two that could have been selected by srs from the population of 12. Table 4-3 shows the distribution of mean income for all of the possible different samples of size two, four, and six that could be selected from a population of 12 families.

Table 4-2   Excerpt from Tables of Random Numbers

```
98 08 62 48 26  45 24 02 84 04  44 99 90 88 96  39 09 47 34 07  35 44 13 18 80
33 18 51 62 32  41 94 15 09 49  89 43 54 85 81  88 69 54 49 94  37 54 87 30 43
80 95 10 04 06  96 38 27 07 74  20 15 12 33 87  25 01 62 52 98  94 62 46 11 74
79 75 24 91 40  71 96 12 82 96  69 86 10 25 91  74 85 22 05 39  00 38 75 95 79
18 63 33 25 37  98 14 50 65 71  31 01 02 46 74  05 45 56 14 27  77 93 89 19 36

74 02 94 39 02  77 55 73 22 70  97 79 01 71 19  52 52 75 80 21  80 81 45 17 48
54 17 84 56 11  80 99 33 71 43  05 33 51 29 69  56 12 71 92 55  36 04 09 03 24
11 66 44 98 83  52 07 98 48 27  59 38 17 15 39  09 97 33 34 40  88 46 12 33 56
48 32 47 79 28  31 24 96 47 10  02 29 53 68 70  32 30 75 75 46  15 02 00 99 94
69 07 49 41 38  87 63 79 19 76  35 58 40 44 01  10 51 82 16 15  01 84 87 69 38

09 18 82 00 97  32 82 53 95 27  04 22 08 63 04  83 38 98 73 74  64 27 85 80 44
90 04 58 54 97  51 98 15 06 54  94 93 88 19 97  91 87 07 61 50  68 47 66 46 59
73 18 95 02 07  47 67 72 52 69  62 29 06 44 64  27 12 46 70 18  41 36 18 27 60
75 76 87 64 90  20 97 18 17 49  90 42 91 22 72  95 37 50 58 71  93 82 34 31 78
54 01 64 40 56  66 28 13 10 03  00 68 22 73 98  20 71 45 32 95  07 70 61 78 13
```

Table 4-3   Distribution of Average Monthly Family Income of All Possible Samples, by Sample Size 2, 4, and 6

Average Monthly Income	n = 2	n = 4	n = 6
Under $300	1	–	–
$ 300–$ 399	7	12	1
$ 400–$ 499	11	67	66
$ 500–$ 599	12	102	193
$ 600–$ 699	9	101	290
$ 700–$ 799	8	103	240
$ 800–$ 899	7	64	122
$ 900–$ 999	6	32	12
$1000–$1099	3	13	–
$1100–$1199	1	1	–
$1200 or more	1	–	–
Number of samples	66	495	924
Average monthly income, all samples	$650	$650	$650

Source: Table 4-1, Monthly Income of 12 Families.

Thus, there are 66 combinations of two elements, 495 of four elements, and 924 combinations of six elements. Notice that the sample means from the smallest sample size cover a much wider range than the others, and that the means from samples of sizes four and six distribute more narrowly, with proportionately more closer to the value of the population mean. This demonstrates one of the fundamental and most useful principles of sampling and estimation, namely that, other things being equal, the distribution of sample means of larger samples varies less than that of smaller samples. The example in Table 4-3 thus provides a small-scale illustration of sampling error.

The mean of all possible sample means of a given size is called the *expected value* for that sampling method. When it is identical with the mean of the population, the sampling procedure is said to be *unbiased*. Remember that *bias* does not refer to the actual difference between the mean of a particular sample and the population value, but to the difference between the expected value for that sample size and method, and the population value.

The utility of a small sample for estimating the population mean is questionable. The possible range of sample means is great and the fact that few fall close to the expected mean is disconcerting. Table 4-3 also shows that the range of sample means is smaller with larger sample sizes, and that the concentration near the population value is also greater with the larger sample sizes. In this example, which shows samples from a very small population, the greater concentration with larger samples results both from the larger absolute sample size and the larger proportion which each successive sample size is of the population. It will be shown later that in sampling from large populations, the *absolute size of the sample* is usually of much greater importance in practical applications than the proportion which the sample is of the population.

Notation and Simple Relationships

Thus far we have reviewed the general notion and the basic concepts of sampling as well as the fundamentals of simple random sampling. To move beyond this elementary level into a working knowledge of sampling, it is necessary to introduce some mathematical tools. Our aim is to present enough definitions and equations to permit a fairly precise discussion of sampling error, estimation, and other key concepts and relationships in sample design. For a more thorough treatment, the reader should consult one of the standard texts, particularly Kish (1965). For practical advice in designing a sample, it is always advisable to seek the assistance of an experienced sampler.

We begin with a system of mathematical shorthand or *notation* which provides a convenient means of presenting relationships. The symbolic representation of a finite population size is indicated by N, while n indicates the sample size. These symbols are used in the equation for the *sampling fraction*, the rate of sample selection, f:

$$f = \frac{n}{N} \tag{1}$$

For example, if we select a sample of 300 from a registered student body of 5400, the sampling fraction would be:

$$f = \frac{300}{5400} = \frac{1}{18}$$

If the population size is known, the equation can also be used to estimate the size of a prospective sample based on a given sampling fraction. For instance, what size sample would be obtained with a sampling fraction of 1:20 and a population of 5400 registered students?

$$\frac{1}{20} = \frac{n}{5400} \qquad 20n = 5400 \qquad n = 270$$

Normally upper case letters refer to the population and lower case letters to the sample. Individual elements in the population are identified by subscripts to N, as with N_1, N_2, . . ., N. Individual sample elements are identified through similar subscripts to n: n_1, n_2, n_3, . . ., n. Because it is awkward to refer to all the individual elements in this way, the subscript i is used as a shorthand. Thus, when i is added to N or n, it refers to each and every element in the population or sample:

$$N_i = N_1, N_2, \ldots, N$$
$$n_i = n_1, n_2, \ldots, n$$

Another useful shorthand device is the capital sigma, Σ, the conventional symbol for summation. Applied below to the population and the sample respectively, these symbols should be read, "The summation (of whatever follows) over all the population cases N, or sample cases n, with i taking all the values from 1 through N, or 1 through n."

$$\sum_{i=1}^{N}(X_i) \quad \text{and} \quad \sum_{i=1}^{n}(x_i)$$

The symbol X or x in the parentheses would usually stand for some attribute or characteristic of the elements, such as height, weight, or income. This notation can be illustrated concretely with one attribute discussed earlier, family income. The monthly income of each of the 12 families comprising the population N can be represented as Y_i and the monthly income of the families in the sample as y_i. The *total* monthly income, over the population and the sample respectively, can be written:

$$\Sigma(Y_i) = Y \quad \text{and} \quad \Sigma(y_i) = y \tag{2}$$

Often the notation above and beneath the capital sigma (i taken from 1 through N or n) is omitted and must be inferred from the context.

One of the most commonly sought statistics in survey research is the *mean* or arithmetic average of an attribute. For example, the mean family income of our sample comprised of families 2 and 10 was obtained by summing the income of these two families ($\$450 + \$1200$) and dividing by the number of sample cases ($\$1650 \div 2 = \$825$). An overbar, $^-$, over the symbol for an attribute denotes a mean. The addition of a prime, $'$, identifies it as an estimate of the population:

$$\bar{x} = (\Sigma x_i)/n = \frac{1}{n}(\Sigma x_i) \tag{3}$$

Population Estimates

An important feature of data from simple random samples is that they permit us to estimate population values. For example, to estimate the mean income of a population, we need only the mean income of a simple random sample from that population. This sample mean is an unbiased estimate of the population mean:

$$\bar{x} = \bar{x}' \tag{4}$$

Thus, even if we do not have income data for the whole population, information from a simple random sample will provide an estimate of it. This is one of the greatest advantages of probability sampling. We saw earlier that population estimates seldom are identical with the population value, but that they cluster about this value and are related to it in predictable fashion. The next section deals with ways of describing how population estimates cluster about the population value.

The Reliability of Estimates

The greatest intellectual hurdle to the acceptance of sampling concerns the reliability of estimates. The potential user of survey research will often ask how one can possibly take a sample of 2000 people in a country of over 200 million people and arrive at a reliable estimate of the number of multiple-car families, potential voters who favor one or another candidate, and so forth. These are good questions which are taken seriously by sampling specialists. We will now show how it is possible to calculate the reliability of means as an example of how statistical information derived from a sample can be used.

We indicated in Equation (4) that the mean of an srs sample is an unbiased estimate of the population mean. To understand how much confidence one should have in such an estimate we need to develop a more precise definition of sampling error. For this it is necessary to introduce the concept of the *normal probability curve,* known more simply as the *normal curve.* This is a bell-shaped distribution which is highest at its center and symmetrical around the mean. Experience has shown that the distribution of many natural events, such as height and weight for individuals, approximates this bell-shaped distribution. In the case of height, for instance, there are many men who measure between 5 ft 8 in and 6 ft 1 in, fewer between 5 ft 3 in and 5 ft 8 in or 6 ft 1 in and 6 ft 6 in, still fewer less than 5 ft 3 in or more than 6 ft 6 in. If we were to do a careful study of height, in almost any country the distribution would probably approach that of the normal curve—many in the middle, fewer as the figures move away from the middle. Further details on the normal curve will be added later. For the moment, it is enough to know that it is a useful theoretical distribution for dealing with the extent to which estimation from samples is reliable.

Even though most surveys use only one sample and, therefore, obtain only one sample mean for a certain attribute, it is important to think of the sample means that might have been obtained by repeated samples. A key principle of probability sampling is that the distribution of all possible srs sample means approximates the normal curve. That is, if we were able to draw repeated samples of the same size from the same population and plot the means obtained on some variable, the distribution of these sample means would resemble the bell-shaped normal curve. This point is illustrated in the example of family income shown in Table 4-3. The last column in this table shows the distribution of average income for the 924 possible samples of six elements from a total population of 12 families. The figures indicate that the largest number of sample means—290—are within the interval containing the population value of $650. The next largest clusters—240 and 193—are for the two categories of income closest to the mean, while the smaller numbers are for categories still further from the mean. While this distribution does not very closely approximate a smooth bell-shaped curve, because of the small sample size and the broad income categories used, similar distributions from real surveys usually come close enough to the normal distribution to permit use of its statistical properties. It is worth noting that, except for small samples of about 30 or less, *the attribute to be estimated* (such as income) *does not have to be normally distributed in the population in order for the distribution of sample means to approximate the normal curve.* In the example cited, this means that it is not necessary for family income to be normally distributed in the population for the distribution of sample means of income to approximate the normal curve.

The relationship between the distribution of an attribute in the population, its distribution in a single sample, and the theoretical distribution of all possible sample means can be illustrated with the attribute of age. Figure 4-1 shows the age distribution of the male labor force in a country (the population), the age distribution in a single sample drawn from that population, and the theoretical distribution of sample means for three sample sizes.

The age distribution of the population (Figure 4-1a) is bimodal, that is, it has two peaks with a dip in the middle. The reason is that the country suffered severe wartime casualties in one cohort of its male population. We include this feature to emphasize that the shape of the distribution of age is largely irrelevant to the shape of the distribution of sample means for age. Thus with few men below 14 or over 65 years in the labor force, the mean population value $(\overline{X})$ is 38 years.

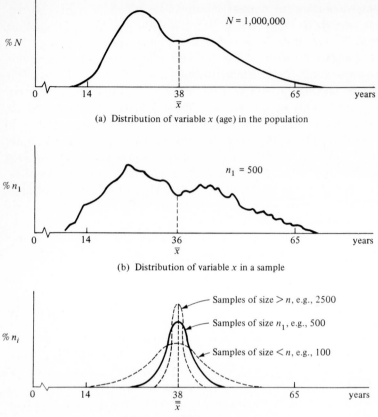

(a) Distribution of variable x (age) in the population

(b) Distribution of variable x in a sample

(c) Distribution of $\bar{x}$ for three sample sizes

Figure 4-1   Age distribution of the male labor force in (a) the population; (b) a single sample; and (c) for three sample sizes.

Figure 4-1b shows the age distribution obtained from an srs sample of
500 from the same population. This distribution closely resembles that of the
population, but also shows chance differences. The mean of the sample $(\bar{x})$ is
36 years, two years lower than the population value. It would be only a
coincidence if they were identical. Although not portrayed, other srs samples
of size 500 from the same population would resemble that portrayed in
Figure 4-1b, but would differ slightly reflecting chance variations.

Figure 4-1c provides a graphic illustration of the influence of sample
size on the reliability of population estimates. The solid line indicates the
distribution that would result if the mean ages from all possible srs samples
of size 500 from the population under study were plotted on the same horizontal axis as the other figures. It is a normal curve with the mean $\bar{\bar{x}}$ (mean

of sample means) equal to the mean of the population, or 38 years. Figure 4-1c also portrays the distribution of mean ages that would be obtained from repeated samples of size 2500 and from repeated samples of 100 cases. Notice that the mean age for these three sampling distributions is the same, and that each has a bell-shaped distribution. However, the variation around the population mean—the width of the curve—*increases* as sample size *decreases* from 2500 to 500 to 100. In other words, the larger the sample size, the smaller the "spread" of the sample means around the population mean of 38 years. This point has critical implications in calculating the reliability of population estimates, for the smaller the expected spread of the sample means, the smaller will be the chance of error between the population estimate and the population value.

Variance and Standard Deviation

Two mathematical expressions of the variability of distributions are used in the estimation process. One is called the *variance;* the other is its square root, the *standard deviation.* Just as the mean or arithmetic average describes the central tendency of a distribution, the variance and standard deviation express the variability or spread of data around the mean. These are described for readers who are not familiar with them.

The variance is the average of the squared differences between the value of the attribute for each of the elements and the mean value of the attribute. Consider the following simple example, a sample of five men from the population indicated in Figure 4-1. We are interested in their mean age and the variance of age around the sample mean. The ages of these five men are: 31, 33, 35, 37, and 39. The mean of this distribution is 35 years. The variance would be calculated as follows:

Age	Difference from mean	Difference from mean, squared
(x_i)	$(x_i - \bar{x})$	$(x_i - \bar{x})^2$
31	-4	16
33	-2	4
35	0	0
37	2	4
39	4	16
		40

Mean $(\bar{x})$ = 35
Sum of squared differences from mean = 40

$$\text{Variance} = \frac{\text{sum of squared differences from mean}}{n} = \frac{40}{5} = 8$$

Separate symbols are used to distinguish among the variance of an attribute in the population and in a sample, and the estimated variance of an attribute in the population. The variance of an attribute in the population, such as the age distribution shown in Figure 4-1a, is defined as:

$$VAR_X = \frac{\Sigma(X_i - \overline{X})^2}{N} = \sigma^2 \tag{5}$$

σ^2 is another common symbol for this variance. The variance of an attribute in a sample, VAR_x, such as the distribution shown in Figure 4-1b, is defined as:

$$VAR_x = \frac{\Sigma(x_i - \overline{x})^2}{n} \tag{6}$$

An equivalent and more convenient formula for computation is:

$$VAR_x = \frac{\Sigma(x_i)^2}{n} - (\overline{x})^2 \tag{6a}$$

The estimate of the population variance, s^2, makes use of the sample variance, VAR_x:

$$s^2 = \frac{n}{n-1} VAR_x \tag{7}$$

The final step is to show the relationship between s^2, the *estimate* of the population variance, and VAR_X, the population variance. This is:

$$s^2 = \frac{N}{N-1} VAR_X \tag{8}$$

For most practical purposes, s^2 can be considered an unbiased estimate of VAR_X. For a population as small as 500, $N/(N-1)$ affects the estimate by only about 0.2 percent, and by considerably less for larger populations.

The *standard deviation* is simply the (positive) square root of the variance. Equations (9) and (10) give mathematical expression to this fact for the population and the sample:

$$\text{Population:} \quad S.D._X = \sqrt{VAR_X} = \sqrt{\frac{\Sigma(X_i - \overline{X})^2}{N}} = \sigma \tag{9}$$

$$\text{Sample:} \quad S.D._x = \sqrt{VAR_x} = \sqrt{\frac{\Sigma(x_i - \overline{x})^2}{n}} = s \tag{10}$$

The discussions and equations thus far have described the variance of a variable in the population and in a sample, their interrelationships, and the estimation of the population variance from the sample variance. The corresponding standard deviations can be readily calculated by taking the square root of the variance in question. Now we will return to the central question of this section: how to determine the reliability of the estimate of population mean based on a single sample. This takes us back to the hypothetical distribution of sample means discussed earlier.

We have seen from Table 4-3 and from the three curves in Figure 4-1c that the larger the sample size, the smaller the variation in the distribution of means from all possible samples of that size. This is the basis for asserting that larger samples are more reliable. In order to be more precise about the reliability of an estimate from any one sample, or to make comparisons of reliability between samples, we must be able to estimate the variance and standard deviation of the means of all possible samples of the same size and type as the one that concerns us. Fortunately, this task is less complicated than it might first appear. Sampling theory provides equations for estimating the variance and standard deviation of these means based on the data from only one sample. The standard deviation of this particular distribution is so basic that it has received a special name, *the standard error.*

The standard error or, more correctly, the standard error of the estimate, is the key to the measurement of the reliability of the estimate of a population statistic—in our example, the *mean* of an attribute such as age. *The standard error of the mean of an attribute is the standard deviation of the distribution of sample means.* Because very few studies are in a position to carry out repeated samples to determine this standard deviation on an empirical basis, it is usually necessary to use the sample itself as the basis for an estimate. The general formula for the standard error and the formula for the estimate of the standard error from sample data are as follows:

$$\text{Standard error} = \sigma_x = \sqrt{\frac{N-n}{N} \frac{\sigma^2}{n}} \tag{11}$$

$$\begin{array}{l}\text{Estimate of the}\\ \text{standard error}\end{array} = s_{\bar{x}} = \sqrt{\frac{N-n}{N} \frac{s^2}{n}} \tag{12}$$

To obtain the corresponding variance it is necessary only to square the standard error.

The Normal Curve

We turn now to consider the normal curve more fully in order to show how to use the estimate of the standard error. A normal curve is a bell-shaped theoretical distribution which describes the expected distribution of sample means and many other chance occurrences. All normal curves share two common features: they are completely identified by their mean and standard deviation, and they can be reduced to a standard form (see Figure 4-2). The mean of a normal curve identifies its highest point and the vertical line about which this bell-shaped curve is symmetrical. The standard deviation establishes the breadth or narrowness of the curve around the mean.

The conventional standard form of the normal curve has three characteristics: (a) the area under the curve is assumed to be 1.000 (with more or fewer zeroes according to the precision desired); (b) the mean is at zero; and (c) the unit of measurement along the x-axis is the standard deviation. These properties also are shown visually in Figure 4-2.

Table 4-4 is a compact version of the typical standardized normal table presented in statistical textbooks with more decimal places. The table shows the proportion of the area under the normal curve corresponding to different ranges along the x-axis stated in standard deviation units from the mean. Since the normal curve is symmetrical around the mean, the table is presented for half the curve, leaving the other half to be inferred.

The row and column heads of Table 4-4 represent distance from the mean in standard deviation units, the rows in whole standard deviation units, the columns in tenths. The body of the table shows the proportion of the area under the normal curve in relation to the distance from the mean expressed by the row-column combination. To determine the area under the curve within one standard deviation to the right of the mean, we would first locate the appropriate row-column combination in Table 4-4. This would be 1.0, the intersection of the unit of 1 and the tenths of 0. The table shows that the area corresponding to the segment of the normal distribution from 0.0

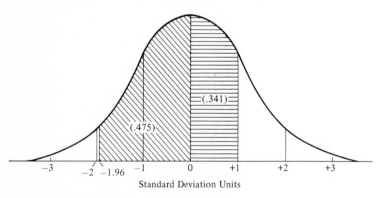

Figure 4-2 The normal frequency distribution or normal curve.

Table 4-4 The Standardized Normal Table

Standard deviations from the mean, in units	Standard deviations from the mean, in tenths of units									
	_0	_.1	_.2	_.3	_.4	_5	_.6	_.7	_.8	_.9
	Area under One-half of the Normal Curve*									
0._	.000	.040	.080	.118	.155	.192	.226	.258	.288	.316
1._	.341	.364	.385	.403	.419	.433	.445	.455	.464	.471
2._	.477	.482	.486	.489	.492	.494	.495	.496	.497	.498
3._	.499	.499	.499	.500†	.500†	.500†	.500†	.500†	.500†	.500†

*Area under the normal curve and over the segment measured in one direction from the mean to the distance indicated in each row-column combination. For example, the table shows that about 68 percent of normally distributed events can be expected to fall within one standard deviation on either side of the mean (.341 times 2), and that about 50 percent will fall within 0.7 standard deviation of the mean. An interval of almost 2.0 standard deviations around the mean will include 95 percent of all cases.

†Less than 0.500 but rounds to 0.500 at three decimal places.

(the mean) to 1.0 is 0.341. This means that about 34 percent of the cases in a normal distribution will fall between the mean and one standard deviation to the right of the mean.

To calculate the area under the curve within one standard deviation on *either* side of the mean, we would double the proportion. Thus, about 68 percent of normally distributed events will fall within one standard deviation of the mean. Similarly, by doubling the proportions presented in the body of the table, we find that about 95 percent of normally distributed events will fall within an interval two standard deviations above and below the mean, and 99 percent within 2.6 standard deviations of the mean.

In probability terms, one could understand from the above that there are 68 chances out of a hundred that a randomly selected event (such as a sample mean) from a normally distributed population of such events would fall within one standard deviation of the mean population value of such events. This property of the normal curve becomes very practical when the researcher wishes to determine the reliability (standard error) of a population estimate based on a single sample.

The earlier example of the age distribution will illustrate how to transform the standard error measured in *years* into the standardized units that appear in the standard table. Suppose we wish to represent in standardized form the distribution of the means of all possible srs sample means of age from samples of size 500, and that the mean (of means) of this distribution is 38 years. This mean is represented in the standard form by the 0 point on the horizontal axis. If the standard deviation of the sample means, the standard error, is 2 years, one standard deviation above and below the mean is 40 and 36 years, respectively.

To find the standardized value of any measurement expressed in original units, subtract the mean from the value to be transformed and divide by the standard error (all expressed in original units). For example, to find the standardized value of 34 years in the example above, subtract the mean (38 years) and divide the remainder by the standard error (2 years) to obtain the result, -2. That is, 34 years in standardized form is -2, or 2 standard errors below the mean. In original units:

$$\frac{\text{(Value, to be transformed)} - \text{(mean)}}{\text{Standard error}} = \text{standardized value}$$

$$\frac{34 \text{ years} - 38 \text{ years}}{2 \text{ years}} = -2$$

The process is reversed to convert a standardized value to original units: multiply the standardized value by the standard error, and add the mean. To illustrate, the standardized value, $+1$, is equivalent to 40 years:

$[(+1)(2 \text{ years})] + [38 \text{ years}] = 40 \text{ years}$
$[(\text{Standardized value}) (\text{standard error})] + [\text{mean}] = \text{value in original}$
units

Population Estimates and Confidence Intervals

We are now ready to apply the basic logic and equations of sampling to a concrete problem. Our task is to estimate the population mean and its reliability for this survey question: "Do you feel that you have adequate contact with your teachers on matters pertaining to your courses?" This question was asked of a 1:20 srs sample drawn from a population of 5400 students. Thus $N = 5400$, $n = 270$. Treating a "yes" response as 1 and a "no" response as 0, the sum of the answers to the question can be represented as $\Sigma(x_i) = 162$; dividing this by n gives a proportion of .60. For the following calculations .60 will be treated as the mean (average) response. Applying the equations reviewed earlier, we arrive at the following figures:

Table 4-5 Student Attitudes toward Contact with Teachers

	(Q. 14. . . . adequate contact with teachers?)			
n	Yes	No	D.K.	Proportion "yes"
270	162	108	–	0.60

Equations to estimate the population mean:

(3) Sample mean $\bar{x} = (1/n)\,\Sigma(x_i) = (1/270)\,(162)$
                                     $= 0.60$

(4) Estimated population mean $\bar{x}' = \bar{x} = 0.60$

Equations to estimate the standard error of the estimate:

(6a) Sample variance $\mathrm{VAR}_x = \dfrac{\Sigma(x_i)^2}{n} - (\bar{x})^2$

$$= \frac{162}{270} - (.60)^2 = 0.60 - 0.36$$
$$= 0.24$$

(7) Estimated population
     variance $s^2 = \dfrac{270}{269}\,(.24) = 0.241$

(12) Estimate of the
     standard error

$$s_{\bar{x}'} = \sqrt{\frac{5400 - 270}{5400}\,\frac{0.241}{270}}$$
$$= \sqrt{0.00085} = 0.03$$

The mean of "yes" answers, more commonly called a proportion, was thus found to be .60 or, if multiplied by 100, 60 percent. The standard error of this estimate was calculated at 0.03, or 3 percentage points. This finding has important implications for the population estimate.

On the basis of an srs sample of 270 cases, the population value was estimated at .60. It would be a coincidence if the population figure were exactly .60, for the estimate from one sample of 270 is only one of many that might have been made from other srs samples of the same size. All these other samples would also yield sample means, and these means would tend to distribute themselves in the form of a normal curve around the population value. This distribution of sample means has a standard deviation, the standard error of the estimate. From our sample it was estimated to be 0.03, or 3 percentage points. With this information we can turn to the normal curve to determine the variability to be expected in these sample means.

According to the normal table, 68 percent of a normally distributed variable—in this case, estimates of "yes" answers from srs samples of 270— can be expected to fall within one standard error of the population value. In our example the interval equivalent to one standard error around the population mean is 0.60 ± 0.03, or the interval of 0.57 to 0.63. Thus, there are about 68 chances in 100 that the population value lies within this range. Put another way, the chances are 68 out of 100 that our estimate of the population value does not differ from the population value by more than 0.03, or 3 percentage points. We thus have a good notion of the reliability of the estimate.

We are now ready to introduce two basic and related terms concerning the reliability of population estimates: the confidence interval and the confidence level. The *confidence interval* is the range around the population value within which estimates from samples can be expected to lie at a given confidence level. The *confidence level* refers to the probability that a given statement is correct. More simply, it has to do with the chances one is willing to take on his estimate. The fewer the chances he is willing to take, the broader the confidence intervals needed. If a researcher works at the 95 percent confidence level, he is willing to accept five chances in 100 of being wrong. At this level it is necessary to use a confidence interval of almost two standard errors around the mean. In the previous illustration this would be 0.60 ± 2 standard errors (0.03) = 0.60 ± .06. Thus there would be 95 chances in 100 (the confidence level) that the population figure of "yes" answers would lie between 0.54 and 0.66 (the confidence interval). For the 99 percent confidence level the confidence interval would stretch to 52.3 − 67.7. The greater the assurance of being correct, the broader the confidence interval. Although there is no strictly logical or mathematical basis for this practice, social scientists frequently use the 95 percent confidence level in presenting confidence intervals for their estimates.

Sample Size and Sampling Error

The next question concerns the factors that contribute to unreliability or sampling error. If we wish to *decrease* the size of confidence intervals for a given estimate, how should we go about it? What changes can be made in the sample design?

Usually the most important factor in reducing the standard error is the *absolute size* of the sample, i.e., *n*. Newcomers to the field sometimes advance the common-sense hypothesis that sampling error depends primarily on the *proportion* of the sample to the total population. They might argue, for example, that a 5 percent sample of a national census will be five times more reliable than a 1 percent sample, a 10 percent sample twice as reliable as a 5 percent sample, and so on. While this argument may seem plausible, it is erroneous.

The relative importance of the absolute sample size and proportionate size in reducing the standard error of the estimate can be demonstrated by a close look at Equation (12). This equation is reproduced as it first appeared, and then rewritten to show its three distinct components:

$$(12) \quad \text{Estimate of standard error } s_{\bar{x}} = \sqrt{\frac{N-n}{N}\frac{s^2}{n}} = \sqrt{\frac{N-n}{N}}\sqrt{\frac{1}{n}}\,(s)$$

The three components are: (1) a factor representing the influence of relative or *proportionate* sample size, $(N\text{-}n)/N$; (2) the factor $1/n$, which represents the influence of *absolute* sample size; and (3) the square root of the sample variance (s). According to Equation (12), the standard error of the estimate is the product of the square root of the sample variance multiplied by the factors representing the proportionate and absolute sample sizes. Since those latter two factors are always less than 1, the standard error will be *less than* the value of the square root of the sample variance. An inspection of this formula will show that increases in the absolute size of the sample do more to reduce the standard error than do comparable increases in the proportion that the sample is to the total population. The practical implication is that in most surveys the absolute size of the sample is of much greater relevance than its proportionate size.

The general principle just stated is illustrated concretely in Table 4-6. The left half of the table shows the values of the corresponding multiplying factors for ascending values of proportionate sample sizes from .01 to .95.

Table 4-6 Effect of the Proportionate and Absolute Size of Sample in the SRS Formula for the Standard Error

Proportionate size			Absolute size		
Sampling fraction f	$\dfrac{N-n}{N}$	$\sqrt{\dfrac{N-n}{N}}$	Sample size n	$\dfrac{1}{n}$	$\sqrt{\dfrac{1}{n}}$
.01	.99	.99	30	.033	.17
.05	.95	.97	50	.02	.14
.10	.90	.94	100	.01	.10
.20	.80	.89	250	.004	.06
.40	.60	.77	500	.002	.04
.50	.50	.70	1,000	.001	.03
.60	.40	.63	2,500	.0004	.02
.80	.20	.44	5,000	.0002	.01
.90	.10	.32	10,000	.0001	.01
.95	.05	.22	25,000	.00004	.01

The right half of the table shows the multiplying factors for absolute sample sizes ranging from 30 to 25,000. The picture that emerges is clear; even with a sampling fraction up to half, the corresponding multiplying factor does not get below .70, while with an absolute sample size as small as 100 cases, the corresponding multiplying factor is .10. In other words, the extremely high relative sample size tends to produce a standard error only about 70 percent of the sample variances, whereas even a small absolute sample size tends to reduce the standard error to 10 percent of the sample variances. The absolute size clearly carries more weight than does the relative sample size.

Another important conclusion can be drawn. Additions to the absolute sample size have a greater payoff in reducing the standard error than increases in the proportion of the sample in the population. For example, increasing the sample from 1 percent to 10 percent of the population reduces the proportionate size factor only about 5 percentage points, from .99 to .94. By contrast, increasing the absolute sample size from 50 to 250 reduces the corresponding multiplying factor by over half, from .14 to .06. Thus, except for samples from a very small population, it is the absolute size together with the sample variance which have the greatest impact on the size of the standard error of the estimate.

Table 4-6 also illustrates another basic precept of probability sampling: the principle of diminishing returns with increases in absolute sample size; that is, the reductions in sampling error decrease as the absolute size of the sample increases. Consider, for instance, the reductions produced by doubling the sample size from 50 to 100, 500 to 1000, and 5000 to 10,000. The multiplying factor $(1\sqrt{n})$ decreases by .04 in the first case, .01 in the second case, and only .004 in the third, despite the fact that the absolute increases in the second and third cases are much larger than in the first.

A final point to be remembered about sample size concerns the subgroups to be analyzed in a given survey. The standard error is determined in good part by the size of the group actually being analyzed, rather than the total sample size. A national survey may be designed with a sample size of 2000. When analyses are carried out on the entire sample, the levels of error are likely to be quite acceptable, for example, .02 or .03 around a percentage. The picture changes, sometimes dramatically, if the analysis shifts to subgroups in the total sample, such as geographic regions, ethnic groups, or occupational categories. Analysts sometimes forget that the standard error of an estimate for one of these subgroups depends on the absolute size of the sample for that subgroup. The more the subgroups to be analyzed, and the smaller the groups, the larger the sample needed to have sufficient cases to keep the sampling error within tolerable limits.

VARIATIONS ON SIMPLE RANDOM SAMPLING

Thus far we have focused heavily on simple random sampling. From both a theoretical and pedagogical standpoint, this is the basic sampling process, and one whose logic is applied to all other methods of probability sampling. Often, however, its stringent requirements cannot be met in a given study. There may be no complete and up-to-date listing of all persons who live in a city, community, or other unit of study. Such lists are rare, especially in the rural and marginal urban areas of less developed countries. Moreover, sometimes information about the population, perhaps from a recent census, can be used to improve on a simple random sampling design, reduce costs, or provide greater convenience in field work. These real-life considerations in survey research often make it necessary or desirable to modify the methods of simple random sampling.

A common reason for introducing variations is to meet the requirements of *area sampling*. This is one of the most frequent and useful applications of sampling for social researchers. Area sampling was developed during the late thirties when survey specialists realized that human populations could be sampled without having huge population lists. An economical alternative was to arrive at a sample of *people* by first sampling *geographic areas* and then subsampling for *dwellings* within the selected areas. The reasoning behind this innovation was straightforward. Except for a minor group of transients, people could normally be identified with a place of residence, and the place of residence with a physical location or area. The individuals who lived at the selected dwellings could then be identified through brief screening interviews at the selected dwellings. An area sampling frame, moreover, could be used repeatedly. Additional samples could be produced by varying only the random selection of dwellings within the sample areas. The method also provided access to widely different populations, such as women of childbearing age, heads of households, school-age children, or persons in the labor force. All that was needed was a change in the procedure for sampling individuals within the selected dwellings. A detailed example involving area sampling of households will be presented in the following chapter. With appropriate modifications, the area sampling technique may also be applied to business establishments, agricultural crops, ground use, and other problems.

We will now consider five modifications of simple random sampling: (1) stratification; (2) clustering; (3) systematic selection; (4) unequal probabilities of selection; and (5) multistage sampling. These modifications, in various combinations, yield an enormous variety of probability sampling methods to meet different survey objectives.

Stratification

In the field of sampling, *stratification is the process of dividing the population into subgroups or strata in order to carry out separate selections in each.* There are two main reasons for using stratification: to control the representativeness of the sample, and to permit the application of different selection procedures in different strata.

Representativeness   We argued earlier that a good sample should represent the differences that exist in the population. In the present context, *representativeness* refers to the similarity between the sample and the population in the proportion of cases falling into each of the different strata. Stratification can often improve representativeness on those variables on which it is based, such as age or sex, and on closely related variables.

To illustrate the uses of stratification, we can return to the hypothetical survey of 270 students from a population of 5400. Assume that the application of simple random sampling gave us a total of 140 lowerclassmen and 130 upperclassmen in the sample of 270. Because of sampling error, it is quite likely that the proportions of lower- and upper-class students in the sample do not correspond exactly to the same proportions in the population. Let us further assume that the registration records making up the sampling list show exactly 2900 lowerclassmen and 2500 upperclassmen. If we had stratified the list by upper- and lower-class standing (in effect, made two lists) and then applied the 1:20 sampling fraction separately to each group, we would have obtained subsamples of 145 lower- and 125 upperclassmen. The differences between these figures and those obtained through simple random sampling could arise from normal sampling error. Stratification of the list by class standing would thus produce some gains in representativeness *for that variable.* It is most unlikely, on the other hand, that this stratification procedure would have much effect on the representativeness of the sex composition of the sample. If, as is the case, the upper- and lowerclassmen were divided about equally among men and women, there would be no relationship between sex and the stratifying variable, class standing. On the other hand, this stratification procedure might well improve the representativeness of the sample with regard to age. Because upperclassmen are generally older than lowerclassmen, there would be a correlation between age and the stratifying variable, and thus a better chance of improving the representativeness of the former.

What could be done to increase the representativeness of the sample on sex distribution? To make improvements here without losing ground on class standing, a double stratification could be used. Four strata in the form of four separate lists could be formed: men and women in the two lower years, and men and women in the two upper years. Srs selection procedures could then be applied separately to each of the strata. The result would be an accurate representation of class standing as well as sexual composition in the sample. Thus, if 22 percent of the population falls into the category of *male, lower-class standing,* the sample also would have 22 percent in the same category. In this way, stratification can improve on the representativeness typically obtained with a straightforward application of srs to the total population list. However, this double stratification procedure would not correct for chance variation in the sample on such qualities as the proportions of freshmen versus sophomores in the lower-class group, or juniors versus seniors in the upper-class category. Nor would it usually have much impact on the representativeness of the students' major fields of interest, political party preference, religion, etc., unless one of these variables was correlated with one or both of the stratifying conditions.

Sometimes a researcher wants to improve representativeness on a certain variable, but can find no direct measure to use in stratification. Suppose, for example, that we wished to improve the precision of our student sample with regard to social class background. The concept of social class, to begin with, is subject to many definitions in sociology and political science. But even if one clear definition were available, it is unlikely that any single empirical condition would serve as an adequate index of social class. Under these conditions, it may be advisable to use *proxy variables* in stratification, i.e., variables which are closely associated with the preferred stratifying variable and on which information is available beforehand. In the student sample, proxy variables for social class background might be chosen from parents' education and occupation, residential areas, or requests for scholarship aid. The closer the proxy variables are to the main stratifying variable, the greater the gains in representativeness on that variable.

Different Procedures The second main reason for the stratification of a sample is to allow different selection or interviewing procedures to be used in the various strata. In many surveys, especially in the developing countries, there are wide variations across sample strata in geography and topography, the availability of maps and materials, customs, language, or simply convenience in administration. Any of these conditions may be sufficient reason to stratify the sample so that different procedures can be applied. In the 1970 national probability sample of Peru, for example, it was convenient and, at times, necessary to develop strata and substrata in which selection was based on a complete listing of the population within them. In other strata, more complex procedures were used, such as multistage area sampling, a process to be discussed later. Similarly, separate strata were formed for different cultural and linguistic areas in Peru, such as the Quechua and Aymara regions.

Separate strata also may be formed to allow small but significant subgroups, such as ethnic minorities, to be oversampled (unequal probabilities of selection). In a survey of Chicago, for example, it may be theoretically or practically important to study the Italian subgroup in that city. In a small sample relying on equal probabilities of selection, the number of selected cases from the Italian subgroup may be too small to permit reliable analysis. Under these conditions, the areas of the city where persons of Italian descent tend to live might be treated as a separate stratum and oversampled. Similarly, it might be practical to form a separate stratum of a large, homogeneous subgroup of the population and then undersample it. The problems involved in using different sampling fractions for different strata will be reviewed in the discussion of unequal probabilities of selection.

We might note in conclusion that elaborate schemes of stratification in survey research are often not worth the effort. Sometimes social researchers are so anxious to improve the representativeness of their samples on key variables that they produce designs that are unworkable in practice. Two observations might be made in this regard. First, stratification is most useful when the stratifying variables are simple to work with, easy to observe, and closely related to the dimension of concern: the greater the number of stratifying variables and the greater the difficulties they pose in the field, the less satisfactory the results for the sample and the greater the headaches during analysis. At best there may be only slight gains in representativeness and at worst the sample may be less representative than a simpler design because of difficulties in complying with overly complex procedures.

Second, the design should recognize an important difference between stratification and analysis. Variables that are crucial to the analytic purposes of a survey need not, and often should not, be used for stratification. Social class might be one of the most critical analytic variables, but the difficulties of classifying sample elements by this criterion may make it a very bad choice for stratification. The information from the survey itself should provide an excellent basis for subsequent classification of individuals on complicated variables such as class. This task can be done well only at the analysis stage when the relevant questionnaire responses are available.

Cluster Sampling

Cluster sampling is a procedure of selection in which the elements for the sample are chosen from the population in groups or clusters rather than singly. The clusters used are often preexisting natural or administrative groupings of the population, such as factories, schools, or political subdivisions. They may also consist of areas and the elements in them, formed by drawing lines around recognizable features of maps or aerial photographs. The basic idea of cluster sampling can be illustrated with a hypothetical study of the health of school children in a certain city. The population in question is made up of all children registered in school districts within the city limits. One approach to the study would be simple random sampling. The researchers could use administrative records to prepare a list of all children attending or registered in public and private schools. A simple random sample could then be drawn from the population list through the techniques indicated earlier. The result, however, would be a sample spread over the entire city. While this sample might present few problems in some types of research, in a study of health it might prove most inconvenient as well as costly to hold medical examinations, laboratory tests, x-rays, etc. in several dozen widely scattered schools.

An alternative would be to take a sample of entire schools, and then measure the health of all students registered in those schools. The schools in this example would be clusters—groupings of sample elements selected together. This sampling procedure would have the advantage of reducing costs and simplifying field work. At the same time, it would probably result in greater sampling error than a simple random sample including the same number of elements (students). Another possibility would be a compromise between simple random sampling and such large clusters as whole schools. The first stage would be to select *schools,* with a larger number chosen than in the previous cluster sampling. The second step would be to carry out a probability sample of *students* within the selected schools to arrive at the desired number of elements. This procedure is known as multistage sampling, and will be treated later as a separate modification of simple random sampling.

The major advantages of cluster sampling are increased convenience and reduced cost. Several specific benefits can be cited. First, the selection of elements in clusters usually simplifies the interviewing process. This is particularly true in studies aiming for state, regional, national, or even international coverage. In such studies it is quite common to have the final selection of elements made within clusters called *primary sampling units* (PSUs). Counties, electoral districts, or similar political subdivisions are widely used for this purpose, with perhaps 60 chosen to represent the many hundreds or thousands in which the total population is found. A sample requiring interviewing in only 60 rather than several hundred areas facilitates the recruitment and training of locally based interviewers and reduces the time and costs involved in travel. The result may be a greater output per day than would be the case in a more scattered sample. Second, the use of clustering also facilitates the supervision of the interviewing and the general administration of field work. We will argue later that the quality of survey data is closely related to the care exercised in field supervision. If several interviewers can work together in a single geographic area, it is much easier for the supervisor to maintain morale, ensure that uniform interpretations are being made, and to control the quality of the interviews. Under these conditions, it is also easier for the central research staff to coordinate the entire survey than if field operations were spread over several hundred disparate sampling points. Third, the greater workload and the simplified administrative structure will often result in reduced costs for salaries, travel, office supplies, postage, telephone calls, and related expenditures. For the same reasons the field work, including the preparation of sampling materials (lists, sketches, etc.), may be completed more quickly than with a simple random sample of approximately the same size.

The principal drawback to cluster sampling, when compared with simple random sampling, is the likelihood of increased sampling error. We noted earlier that, in general, as the size of the sample increases, the size of the standard error decreases. This rule applies, however, only when each sample element is selected independently of every other element, as in simple random sampling. In cluster sampling the elements are, by definition, selected in a group rather than independently. The effect of clustered selection on the standard error will depend on the similarity between the elements in the cluster and those in the population. In many cases, sample elements selected in clusters will not show the same variation as an equivalent number selected independently. People who live in the same block or the same apartment complex may be more like each other with regard to a certain characteristic, such as income, than people in the population at large. As a result, sample cases selected in clusters may not have the same effect in reducing sampling error as the same number of cases selected by simple random sampling. Thus, a study might need a clustered sample of 2000 cases to produce the same standard error as an unclustered sample of 1000. This point has important practical implications when the researcher is calculating the costs and benefits of one sampling method versus another.

The relationship between clustering and sampling error may be summarized as follows. If all the elements in the cluster (e.g., a city block) were identical with regard to a certain characteristic (e.g., income), and totally different from the elements in other clusters, the sampling error would be extremely high. Clustering, in this case, would tend to make the clustered sample equivalent in size to a simple random sample with as many cases as there are *clusters,* rather than *elements.* Hence, a sample made up of 60 clusters might be equivalent to a simple random sample of 60 individuals. This is obviously an extreme that is never seen in practice. At the opposite extreme would be a series of clusters showing the same variation within each cluster as simple random samples of the same size. In this case, each cluster would represent the entire population, another condition rarely met in practice. Most sampling situations fall between these extremes, tending toward one or the other according to the characteristic being studied. In general, experience has shown that well-designed cluster samples will produce standard errors that often are about one and one-half times as large as the standard errors from srs samples of the same size. The mathematical procedures used in calculating the standard errors of clustered samples are beyond the scope of this discussion. These procedures are well treated, however, in standard works on sampling, especially Kish (1965).

Finally, several practical considerations might be mentioned. Clustered samples may be designed to include all of the elements in a given cluster, or only part. The results are called *take-all* or *take-part* clusters, with the latter requiring an additional stage of sampling. In either case, the clusters of final elements (the individuals, households, etc. actually selected for interviews) should be kept small if they are likely to produce little variation on the main variables of the study. For example, numerous studies have shown that such characteristics as race, social class, and ownership of automobiles or durable consumer goods tend to be similar among neighbors in the United States. If these characteristics are critical to the analysis plans of the study, it would be advisable to use relatively small clusters of households in the sample. On the other hand, some variables, such as fertility, show considerable variation among neighbors, so that the effects of clustered household units may be slight. In general, where it is possible to choose the design of final element clusters, it is safest to have more rather than less heterogeneity within clusters. This suggestion is particularly germane when it is difficult to obtain advance information about the effects of clustering on the variability of key measures in the study. In this situation the final clusters of dwelling units should usually be around 4 to 5 units, and should rarely go above eight. If the use of large clusters cannot be avoided, a selection procedure should be devised to reduce their size, such as making them *take-part* clusters.

Systematic Selection

A third modification of simple random sampling is *systematic selection*. This is *a method of selecting units from a list through the application of a selection interval, I, so that every Ith unit on the list, following a random start, is included in the sample.* The interval, *I,* is readily determined by dividing the population size (N) by the desired sample size (n). The result is the inverse of the sampling fraction, *f.*

$$I = \frac{N}{n} = \frac{1}{f} \tag{13}$$

Suppose we wished to select a systematic sample of 270 students from the registration list of 5400 students in the earlier example. The first step would be to calculate the interval, $I = 5400/270 = 20$. Then we would choose a random start from the first segment of length I on the list, that is, 1–20. A random number within this interval could easily be obtained from a table of random numbers. If the number 16 was drawn, the 16th student on the registration list would be the first selection. The final step would involve adding the value of the interval (20) to the random start (16) and to the succeeding numbers obtained in this way. In our example, the sample of registration numbers would be:

$$16, 36, 56, 76, 96, 116, \ldots , 5356, 5376, 5396$$

The students corresponding to the selected numbers would comprise the sample elements.

The lists to which systematic selection is applied may be either *written lists,* such as a registration list of students at a university or lists of dwellings, or *proxy lists,* such as rows of houses found side-by-side along a street or around a block, or individual case records in a file. The only requirement for such unwritten lists is that the principle of ordering, such as physical proximity in the cases of houses and files, be unambiguous. Much of the art and hard work of survey sampling lies precisely in discovering lists through which sampling may be carried out. Once the lists are ready, the selection process itself is often simple and swift.

The main advantage of systematic selection is simplicity and ease of administration. In the example cited, it is much more convenient to draw one random number and then proceed by intervals of 20 than to draw 270 random numbers. Another advantage arises when the lists from which the selections are made are implicitly stratified on a variable important to the study. An example would be a list of factory employees organized by the date of hiring. If length of service at the factory were a significant control variable in the study, the use of systematic selection from this list would have the side effect of stratifying along that variable. Sometimes an element list is intentionally ordered with this benefit in mind.

In contrast with simple random sampling, units are not selected independently in systematic selection. Depending on the selection interval and the order of the sampling units on the list, only certain combinations of units can be chosen. This condition does not prevent each sampling unit from having a known or even identical chance of selection, but it does violate one of the formal assumptions of srs sampling, which is independence in the selection of elements. Whether this departure brings advantages or disadvantages in reducing sampling error depends on the specific circumstances involved. In the case of the employee list ordered by length of service at the factory, the gains are likely to outweigh the drawbacks. In other cases, bias may be introduced.

The greatest danger of bias lies in a coincidence between the selection interval and a cyclical repetition of some characteristic along the list. An extreme example would be a population list made up of husbands and wives who appear alternately on successive lines. An even-numbered selection interval (e.g., 4) would yield a sample made up of either all men or all women, despite the fact that the list is evenly divided between the two. To protect against less obvious occurrences of this type, a long list can be broken up into several sections with separate selection procedures, including a different random start, applied to each. Another source of bias would arise in a list ordered in steadily increasing values of a variable to be measured. A survey of credit union borrowers, for example, might use a list organized by the size of loans received by each borrower. In this case, the point selected for the random start could influence the average loan size found in the sample. If the interval was fairly large and the random start was near the beginning, the average amounts would be lower than if the start was at the other end of the interval. The importance of this bias would depend on the range of differences across the interval. If the range is small relative to the mean and median, the resulting bias could probably be ignored.

Unequal Probabilities of Selection

In unequal probabilities of selection, sampling units are chosen by a procedure giving some elements a higher or lower chance of selection than others. As a result, the sample includes proportionally more of the oversampled cases. One reason for adopting unequal probabilities of selection (UPS) was mentioned in connection with stratification. A subgroup of the population, such as the wealthy or a minority group, may be extremely important in meeting analysis objectives and yet very small relative to the total population. A sampling fraction suitable for producing the desired overall sample size might yield too few cases to permit separate analysis of the subgroup in question. With very few sample cases, the standard error of an estimate based on data from the subgroup would be unacceptably high. Under these conditions, it may be advisable to form separate strata for the subgroups (e.g., the wealthy or the minority group) and then sample them separately at a higher rate. Through procedures to be discussed shortly, the same result may be achieved by using unequal probabilities of selection without explicitly forming separate strata.

Another situation in which the use of UPS is common involves sampling units of widely different size. In area sampling, for example, the first-stage or primary sampling units (PSUs) are often defined by natural or administrative units such as blocks, political subdivisions, or segments marked out on maps and sketches. Such units will usually show marked differences in the number of elements within their boundaries. Rather than forcing the formation of sampling units with approximately equal numbers of elements, the researcher can control for size differences by adjusting the selection probabilities to each unit's size. One method of doing so, known as *probabilities proportional to size* (PPS), will be explained later.

A third reason for using UPS is to reduce the costs of sampling and/or interviewing. In some surveys, particularly in the developing nations, the expenses involved in preparing a sample and collecting data will differ substantially between areas or population subgroups. Other "costs" might include danger to the lives or health of the sampling team or the interviewers, a legitimate concern in many countries. In such cases, it may be worthwhile to use lower sampling rates for the more expensive or difficult segments of the population.

A further consideration is the variance of important data across the several strata. Sample cases from population subgroups showing high variance on a given characteristic (e.g., income) will tend to reduce the overall sample variance more than the same number of cases from low variance subgroups. In such circumstances there are advantages to oversampling the high variance subgroups. Further consideration of these and related factors are found in sampling textbooks under the heading of *optimization.*

Offsetting these benefits is one overriding drawback: the need to assign compensating weights to sample elements, and to apply them during analysis. Used alone, a sample drawn according to UPS procedures would give a distorted picture of the population. The oversampled strata or elements would carry more weight than they deserve, while the rest would be underrepresented. For example, if the wealthy were sampled at three times the rate used for the other strata, and if no adjustments were made during analysis, this subgroup would carry three times its proper weight in any tabulation based on total sample data. It is usually necessary, therefore, to assign weights which will either *decrease* the overall representation of oversampled elements, or *increase* the representation of those that were undersampled. Thus, if the wealthy were oversampled by a 3:1 ratio, their weights in analyses carried out on the total sample should be one-third that of elements sampled at the regular rate.

Other problems in using UPS procedures may arise from the researcher's inexperience in working with weighted data or the lack of appropriate data processing equipment. Any tabulation or calculation based on weighted data is more complicated than the same operation carried out with unweighted data. The more intricate the analysis, the greater the complexities. Sometimes the proper balance can easily be restored by duplicating the punch cards for those categories of sample cases that are underrepresented because of UPS. For example, if the stratum corresponding to the more wealthy respondents is sampled at twice the rate of the remainder of the population, the problems of weighting can be dealt with rather simply by duplicating the cards for all of the samples except those in the wealthy stratum, and then merging the two sets of cards for data processing. With more complex UPS samples, this procedure may be inadequate. While most of the mechanics of weighting can be handled semiautomatically by modern computers and available data processing "packages," the analyst must be well trained in the use of weighted data and constantly alert to their implications. If he has little experience with survey analysis, or if the data processing equipment has very limited capacity for handling weighted data, UPS would be inadvisable.

Beyond these drawbacks, there are other costs to be considered with UPS. Careful records must be maintained on the sampling fractions used with different strata or elements, the preparation of tabulation requests must give explicit attention to compensating weights, the description of the sampling procedure is lengthened, and the increased possibility of error requires extra attention to the details of data processing and analysis.

Procedures and Examples   UPS can be built into a sample design in at least three ways: (1) by dividing the population into strata and then applying different sampling fractions to the strata; (2) by deliberately assigning higher (or lower) probabilities of selection to special types of elements on a list without forming separate strata; and (3) by assigning probabilities proportional to the size of sampling units, such as city blocks or clusters of dwellings. The first two can be illustrated in a hypothetical study of the financing of student expenses of higher education. To determine how students pay for their university training, a team of social scientists plans to conduct a survey of students from the same population. In anticipation of the survey, each student was asked at registration if he or she expected to work for pay during the school year. The analysis plan calls for comparisons between those who expected to work and those who did not. A total of 1800 students fell into the former category, 3600 into the latter. The task now is to draw a sample of 400 students divided equally between these groups, that is, with 200 from each.

The first possibility is to form distinct strata consisting of those who expect to work and those who do not, and then carry out separate selections in each to obtain 200 students. The sampling fractions required would be 1:9 for those who expect to work (1800) and 1:18 for those who do not (3600), as shown in Table 4.7.

Table 4-7 Sampling Fractions to Obtain Equal Numbers of Sample Cases of Workers and Nonworkers from the Student Population

| | Subgroup | |
Item	Planning to work (workers)	Not planning to work (Nonworkers)
Desired n	200	200
N	1800	3600
$f = n/N$	1:9	1:18

The second option would produce the same result without explicit stratification. The procedure would be to assign two selection chances to each nonworker on the list compared to one chance for each worker. The chances would then be cumulated for workers and nonworkers in a column called *selection possibilities,* as in Table 4-8. Systematic selection would next be applied to the list of cumulated chances. The appropriate interval, $I,$ would be calculated by dividing the total number of selection possibilities, 7200 (1800 students with two chances each, plus 3600 students with one chance each), by the desired sample size of 400. Thus $I = 7200/400 = 18.$ This interval would be applied to the random start, 7, and successively to the numbers that follow (25, 43, 61, 79, etc.). The student corresponding to line number 4 (selection number 7) would be the first selected. By chance he is a person who is expected to work. The probability of selection for a worker in this method would be $2:18 = 1:9.$ The worker has two chances that the selection number chosen from every interval of 18 numbers will coincide with his own. The nonworker has only one chance in 18 of being selected, or half as much as for a worker. Though the selection procedures are somewhat

Table 4-8 Excerpt from List of Registered Students with Modifications Providing for Unequal Probabilities of Selection Based on Students' Plans to Work

Line number	Student identification	Classification	Selection weight	Selection possibilities
1	(Name, or other	worker	2	1,2
2	unambiguous identifi-	non-worker	1	3
3	cation)	worker	2	4,5
4		worker	2	6,7
5		non-worker	1	8
6		non-worker	1	9
etc.		etc.	etc.	etc.
5399		non-worker	1	7199
5400		non-worker	1	7200

different, the result produced by this method is almost identical with that obtained from the approach shown in Table 4-7.

Selection with *probabilities proportional to size* (PPS) is a special case of unequal probabilities of selection. Here the selection probability assigned to a unit, such as a city block or an apartment complex, is set proportional to a measure of its size. For example, when the primary sampling unit (PSU) for an area sample is the county, probabilities of selection could be based on the size of each county in the last census or on an updated estimate of this figure. Of course, the measure of size should be in units related to the population under study. In the case of counties it should be in units of population or dwelling units rather than land area. Because of wide variations in the densities of population by counties, a measure of land area might bear little relation to population size. The procedures for applying PPS will be shown in the example presented in Chapter 5.

Adjusting for Unequal Probabilities   Two basic techniques are used to compensate for the overrepresentation of sample elements produced by UPS. The first is *weighting* of sample elements. This is a procedure for restoring to the sample the same proportion of elements of different types that are found in the population. Higher weights are assigned to individual sample cases of the type that are underrepresented, and lower weights to those that are overrepresented. In the illustrations shown in Tables 4-7 and 4-8, a weight of two could be assigned to data obtained from nonworkers in the sample, and a weight of one to workers. The sample data are then tabulated using the respective weights of the sample cases. Usually the simplest method of assigning weights is to use the inverse of the sampling fractions by which elements were selected. As in the previous example, a sample element selected at 1/4 of the highest rate would have a weight four times as large as elements selected at the maximum rate. The calculation of most statistics from the sample, such as means, medians, and percentages, should be carried out from weighted tabulations. However, in calculating sampling errors, the actual sample size rather than the sum of the weights should be used. No increase in the effective sample size is brought about by adding weights.

The second technique for dealing with unequal selection probabilities is that of *compensating selection probabilities in sampling.* Equal overall probabilities may be restored by using a multistage design in which selection probabilities are unequal within each stage, but with a chain of selection ultimately producing the same selection probability for each selected case. For example, some dwellings might be within blocks selected at 1:5, with the dwellings subselected at 1:20. The overall sampling fraction is thus 1:100. Other sample dwellings may be in blocks selected at 1:25, with dwellings subselected at 1:4. The overall probability of selection would be the same as in the previous case, 1:100. Other combinations could be devised to yield the same rate.

Multistage Sampling

The final variation on simple random sampling is *multistage sampling*. This is *a process of selecting a sample in two or more successive, contingent stages.* The basic procedure will be outlined for two stages, although the logic can be extended to others. The first stage consists of two fundamental operations: dividing the target population into many large groups or clusters of elements (primary sampling units—PSUs), and then using chance procedures to select a large number of these to represent the entire set. The first stage is intentionally designed so that the selected PSUs, such as city blocks or counties, will contain more elements than are needed for the final sample. The second (or final) stage involves subsampling within the selected PSUs to draw the sample of elements, such as individuals or dwelling units.

Compared with simple random sampling, the major advantages of multistage sampling are convenience and economy. Costs of sample preparation are greatly reduced by the fact that the multistage approach does not require complete lists of individual elements in the population. Consider, for example, a two-stage sample of dwellings in a city in which the PSUs are blocks or other small geographic areas. In this case, it would be necessary to have lists of individual dwellings only for the *selected* PSUs rather than for the entire city. To carry out the first-stage selections, all that is needed is a list of the PSUs and a measure of size for each. Similarly, for a sample of school children, a list of schools and homerooms would suffice for the first-stage selections. Lists of the individual children would be needed only for the selected homerooms. Other advantages were discussed in the section on clustering where we pointed to possible economies in interviewing and field administration. Multistage sampling may also contribute to better quality in carrying out the final selections, especially by permitting greater flexibility and control in developing lists of population elements.

The major disadvantage of multistage sampling is the same as with cluster sampling: increased sampling error arising from the selection of sample cases in groups rather than independently. This drawback can usually be kept to tolerable proportions by selecting a relatively large number of PSUs in the first stage. In cases where the possible number of PSUs ranges from several hundred to a thousand or more, it is often advisable to select from 50 to 100 PSUs to keep sampling error within reasonable bounds. This is a very imprecise rule of thumb introduced to illustrate a point rather than to suggest a sample design. In any concrete case, the best policy is to consult an experienced sampling statistician.

Finally, several practical considerations should be taken into account in developing a multistage sample. First, the primary sampling units should cover all the elements in the population to be sampled. Every element must be identified with one or another PSU, but none should appear in more than one. Observance of this rule ensures that every element will have a known chance of being selected into the sample at the first stage. A second and related point is that the definitions and boundaries of PSUs, as well as associated sampling instructions, should be unambiguous. If these first-stage units consist of geographic areas, it is not enough to list the name of a certain suburb or area of the city. The instructions must give exact delimitations of its boundaries, stating precisely what is and is not included in this area. In many cases, this will mean walking around the area or consulting an enlarged aerial photograph to determine the exact limits. In area sampling it is common practice to use clearly defined natural boundaries, such as rivers, or unambiguous manmade limits, such as roads, in an effort to provide sharply delineated PSUs.

Third, the success of a probability sample depends on the observance of predetermined rules for the last stage of sampling, the selection of respondents. Only rarely will sampling requirements be met by interviewing whomever happens to appear at the door. Usually the target population is defined as a specific set of individuals living within the dwelling units, such as heads of household, all women fifteen to forty-nine years of age, employed males, or female registered voters. The instructions to the interviewers should specify exactly who is to be interviewed and how they are to be chosen. A brief filter interview is often used to identify the habitual residents in the selected dwellings. This information can then be used to make the final stage of selection. It is worth emphasizing that all of the elaborate work of preparing a multistage sample can come to naught if proper procedures are not followed in selecting respondents. Interviewers should have extensive training and practice before being assigned to this task.

Fourth, with a relatively modest additional investment, a multistage sample may be converted into a *master frame* for subsequent samples. The administrative overhead and the field work required to develop a national sample, for instance, may provide handsome dividends for other studies requiring probability sampling. The key to this benefit is the fact that the selected PSUs usually contain many more elements than are needed for one sample. It is thus possible to draw new samples from the same population by repeating the final stage of the selection process with different random numbers. In most cases, the first-stage selections of PSUs and the listing of sample units, such as dwellings, within these PSUs are the only raw materials needed. Additional work may be needed, however, if there will be a time lag of more than a few months between the first and later samples. In this case, it is often necessary to review the listings to determine the extent of new construction, demolition of buildings, and similar changes affecting the sample. The costs involved in updating the master frame are usually small. A detailed illustration of a multistage probability sample follows in the next chapter.

CONCLUSION

This chapter has dealt with the basic principles of sampling and estimation largely at the conceptual level. But there is much more to designing a good sample than having a grasp of concepts and principles. With area samples, there is considerable art to making the best use of available materials (maps, census data, etc.) and manpower. Training and careful supervision during sampling field work are usually as important for high quality as an adequate theoretical design. Unfortunately, slipshod work in the practical procedures of sampling often escapes notice when the data are reported. Statistical tables report findings for the various groups and subgroups under study, but they typically provide no means for evaluating the quality of the information. Here, as elsewhere in survey research, theory and practice are inseparable in producing reliable information. Study directors who feel that the basic task is done when calculations of standard error are completed could be sadly disappointed by a careful scrutiny of the field operations in sampling. Thus, we would strongly recommend that in any complex sample, including almost any large area sample, the study directors consult an experienced sampling statistician in the early stages of planning. The great advantage of such consultants is that they can effectively blend theory and practical considerations to recommend an optimum design for the purposes of the research.

While careful sampling is extremely important in survey research, it is but one part of an integral process. It should be remembered, in particular, that sampling is only one of several sources of error. The quality of the questionnaire, the overall response rate in interviewing, and the interviewing and reliability of the coding are other factors which influence the accuracy of the data. Thus, in planning the survey the study directors should aim for an adequate balance of precision and quality at all stages. There is little to be gained from a highly precise sample which is undercut by a low response rate, or from a top-flight interviewing staff which works with a third-rate sample. Each incomplete interview with a selected respondent is potentially damaging to the sample, as are substitutions of respondents or other failures to follow instructions. In short, while it is often tempting to see sampling as an end in itself, partly because the results are more easily measurable than in some other parts of survey research, the proper perspective is to set sampling in the total context of the research. In the end what counts is overall error, rather than just sampling error.

FURTHER READINGS

Kish, L., *Survey Sampling* (New York: Wiley, 1965).
Stephan, F. F., and P. J. McCarthy, *Sampling Opinions: An Analysis of Survey Procedure* (New York: Wiley, 1963).

Chapter 5

Illustration: A Multistage Area Sample

The sampling concepts introduced in Chapter 4 can be illustrated with an example of multistage area sampling. The aim of the exercise is to develop a sample of heads of households for the national urban survey that serves as the principal example in this book. The guiding questions for the survey were outlined in Chapter 2, the questionnaire is presented at the end of Chapter 6, and a specimen code for the information is supplied in Chapter 9. In line with the objectives of this study, the sample design covers only the largest cities in the country of Pacifica. These are defined as cities whose population is 50,000 or over. Eight cities in Pacifica qualify, as is shown in Table 5-1. The immediate task is to design and select a probability sample which will yield about 2000 interviews with household heads from this urban population.

ESTABLISHING THE SAMPLING FRACTION

To obtain 2000 completed interviews, it will be necessary to select more than that number of dwelling units. Some dwellings in the population will be

Table 5-1   Cities in Pacifica with 50,000 or More Inhabitants

City	1970 population	Estimated population in 1974	Estimated number of dwellings in 1974
San Pedro	1,000,000	1,200,000	324,000
Capital City	300,000	360,000	98,000
New Britain	190,000	228,000	62,000
Venice	140,000	168,000	45,000
Bangla	120,000	144,000	39,000
Haneda	110,000	132,000	35,000
Nairobi	80,000	96,000	26,000
Praha	60,000	72,000	20,000
Total	**2,000,000**	**2,400,000**	**649,000**

vacant or uninhabitable, and others will not be used as residences. As these appear among the selections, they can be set aside without replacement, for they are nonsample addresses, those currently without a head of household. Among the occupied dwellings, some heads of households will be temporarily away and others will fail to keep appointments or refuse to be interviewed, and will thus be classified as noninterviews. The selection target (x) must thus be increased to take account of these and related contingencies. Estimates of the frequency of vacant dwellings, refusals, etc. may be based on past experiences in the country, if there are any, or simply on educated guesses by an experienced sampler.

Given a target of 2000 completed interviews, the number of dwelling units to be selected can be determined with an equation involving the rates of occupancy and completion of interviews. In our example we will assume that 96 percent of the dwellings will be occupied, and that 85 percent of the household heads of occupied dwellings will be successfully interviewed. With this information we can solve for x, the number of dwellings to be selected:

$$\frac{\text{Completed interviews}}{2000} = \frac{\text{completed interview rate}}{(.85)} \times \frac{\text{dwelling use rate}}{(.96)} \times \frac{\text{dwellings to be selected}}{(x)}$$

$$(x) = \frac{2000}{(.85)\,(.96)} = \frac{2000}{.816} = 2450$$

This calculation shows that it will require a sample of 2450 dwellings to yield about 2000 completed interviews.

The next ingredient needed for the sampling fraction is an estimate of the total number of dwellings in 1974, the period of the study. The last available data on the population of cities in Pacifica comes from the 1970 census. These figures would have to be adjusted, however, to take account of an urban growth rate of about 4 1/2 percent per year. With this growth rate, the population of 2,000,000 in 1970 would be about 20 percent larger, or 2,400,000 in 1974.[1] Data on the number of dwellings in 1970 may be available from the census, or they may have to be estimated from population data through figures on average household size. In Table 5-1, the estimated number of dwellings in 1974 is obtained by dividing the estimated population for each city by average household size for that city (roughly 3.6 to 3.7). Let us assume that the figures in the last column are reasonable estimates and proceed with the selections on that basis.

Using Equation (1) from Chapter 4, we can now calculate the overall sampling fraction that would produce the desired size of 2450 dwellings.

$$f = \frac{n}{N} = \frac{\text{desired number of dwellings}}{\text{total number of dwellings}} = \frac{2,450}{649,000} = \frac{1}{265}$$

Assuming equal overall probabilities of selection, the rate is thus 1:265.

FIRST-STAGE SELECTIONS

Different strategies could be followed to select the 2450 dwellings. For example, a proportionate number could be selected from each city. This procedure would mean that sampling field work and interviewing would have to be carried out in all of the eight major cities. Another option is to cluster the sample by selecting only some of the cities. This procedure would reduce the costs and complexity of field work, and allow for greater quality control in the interviews. At the same time, it would result in higher sampling errors than the first method. Let us assume that considerations of cost and convenience ultimately lead us to a design involving multistage cluster sampling, stratification, and systematic selection with probabilities proportional to size (PPS). The first task is then to divide the eight cities into different strata.

San Pedro, with half of the target population in Pacifica, is so large that it must clearly enter the sample. It is thus included as a separate stratum with a probability of selection of 1:1, or certainty. Under these conditions, it is said to be self-representing. That is, when the overall selection rate for the elements (dwellings) is applied to San Pedro, the resulting sample cases will represent this city alone.

How many of the remaining seven cities should be chosen, and how should the selections be made? We argued earlier that strata should generally be composed of sampling units that are similar to each other and that include about the same range of diversity. We are now confronted by Capi-

[1] The actual estimate is 2,385,000, but it is rounded for purposes of simplicity. This is obtained by multiplying 2,000,000 × 1.045, the product of these figures (2,090,000) by 1.045, and so on for four years.

tal City, the national capital, which has an economic base unlike another city in Pacifica. Fortunately, it is the largest of the remaining cities. In fact, the other six cities combined have only a little more than twice the number of its dwellings. This situation argues for making Capital City another self-representing stratum, and for placing the other six cities in two additional strata.

The two next largest cities are New Britain and Venice. Both are seaports, with fishing and international trade playing an important role in their economies. The four remaining cities are inland, serving in large part as commercial and trading centers for the areas in which they are located. By coincidence this division between coastal and inland cities produces two groups of roughly equal size, and makes a reasonable basis for two strata. We thus have four strata: two self-representing ones; another one made up of two coastal cities; and the last including the four inland cities of Bangla, Haneda, Nairobi, and Praha. The sample design thus calls for the inclusion of the cities that make up the two self-representing strata, and for the selection of one city from each of the other two strata.

To conclude the first stage of sampling, it is necessary to select the cities that will be in the sample. The four strata and related population data are shown in Table 5-2. To simplify the presentation, we will call these first strata *domains* and the first round of sampling units chosen within each city the *primary sampling units* (PSUs).

The selections from the first two domains have been made by definition, i.e., both San Pedro and Capital City are automatically included because of

Table 5-2 Population Estimates in 1974 for Eight Largest Cities of Pacifica, Selection Probabilities and Selection Ranges, by Domains

City	Estimated population in 1974		First-stage selection probabilities	Selection ranges
	Inhabitants	Dwellings		
Domain 1				
San Pedro	1,200,000	324,000	1:1	Certainty
Domain 2				
Capital City	360,000	98,000	1:1	Certainty
Domain 3	396,000	107,000		
New Britain	228,000	62,000	62:107	001–062
Venice	168,000	45,000	45:107	063–107
Domain 4	444,000	120,000		
Bangla	144,000	39,000	39:120	001–039
Haneda	132,000	35,000	35:120	040–074
Nairobi	96,000	26,000	26:120	075–100
Praha	72,000	20,000	20:120	101–120

the sample design. The dwellings within these cities will be sampled at the overall rate for the study, 1:265.

The task remaining is to select one city each from Domains 3 and 4. The selection process outlined in Table 5-2 uses the procedure of selection with *probabilities proportional to size* (PPS). That is, each city within the domain will have a chance of being selected proportional to the size of its population. Domain 3, for example, has a total of 107,000 dwellings, 62,000 of which are in New Britain and 45,000 of which are in Venice. According to PPS procedures, New Britain will have 62:107 chances of being selected, while Venice will have 45:107. The same principle is applied to the cities in Domain 4.

To carry out the selection for Domain 3, we must first select a three-digit random number corresponding to the range of 001 to 107. If this number is from 001 to 062, New Britain will be selected for the sample; if it is between 063 and 107, Venice will be the choice. Similarly, for Domain 4, the random number would be chosen from the range of 001 to 120. The probabilities of selection are again determined by the size of the city, i.e., the proportion of dwelling units in that city to the total number of dwelling units in the domain, 120,000. Bangla would be chosen if the random number fell in 001 to 039, Haneda if it were 040 to 074, and so on as in the last column of Table 5-2. The random number selected for Domain 3 is 071. This falls into the selection range for Venice and selects that city. The random number for Domain 4 turns out to be 113, within the selection range for Praha. The four cities falling into the sample, therefore, are San Pedro, Capital City, Venice, and Praha. The first two will represent themselves, and the last two will represent themselves *and* the unselected cities from their strata.

LATER STAGES

An immediate problem created by the selection of Pacifica cities is the unequal first-stage selection probabilities between the domains. San Pedro and Capital City were selected at a rate of 1:1, Venice at 45:107, and Praha at 1:6. Earlier we set as an ultimate target an overall sampling rate of 1:265 dwellings. To move from the first-stage probabilities just indicated to the same overall rate, compensating selection rates must be made in subsequent stages of sampling. That is, it will not be correct to sample *within* all four of the selected cities at the rate of 1:265 which is applicable to San Pedro and Capital City. To select at this rate within the other two cities would give the dwellings from those domains a lesser chance of being selected into the sample.

Table 5-3 presents two basic items of information about the first-stage sample: (1) the first-stage selection probabilities for the selected cities, and

Table 5-3   Selection Probabilities for Cities Chosen in the First-stage Selections

Domain number	City	Selection probabilities		
		1st stage $\times$	Subsequent stage(s) =	Overall
1	San Pedro	1:1	1:265	1:265
2	Capital City	1:1	1:265	1:265
3	Venice	45:107	1:111.4	1:265
4	Praha	1:6	1:44.16	1:265

(2) the selection rates required in subsequent stages to provide for equal *overall* rates of selection for the sample dwellings in the four domains. For example, with a first-stage selection probability of 45:107, the dwellings within Venice will have to be selected at a rate of 1:111.4 in later stages to produce an overall selection rate of 1:265. By contrast, San Pedro and Capital City, which were selected at 1:1, will have an overall selection rate of 1:265 by using that rate in later stages. The basic point is that the product of the first stage and subsequent stage selection probabilities should result in the same overall fraction (1:265) for each domain. If the sample design calls for *unequal probabilities of selection* across the domains, the figures could be adjusted to produce the desired yield. For example, if we wished to halve the sample fraction for Domain 4 from 1:265 to 1:530, the selection rate in the subsequent stage would be 1:88.32 rather than 1:44.16.

Forming PSUs

The next step is to divide each of the four selected cities into *primary sampling units* (PSUs), area units which will themselves be sampled at the next stage of selection. They may be formed by city blocks or combinations of blocks, or other land areas with clearly identified boundaries on a map, sketch, or aerial photograph. An important requirement for PSUs is that each unit have an associated measure of size representing its total number of dwelling units. Measures of size usually come from two sources: information, including detailed maps, provided by the national or local census office, and *quick counts* of dwellings carried out by field workers in the survey. For example, in the present study we find that block maps are unavailable for parts of San Pedro, especially newer sections made up of improvised dwellings. To provide raw materials from which to develop PSUs, field workers should be sent to make map sketches of those areas. The areas should then be divided into segments with clearly recognizable boundaries, and the number of dwellings in each segment estimated by a quick count. Whatever the method, it is essential to arrive at a list of PSUs covering each selected city in its entirety, including commercial areas, parks, and empty lands as part of nearby PSUs. The reason for including *all* lands is that new construction may take place between the time of the initial sampling field work and the time of the survey.

In planning this step, it will be necessary to refer to the definition of the city which was adopted in the study, and on which the population estimates were based. Does the exact territory covered by the city include all of the suburbs and communities in the immediate area around the city, or is it confined to the municipality itself? For this example, we decided on the latter rather than the greater metropolitan area and its attendant complications.

Once the list of PSUs is ready it can be sampled, using PPS to draw the desired number of PSUs for each domain in the sample. Then careful and complete lists of dwelling units will need to be compiled for the *selected* PSUs. These will permit further subsampling to select individual dwelling units. The sampling procedure is thus multistage, with each stage intimately linked to the others.

In the Pacifica study, sampling error and related considerations suggest that the full urban sample should be drawn from a total of at least 60 PSUs, and the PSUs should be selected in pairs from broad strata to facilitate calculations of sampling error.[2] A total of at least 60 PSUs would mean one pair of PSUs for every 21,600 dwellings in the population. Given the differences in size between domains, the 60 PSUs would be distributed as follows: 15 pairs of selected PSUs would be drawn in San Pedro, 5 pairs in Capital City, and 5 and 6 pairs respectively in Venice and Praha. These sets of pairs add up to a total sample, distributed roughly in proportion to the sizes of the domains. In addition, the plan provides that San Pedro and Capital City will be self-representing.

At this point, the study directors want to know if the sample of Capital City will provide enough cases to permit reliable generalizations about that city taken alone. The answer to this question depends basically on the total number of completed interviews and the standard errors of estimates arising from this particular sample design. On the question of numbers, the overall sampling rate will produce about 370 selected dwellings in Capital City, provided that the estimated number of dwellings in 1974 is roughly correct. Losses from vacancies and noninterviews are likely to reduce this figure to about 300 completed interviews.

With regard to standard errors of estimates, it must be remembered that errors will be greater in this design, which relies on cluster sampling, than in a straight srs procedure. On the basis of his calculations, the sampling advi-

---

[2] In complex designs such as this, sampling error calculations frequently depend on paired selections. The point is too technical to pursue in this work, but the potential user of sampling should be alerted to its importance and should discuss it with a qualified sampling advisor.

sor concludes that much more reliable generalizations based on Capital City alone would result if the 370 selections were obtained from 30 pairs of PSUs, instead of the 5 pairs called for in the design of the National Urban Sample.

The study directors decide that, in the interests of attaining reliable generalizations about Capital City, the sample design should be modified to spread the 370 selections across 30 pairs of PSUs. This example illustrates the importance of anticipating the calculation of standard errors of estimates well in advance of the analysis stage. Then it is too late to make changes which would increase reliability.

Selecting PSUs: Capital City

The concrete procedures followed in forming and then selecting PSUs can be illustrated by following the example of Capital City. We just noted that the modified sample design for this city calls for a total of 30 pairs of selected PSUs to provide a self-representing sample with an overall selection rate of 1:265 and an expected yield of 370 sample addresses. The following steps are involved in forming and selecting the PSUs: (1) stratification, (2) formation of the PSUs, (3) preparing a list of PSUs, and (4) applying the selection procedure.

Stratification   It is advisable to group the PSUs from a selected city into broad strata before carrying out the selection process. Chapter 4 showed that stratification can be carried out in a variety of ways. The method used in this example is to form the PSUs into strata by socioeconomic level. Using information from the 1970 census, it is possible to organize all the various blocks and segments of Capital City into groups ranging from high to low socioeconomic status. The immediate use of this classification of PSUs is to order the PSUs on a single list beginning with those at the bottom of the socioeconomic scale and moving to those at the top. The list will also contain information on PSU size, i.e., the number of dwelling units corresponding to each PSU. A *stratum* will be defined implicitly as every 3332 dwelling units on the resulting list. This figure is obtained by dividing the total number of dwelling units for the city, 99,960,[3] by the desired number of strata, 30. The latter figure represents the total number of *pairs* of PSUs to be selected. In this sample a stratum thus becomes a group of 3332 dwelling units showing rough similarities on socioeconomic status. The procedures involved in stratification should become more clear as the example develops.

---

[3] The estimated number of dwellings shown for Capital City in Table 5-1 is 98,000. However, information gathered in the city through quick counts and similar procedures suggests that the city grew faster than expected. The revised estimate is therefore 99,960. The discrepancy between the two figures is not large enough to argue for a return to the first stage of sampling, where 98,000 was used as a measure of size. The sampling can thus proceed with the new estimate, which, if correct, will automatically increase the number of selected dwellings by about seven.

Forming the PSUs   Blocks and other well-defined land areas typically are the area units from which urban PSUs are formed. Before these can actually be combined or subdivided to form PSUs, it is necessary to set a target for PSU size. Three main factors enter into this decision. The first is the expected size of the final clusters of dwelling units to be drawn from the selected PSUs. Should we take many or few dwellings? To keep sampling error within bounds, we decided to select about six dwelling units per PSU (see the discussion of cluster sampling in Chapter 4). The second factor is the number of additional samples that may be wanted for future studies. Let us say that there is a good chance that five other studies will be carried out in Capital City. Thus, the PSUs should be large enough to allow for six selections of six dwelling units per sample without overlap. These considerations set a minimum size for PSUs of 36 dwelling units (DUs). A third consideration is the size of the *stratum*. The *maximum size* of a PSU should be considerably less than half a stratum, otherwise it would not be possible to sample *within* the stratum, as in the present case where we want a pair of PSUs from each stratum. Since the stratum size for Capital City is 3332 dwellings, no PSU should be larger than 1666 DUs. In practice, the ideal size of PSUs falls between these extremes (36 and 1666), and tends toward the small side since all the dwellings within the selected PSUs will have to be listed. A common rule of thumb is to make the PSUs small enough to minimize the amount of work involved in listing, and large enough so that few, if any, blocks are larger than a single PSU.

Applied in Capital City, these guidelines point to 100 to 300 DUs as the ideal size of a PSU. This figure will also give us considerable flexibility in combining blocks.

The next step is to combine blocks and other segments in Capital City to form PSUs with between 100 and 300 DUs. This is usually done by combining neighboring blocks with each other to reach the desired size. One constraint following from the intention to stratify PSUs is that the groupings should not mix blocks which fall into widely different socioeconomic strata. Smaller but more homogeneous PSUs would be preferable to odd combinations of scattered blocks. The process of PSU formation continues until each and every part of the city is satisfactorily included in one or another PSU without duplication or overlap.

Preparing the List   Once the entire municipality of Capital City is included in one or another PSU, the PSUs can be put in order on the basis of the stratifying variable, in this case socioeconomic status. As noted earlier, information from the 1970 census supplemented by observations during the sampling field work allow us to group the various PSUs into ten groups ranging from low to high socioeconomic status. To the extent possible the PSUs are also ordered on this variable within each group so that the entire list progresses from low to high. With this information, we are ready to prepare the *master list* of PSUs.

The master list for Capital City contains four kinds of information, in addition to the serial numbering of lines:

1 *PSU identification:* a unique number or some other unambiguous description of each PSU, the sampling unit of this stage.

2 *Measure of Size (MOS):* in Capital City, and in most area samples, this consists of the total number of dwelling units for each PSU.

3 *Cumulative Measures of Size (Cum MOS):* the cumulative total of dwelling units beginning with the first PSU and continuing through the last.

4 *Selection range:* the range of chances of being selected corresponding to each PSU. The selection range for any line (PSU) is *anything larger than the Cumulative Measures of Size (Cum MOS) of the previous line, up to and including the Cum MOS for the line in question.*

Table 5-4 presents an excerpt from the master list for Capital City. The first column on the left is simply the line number for each PSU. The PSU identification for the first three lines is given by zone and block numbers. This information is unmistakable since the zones and blocks are clearly demarcated in the census maps. PSUs formed in the newer marginal areas of Capital City would have to be given brief descriptions which were also unambiguous. The third column contains the Measure of Size (MOS) for each PSU. The fourth column is the Cumulative Measures of Size (Cum MOS). As the table shows, this column is prepared simply by adding the number of dwelling units for a given line to the total MOS of all the lines before it. Hence, the Cum MOS for the first line is 220, the same as the MOS for that line. The Cum MOS for the second line is the sum of the MOS for the second line (260) and that of the first line (220), or 480. The Cum MOS for the third line, 620, is the sum of the MOS for that line, 140, and the total for the previous two lines, 480, and so on. The final column is the selection range. The range is based directly on the Cum MOS, but with enough leading zeroes to cover the total range of figures. Because the *highest* Cum MOS is five digits (99960), the numbers in the lower ranges should also be in five digits to allow for the proper application of systematic selection. Thus the selection range for Line 1 is 00001 through 00220. The range for Line 2 begins with 00221, which is the next number after the highest number in the previous line, and continues to the Cum MOS for that line (00480). Any selection number falling within the selection range would pick the corresponding PSU for inclusion in the sample.

Applying the Selection Procedure The selection of PSUs in Capital City is very much like the procedure followed in selecting cities from Do-

Table 5-4   Excerpt from Stratified List of Capital City PSUs, with Their Associated Measures of Size

Line number	PSU identification	MOS	Cum MOS	Selection range
1	Zone 1 Block 6	220	220	00001 thru 00220
2	Zone 1 Block 4	260	480	00221 thru 00480
3	Zone 6 Block 7	140	620	00481 thru 00620
4		190	810	00621 thru 00810
5	(Identification as	220	1030	00811 thru 01030
6	appropriate)	160	1190	01031 thru 01190
7		120	1310	etc.
8		280	1590	
9		200	1790	
10		220	2010	
11		160	2170	
12		210	2380	
13		200	2580	
14		220	2800	
15		190	2990	
16		180	3170	
17		250	3420	
18		150	3570	
19		200	3770	
20		190	3960	
21		200	4160	
22		210	4370	
23		250	4620	
24		180	4800	
.		.	.	
.		.	.	
.		.	.	
488		210	96920	
489		180	97100	
490		250	97350	
491		170	97520	
492		200	97720	
493		200	97920	
494		170	98090	
495		190	98280	
496		210	98490	
497		200	98690	
498		180	98870	
499		270	99140	
500		220	99360	
501		160	99520	
502		210	99730	99521 thru 99730
503		230	99960	99731 thru 99960

mains 3 and 4. The difference is that the strata are not explicitly separated in the present case. The master list for the sample is ordered by the socioeconomic status of the PSUs, but the list itself does not show strata organized on this basis. The definition of strata in Capital City, as we have seen, is essentially *statistical.* That is, a stratum is defined as every 3332 dwellings on the list. It could happen that the PSU after the one corresponding to the 3332d dwelling shows exactly the same socioeconomic status as the one before it. This fact makes no difference in the present sampling procedure. The strata are defined by ordering PSUs on socioeconomic status rather than by distinct categories of socioeconomic status.

The sample design calls for the selection of a *pair* of PSUs from each stratum of 3332 dwellings for a total of 30 pairs. The first in the pair of PSUs for the first stratum is selected by choosing a random number from the range equal to *half* the stratum size, i.e., half of 3332 = 1666, or calculated another way, by dividing the total estimated number of dwellings, 99960, by the desired number of selections, 60 = 1666. The number 1666 would thus form the sampling interval, I, for systematic selection. Using a table of random numbers, we obtain the number 0575 to select the first PSU in the first stratum. Table 5-4 shows that the number 0575 falls within the selection range corresponding to Line 3 (00481 through 00620). The PSU described in Line 3 thereby enters the sample. The next selection is made by adding the selection interval, 1666, to the randomly chosen start, 0575. This brings us to 2241, the number corresponding to Line 12. The pair of PSUs selected from the first stratum is, thus, Line 3 and Line 12. The procedure continues by again adding the selection interval, $I,$ to the previous selection number. According to this procedure, the first PSU selected from the second stratum would be Line 20, from the number 3907 (the last selection number, 2241, plus 1666). The selection process would continue in this way throughout the list (575, 2,241, 3,907, 5,573, 7,239, . . . , 93,871, 95,537, 97,203, and 98,869). The thirtieth pair of PSUs selected from the list corresponds to Lines 490 and 498. This brings us to the desired number of PSUs and completes the selection of PSUs, for the next selection number exceeds the highest number on the list.

Selection of Dwellings

In the next stage, we move inside the 30 pairs of selected PSUs. Each selected PSU be located in order to work on the main task, the preparation of a comprehensive *listing* of all the dwelling units in every *selected* PSU. The usual procedure is to have one or more field workers go to each PSU to exactly identify its boundaries on site using the maps, sketches, and other indications, and to list all the dwellings within the boundaries on a special listing sheet. These sheets are designed with separate numbered lines, with one line for each dwelling unit. Field workers should be instructed to begin at a predetermined point in the PSU and to follow a predetermined route or other convention for covering the structures within the area. The aim of such exact procedures is twofold: first, to ensure comprehensive coverage of the PSU; and, second, to assist the interviewers in the subsequent location and identification of the dwelling units. By following some describable, and clearly marked path, the field worker preparing a listing can establish markers and reference points for the interviewers. These procedures will make it easier to check the quality and the completeness of the listings, to identify selected addresses, and later to identify missed dwellings or new construction.

When the listing of a PSU is complete, we can begin the final selection of dwellings. Before doing so, however, it is usually a good practice to set up a file folder for each selected PSU containing essential maps and sketches, the listing of dwellings for that PSU, special notes or instructions, and related materials. To prevent confusion in using the materials, the selection probabilities for both the PSU and subselections should be written on the outside of the folder.

The next step is to determine the selection probabilities required in the final stage to bring the *overall* selection rate for dwellings to the desired figure of 1:265. Table 5-5 shows the selection probabilities at the various stages for several PSUs in Capital City.

Table 5-5 Selection Probabilities by Stages and Expected Yield for Selected Sampling Units of Capital City

			Selection probabilities				
PSU number	MOS	City	PSU	×	Final stage	= Overall	Expected yield
3	140	1:1	14:1666		1:22.27	1:265	6 or 7
12	210	1:1	210:1666		1:33.40	1:265	6 or 7
20	190	1:1	190:1666		1:30.22	1:265	6 or 7
29	200	1:1	200:1666		1:31.81	1:265	6 or 7
.							
.							
.							
490	250	1:1	250:1666		1:39.77	1:265	6 or 7
498	180	1:1	180:1666		1:28.63	1:265	6 or 7

For example, in the case of PSU 3 (and all others in the table) the *city* was selected at a rate of 1:1, or certainty. PSU 3 itself was selected at a rate of 140:1666, the Measure of Size (MOS) over the sampling interval. With these two sampling rates and the overall desired rate of 1:265, we can solve for the sampling rate in the final stage needed to bring the overall rate to 1:265. Let us call the rate for the final stage $(Y) = 1/I_f$, where I_f is the selection interval to apply within the selected PSU.

$$\left(\frac{1}{1}\right) \left(\frac{140}{1666}\right) (Y) = \frac{1}{265}$$

$$(Y) = \frac{1}{I_f} = \left(\frac{1}{1}\right) \left(\frac{1666}{140}\right) \frac{1}{265} = (11.9) \left(\frac{1}{265}\right) = \frac{1}{22.27}$$

Thus, the selection probability at the final stage for PSU 3 must be 1:22.27 in order for the overall rate to be 1:265. Similar calculations have been carried out for the other PSUs shown in Table 5-5.

The last column in Table 5-5 shows that the expected yield of dwellings in each PSU is approximately the same: 6 or 7. This outcome is the result of selection with probabilities proportional to size (PPS). Specifically, it occurs because the chains of calculations just shown lead to equal overall probabilities of selection from strata of about the same size, in this case every 1666 dwelling units on the master list. If we were to double the overall selection rate for a stratum (2:265) or halve it (1:530), the expected yield would likewise be double or half. Once the selection rate is determined, the only remaining step in this stage is to apply the resulting selection interval (e.g., 22.27 in PSU 3) to the list of dwellings for the PSU in question. The principles of systematic selection, including a random start within the interval, would be applied to produce the sample addresses scattered throughout the PSU.

The procedure just described contains a built-in correction factor to compensate for errors in the measures of the size of the PSUs. If the number of listed dwellings is actually *less* than the corresponding MOS, systematic selection will stop sooner and the total number of dwellings selected will be fewer than anticipated. The main cost of this is a sample somewhat smaller than expected. If, on the other hand, the number of listed dwellings is greater than the MOS, more sample cases will be selected, and the main cost will be the extra field work and related expenses necessary to interview and process the larger sample. In either case, the selection procedure corrects for errors that might have occurred in earlier estimates of size. We might note that this advantage does not hold with quota samples. Apart from its other limitations, the representativeness of the quota method depends on the accuracy of the original estimates of population subgroups. If an area or subgroup size was estimated incorrectly, or if it has changed substantially between the period covered by the original figures and the time of the study, as when there is a great deal of new construction, the quotas will not accurately reflect the characteristics of the current population and its subgroups.

Selecting the Respondents

To maintain the representativeness of the sample, probability methods also must be applied to the selection of respondents. In Chapter 4 we presented an example showing how much of the work involved in reaching the stage of selecting dwellings can be undercut by carelessness at the door of the selected dwelling unit. The population to be studied must be clearly defined before the interviews, and the interviewers must be given clear and exact instructions as to the persons to be interviewed, and on how to select respondents in the selected households. For example, if the population under study is all adults twenty-one years of age and over, the instructions should indicate whether interviews should be carried out with *all* individuals in the household who meet this criterion, or only one. If only one is to be interviewed, the study should use a probability-based method for selecting among available adults when there are more than one. A common practice is to include a random selection procedure on the interview schedule, but with different selection numbers on different schedules. The interview might begin by having the respondent who answers the door provide a list of all members regularly living in the household. These individuals would be listed in a prescribed order (e.g., head of household, wife or husband of head, eldest son, etc.) on a numbered form. When more than one individual qualifies under the definition of the population to be studied, the random selection procedure would indicate which one should be interviewed. The instructions on one schedule might say, for instance: If 4, interview number 3; if 3, interview number 1; if 2, interview number 2, etc. Kish (1965) provides a very helpful review of practical procedures for developing a random selection method for respondents.

This chapter shows the intimate connection between the various stages of a multistage sample. Not only do the probabilities of selection at one stage affect those at later stages, but the representativeness of the sample can be destroyed or seriously jeopardized by the introduction of nonprobability methods at *any* stage. In this sense the methods used to choose a *city* are just as important as those followed in selecting *dwellings*. The discussion also underscores the vital role of accurate field work in sampling, and of locating the selected dwellings for interviewing. A good survey sample involves much more than elaborate armchair calculations of sampling error, variances, and confidence limits. These statistical procedures are basic, but their worth ultimately rests on such mundane factors as the accuracy and comprehensiveness of maps, the location of boundaries in the field, the quality of listings within PSUs, and similar practical considerations. Theory and practice are thus inseparable in the development of a good survey sample.

Questionnaire Design

Survey research is marked by an unevenness of development in its various subfields. On the one hand, the science of survey sampling is so advanced that discussions of error often deal with fractions of percentage points. By contrast, the principles of questionnaire design and interviewing are much less precise and systematic. Experiments suggest that the potential range of error involved in sensitive or vague opinion questions may be twenty or thirty rather than two or three percentage points (Payne, 1951). This situation does not mean that we should abandon survey research or throw up our hands in despair when designing the questionnaire. It does mean, however, that in most surveys based on probability samples there will be more room for improvement in the questionnaire than in the sample. It makes little sense to design one with a scalpel and the other with a broad axe.

The newcomer to the survey often correctly perceives that there are no fixed rules for designing a questionnaire, but mistakenly concludes that each man can be his own expert by applying common sense and good grammar. There *are* sound principles of question-writing drawn from controlled experiments as well as codified experience, but these must always be adapted to the peculiar circumstances of the study. Common sense is essential to the latter task, but it is not enough. Our aim in this chapter is to set forth both general principles and specific suggestions for designing a *survey* questionnaire.[1] Our implicit model here as elsewhere is the household or organizational survey rather than the psychological clinic or the employment interview.

GOALS AND CONSTRAINTS

There are two basic goals in questionnaire design: (1) to obtain information relevant to the purposes of the survey and (2) to collect this information with maximal reliability and validity. These goals can be called *relevance* and *accuracy*. To ensure relevance, the researcher must be clear about the exact kinds of data required in the study. Specifically, he or she should have an explicit rationale for each item in the questionnaire covering not only why the question will be asked but what will be done with the information. As noted earlier, this requires decisions about coding and analysis. The preparation of "dummy tables" of the type to be included in the final report may show that the response options on a given question, such as income or education, are too gross, or that those on another are unnecessarily detailed.

Accuracy is enhanced when the wording and sequence of the questions are designed to motivate the respondent and to facilitate recall. Cooperation will be highest and distortion lowest when the questionnaire is interesting and when it avoids items which are difficult to answer, time-consuming, embarrassing, or personally threatening.

A major constraint in questionnaire design is respect for the dignity and privacy of the respondent. This is related to accuracy, for a prime source of distortion is the respondent's feeling that the survey has trespassed on his dignity. But beyond such expedient concerns as accuracy and "public relations" is the researcher's obligation as a professional to respect the rights of those with whom he deals. As Frederick Stephan notes:

> The greater part of opinion research rests on kindness and confidence: kindness in the willingness of respondents to give time to the interview and to do what is requested, confidence in accepting the implicit assurance of the interviewer that he will not take advantage of the respondent and that the survey will in no way harm his interests. Usually nothing is offered respondents except an opportunity to help a stranger and his organization to make a study that is important to them and, in some instances, to the larger community [1964, p. 118].

[1] In some works a distinction is drawn between a *questionnaire* and an *interview schedule,* with the first referring to a series of questions read and answered by the respondent himself, and the second to a form administered by an interviewer. Here we use the term *questionnaire* to cover both possibilities.

In this sense, the survey is very different from the clinical or psychiatric interview in which the patient has come to the office seeking help. Here there is some justification for projective tests and intensive probing into personal affairs, for the resulting information may be necessary for counseling or psychotherapy.

While there is still no general agreement or code of ethics covering the limits of inquiry in the survey, the researcher certainly has no right to engage in deception or prowl unchecked into the individual's private life.[2] Respondents will sometimes complain, in fact, that they were tricked into providing information that, in retrospect, they wish they had not revealed. If the issue is not well handled at the time, they may also leave the interview wondering about the uses to which the information will be put. Positively, respect can be shown by candor, clear guarantees of confidentiality, an attractive and interesting questionnaire, and by following the rules of common courtesy in manners, speech, and dress, as well as by maintaining a generally pleasant atmosphere during the interview. We might note that there is clear evidence that respondents in the United States, particularly in central city areas, are becoming more suspicious of household interviews, and more resistant to providing data without some assurance of direct personal benefit. This shift in attitude may necessitate major changes in the methodology of survey research, such as increased reliance on telephone interviews. In any event, it is a further argument for courtesy, respect, and the highest ethical standards in questionnaire design as well as interviewing.

PRELIMINARY STRATEGIC DECISIONS

Before turning to the actual wording of questions, the researcher must resolve two basic questions of research strategy: whether or not to use interviewers and the problem of open versus closed questions. In exploring new terrain, it is difficult to reach a decision on these issues without considerable exploration and pretesting. Nevertheless, it is helpful to have a sense of the advantages and limitations of the various options.

Interviewers versus Self-administered Questionnaires

The four most common methods of data collection in the sample survey include a questionnaire administered by an interviewer in the presence of the respondent, the telephone interview, the mail questionnaire, and the self-administered questionnaire completed in group sessions, as in a classroom or

---

[2] There have been instances, unfortunately, in which individuals have misrepresented themselves or the purposes of the study in survey research. For a case study of misrepresentation and a discussion of the general issues which it raises, see Warwick (1973).

office. Let us first consider the general advantages of using an interviewer versus using a questionnaire without an interviewer being present.

In most surveys there are powerful advantages in having the information gathered by a well-trained interviewer, whether in person or over the telephone. The two most compelling advantages concern the motivation of the respondent. First, by arousing his or her initial interest, the interviewer increases the chances that the individual will participate in the study. Completion rates on many mail questionnaires are notoriously low, with figures of 40 or 50 percent considered good. By contrast, a response rate of 75 percent, achieved only rarely and under optimal conditions with a mail questionnaire, is often the minimal acceptable level in surveys using household or telephone interviews. Group sessions typically fare much better than the mail survey, again because someone is present to provide the motivational impetus. Second, by creating a permissive atmosphere for discussion, the interviewer frequently increases the respondent's motivation to provide complete and accurate answers. This factor can become extremely important in studies using a long questionnaire and/or complex questions. For this reason, mail surveys must typically be brief and confine themselves to fairly simple and straightforward questions. Also, in the telephone interview this advantage is reduced by the lack of personal contact and by the related pressures to make the interview brief.

Third, the face-to-face interview and, to a lesser extent, the telephone interview permit greater flexibility than self-administered questionnaires in asking questions and in clarifying ambiguous answers. When an item is not understood the interviewer can repeat it, while in the self-administered form it might simply be skipped. More importantly, the interviewer can improve the quality of the data by probing for added detail when a response is incomplete or seemingly irrelevant. In the face-to-face interview, he or she can also show cards or pictures and complete ratings on the respondent's appearance, home, attitude toward the study, and similar characteristics.

Fourth, unlike the self-administered questionnaire, the personal interview is not highly dependent on literacy, educational level, or visual acuity. Though much depends on the nature of the questions and the level of vocabulary, in most countries a well-designed schedule can be used in interviewing members of all but the most isolated social strata. Finally, the use of an interviewer ensures control over the sequence of the questions and other aspects of the data-gathering process. One of the limitations of self-administered questionnaires is that respondents can and do look ahead, skip around, or compare their answers with those of others. This lack of control is particularly serious when the responses to later questions affect those coming before, or when it is especially important to have the person's *own* view. In short, the personal interview is usually more appropriate than self-administered questionnaires when the survey covers highly diverse individuals with little intrinsic interest in the subject matter, when the questionnaire is likely to be long or the items complex, and when the study design calls for more than one or two open-end questions. In general, the telephone interview has fewer advantages than the face-to-face interview on these dimensions, but it still comes out ahead of the self-administered questionnaire.

The limitations of the personal interview are identical with the strengths of the self-administered questionnaire. The greatest appeal of the latter is that it is less expensive than the face-to-face interview and, usually, than the telephone survey. With household interviews, the survey budget must allow for the interviewers' salary, travel, per diem, and the costs of field supervisors. And, as Oppenheim notes, the interviewers "leave or get stale, so that replacement is a constant problem" (1966, p. 32). The self-administered form may be preferable when the study deals with highly personal or embarrassing topics. Under these conditions, the respondent may be more willing to answer questions if he or she does not have to discuss them, and may have greater trust in the confidentiality of the responses with a paper-and-pencil questionnaire. This approach also allows for greater use of rating scales, checklists, and other forms of measurement that are either too unwieldy or too time-consuming when read aloud by the interviewer.

The Telephone Interview In recent years, particularly in the United States, survey researchers have turned increasingly to the telephone interview, either as the sole source of information or as a supplement to other methods. This method shares many of the advantages of the face-to-face interview, and yet is considerably less expensive. Another reason for its rising popularity in the United States is both the mounting danger and the high refusal rates in some central city areas. Some respondents who would be fearful of an interviewer on their doorstep might be perfectly willing to talk with the same individual on the telephone.

These advantages, however, should not obscure the real limitations of telephone surveys. The greatest drawback, which has direct implications for questionnaire design, is the difficulty of establishing the same rapport that is possible in a face-to-face situation. With the telephone, the parties are linked only by voice and technology, and thus lack the opportunity for communication through facial expressions, appearance, and sheer physical presence. It is also much easier for the respondent to end the interview by hanging up than it would be to dismiss the interviewer from one's porch, doorstep, or front hall. As a result, the questionnaire used in a telephone interview must usually be shorter and less demanding than in the face-to-face interview. Moreover, there are greater risks attached to asking sensitive questions. With the only assurance of confidentiality being a voice on the other end of the line, the respondent will often vacillate between trust and suspicion. If the balance is tipped toward the latter by too many personal questions, too much insistence by the interviewer, or other negative factors, the questioning process can be over in a second's time. In North America the situation is not helped by the fact that telephone surveys are sometimes used by magazine and other salesmen as a lead-in for a sales presentation, and by the frequency of other types of phone solicitation. Nevertheless, several survey organizations, including the Michigan Survey Research Center, have obtained good results from telephone interviews conducted with prudence and skill.

Perhaps the greatest success with the telephone interview is seen when it is used as a follow-up to a face-to-face interview. This combination was adopted in a study of the socioeconomic correlates of fertility in the Detroit metropolitan area.

> At the close of the first interview each respondent was told that the study was to be a continuing one and that she would be called again to discuss any changes in her family situation or plans. As an aid in contacting her later, each woman was asked to give the names and addresses and telephone numbers (if possible) of three persons who would know where to reach her if she moved. This information was obtained from almost the entire sample [Coombs and Freedman, 1964, p. 112].

The response rate obtained on the follow-up survey was 97.6 percent, an extraordinarily high rate.

The Mail Questionnaire   The boon of mail questionnaires is their low cost and the bane is their low response rate. We might note, in passing, some of the devices that have been used, with varying degrees of success, to increase response.[3] In North America one of the most effective devices seems to be stamped, addressed, return envelopes, rather than a conventional business reply envelope or none at all. Higher returns have also been noted when questionnaires are sent out by first-class rather than third-class mail. In one study, respondents seemed to react favorably when multicolored, small-denomination stamps were placed on the envelope, and when the packet included a personally typed letter. Other useful techniques include follow-up phone calls, suggested deadline dates, green questionnaires rather than white, and postcards sent as reminders. The postcard follow-up has been evaluated more than the other suggestions, with rather mixed results. However, the general conclusions emerging from these studies, and the authors'

---

[3] For a good summary of this literature see Roeher (1963). Gullahorn and Gullahorn (1963) also offer useful observations on the mail questionnaire.

own observations, are that: (1) the greater the amount of work required of the respondent, whether in the form of a long questionnaire or in the need to search for an envelope, the lower the response rate; (2) some type of "personal touch" may be helpful, whether in the form of a supporting phone call, the use of first-class mail, unusual stamps, or an introductory letter; and (3) the greater the intrinsic interest of the subject matter to the respondents (e.g., a survey within an organization on suggestions for promotion and salary increases), or the greater the links between the researchers and the respondents, the better the chances of a high response rate. With regard to the last point, a survey of about 100,000 members of women's religious congregations in the United States conducted by Sister Marie Augusta Neal of Emmanuel College attained an almost perfect response rate. The conditions in this survey were optimal: it was sponsored by the religious congregations and had their full support, the subject matter was of high intrinsic interest to the respondents, and the research was conducted by a visible member of one of the religious orders. Needless to say, most surveys fall far short of this ideal situation. Also, in many developing countries, where the mails are unreliable, literacy low, and the very notion of a mail survey almost unheard of, this approach will be practically useless.

Open versus Closed Responses

How much freedom should respondents be given in answering questions? Should we let them tell their own story in their own words while we probe only for more information, or should we have them comment on or rate the possibilities *we* have in mind? Much depends on the aims of the study, the type of respondent, and, of course, the purpose of the specific question. Three common possibilities are (1) free response with no classification, (2) free response with some classification by the interviewer, and (3) the closed or structured response in which a question or supplementary card presents the answers to be considered by the respondent. These options are illustrated by the following questions on experience with crime.[4]

    Free response, no classification
        Q. 47   There has been a lot of concern about crime lately. During the last year have you or has anyone living with you had any crime happen to them? (Anything else?)

            _____

            _____

            _____

[4] The topic was included in the Boston Area Study, 1969, Joint Center for Urban Studies of Harvard University and the Massachusetts Institute of Technology. The third version was the item actually used in the survey.

Free response, some classification

Q. 47 There has been a lot of concern about crime lately. During the last year have you or has anyone living with you had any crime happen to them? (Anything else?) INTERVIEWER: CHECK THOSE CATEGORIES MENTIONED BY R
[ ] HOUSE BROKEN INTO OR ROBBED
[ ] POCKET PICKED OR PURSE SNATCHED
[ ] MAILBOX ROBBED
[ ] PROPERTY DAMAGED OR DESTROYED
[ ] PEOPLE ATTACKED OR BEATEN UP
[ ] CAR STOLEN
[ ] OTHER (SPECIFY): _____

_____
[ ] NOTHING (SKIP TO Q. 50)

Closed response

Q. 47 There has been a lot of concern about crime lately. Here is a list of some of the crimes that happen to people. In the past year has anything like this happened to you or to anyone living with you? (Anything not on the list?)
(Anything else?)
[ ] HOUSE BROKEN INTO OR ROBBED
[ ] POCKET PICKED OR PURSE SNATCHED
[ ] MAILBOX ROBBED
CARD C [ ] PROPERTY DAMAGED OR DESTROYED
[ ] PEOPLE ATTACKED OR BEATEN UP
[ ] CAR STOLEN
[ ] OTHER (SPECIFY) : _____

_____
[ ] NOTHING (SKIP TO Q. 50)

In the first example, the interviewer reads the question aloud and records the answer verbatim.[5] He may have to probe for greater detail, but the goal remains that of achieving an exact transcription. In the second case he reads only the question and then tries to fit the resulting information into one or more of the categories provided. Some studies require the interviewer to check the boxes *and* fill in the verbatim response, but this usually imposes a heavy burden on his attention and may lead to poor quality in both places. In the third situation, the interviewer reads the question and then shows the respondent a card containing the various response options. What, then, are the relative advantages of the open and closed response forms?

_____

[5] We assume here that the questions are being administered by an interviewer. The second alternative would not be applicable to the self-administered questionnaire.

Open Response   The beauty of the open-end question lies in its freedom and spontaneity. The respondent can follow his own logic and chains of association, free from the constraints of an imposed scheme. There are also more opportunities for self-expression and verbal catharsis, both incentives for continued interest in the interview. As Stanley Payne points out in his small classic on question-writing, this approach gives the respondent

> . . . the chance to have his own say-so with ideas which more restrictive types of questions would not permit him to express. Courtesy may require that when we ask a person's opinion we should at least give him the opportunity to state the ideas on the subject that are uppermost in his thinking, even though they may not be important for the purpose of the survey. The respondent should be satisfied that the interviewer asks the right questions or else he might think we questioners are stupid—and in such cases he might be right [1951, p. 50].

The resulting material, moreover, is often helpful as a source of hypotheses and "quotable quotes" which lend color and authenticity to the research reports.

In addition to these general advantages, the open response may serve several specific functions in the research process. First, it is usually valuable and often essential in exploring certain qualitative aspects of a problem. These may include the respondent's frame of reference in answering a question; the intensity of his attitudes, opinions, aspirations, or intentions; the average level of information reflected in the answers; the "natural logic" followed by individuals in structuring their responses; and the special vocabulary used in the various research sites (cf. Gorden, 1969, Chap. 2). Such information may be valuable in itself and as a means of developing structured items for subsequent interviews. Second, open questions are often useful in determining the range of responses that will be elicited by a given question. These may be quantitative, as in the case of questions about rent and income, or qualitative, as in many opinion questions. Third, the free response form is advisable when it is likely that some of the respondents will have no information or experience in the area under study or, in the case of opinion surveys, no opinion. The use of structured items under these conditions may produce answers where none really exist. Finally, open responses are helpful in allowing the respondent to "warm up" at the beginning of an interview and as a pleasant change of pace after a series of closed questions.

With this impressive roster of qualifications one might conclude that the open response form would be wholeheartedly recommended for general consumption, but this is far from the case. As Payne laments, "How can anything so good be so bad?" [1951, p. 51]. Its major drawbacks have to do with the coding and analysis of the data. When an open question is set within a clear frame of reference, its freedom and spontaneity can generate an enormous variety of responses. The problem, however, lies in developing a coding scheme which both encompasses the full range of answers and provides enough cases in each category to permit statistical analysis. To complicate matters, respondents vary greatly on the length of their responses to open questions, and interviewers differ on the extent to which they probe for more information. This means that some questions may produce one sentence while others yield ten or more. In some studies the problem is handled by coding the first three points, but this policy rests on the questionable assumption that the points mentioned first are the most salient or important.

There are other limitations to the open response. For one, it makes more demands on the time and energy of both the respondent and the interviewer. A single open question may be good for a "warm up," but ten can produce overheating. Moreover, since most interviewers are not adept at verbatim recording, they can introduce bias by emphasizing those aspects of the response which *they* consider interesting or relevant. Selectivity is also a problem in the case of the respondent whose frame of reference in answering today may be quite different from what it was yesterday, or who may feel that some points fundamental to the survey are so obvious as to require no comment. In question design, as in politics, freedom is often achieved at the expense of order, and vice versa.

Closed Response   The advantages of the closed or structured response were implied in the previous discussion. This type of question is easier to answer (though it may be more difficult to ask), easier to code and analyze, requires less skill and effort on the part of the interviewers, shortens the interview, and may make it easier for the individual to comment on sensitive or unpleasant subjects. People are often more willing to reveal both their virtues and vices by checking "those that apply" on a standard list than by answering an open question on the same topics.

When the time arrives to write the final report, the greatest advantage of the closed response is that the answers are *comparable* from individual to individual and limited in number. Consider the following question used in a study of managers in the U.S. Department of State:[6]

> **6** How much does your job as a program manager *challenge you*—in the sense of demanding your skills and abilities?

To begin with, this question would present a challenge to even the most articulate respondents who may not be accustomed to thinking of their jobs in these terms. The first problem in answering, then, is to select an appropriate frame of reference. A manager might properly speak of the *number* of his abilities challenged, or the *frequency* with which he feels challenged, or the *intensity* of the challenge experience ("It's so challenging I go home ex-

---

[6] This study was conducted by the Institute for Social Research of the University of Michigan in 1966-1967.

hausted each night"). He might also launch into a discussion of *why* the job is or is not challenging. All in all, one could easily think of five broad frames of reference yielding between five and ten possibilities each for a total of 25 to 50 options. Since the number of managers totaled only 40, the free-response approach did not seem appropriate to the goals of the study, which included statistical comparisons. Therefore, the study directors decided to structure the response options in the following way:

6 How much does your job as program manager *challenge you*—in the sense of demanding your skills and abilities?
[ ] Completely demands my abilities
[ ] Demands most of them
[ ] Demands about half
[ ] Demands some of my abilities
[ ] Demands very few of them

The respondents were thus given a constant frame of reference in commenting on the degree of job challenge. Some information was lost in the process, but the task of analysis was greatly facilitated by having comparable answers from all 40 managers.

The most serious limitation of the closed response is that it may put words in the respondent's mouth (or on his pen), especially by providing acceptable or face-saving answers which he would not have considered if left to his own resources. Most respondents are reluctant to admit their ignorance of an issue or problem, and can easily escape by selecting a reasonable answer from among those provided. They may also be tempted to avoid work by selecting the easiest alternative, such as "Don't know." And, as suggested earlier, the structured question permits less spontaneity, provides fewer opportunities for self-expression, and may leave the respondent feeling that the study imposed an artificial view of reality.

Combining the Open and the Closed Response With proper planning and a dash of creativity, the survey researcher can often work out a compromise design capitalizing on the advantages of both response forms. Several possibilities are open. First, one can begin with a series of unstructured exploratory interviews and then, as the issues become more clear, use the preliminary results as the basis for developing a more structured questionnaire suitable for the entire sample. This approach is illustrated in a study of attitude and value change among students at Bennington College (Newcomb, Koenig, Flacks, and Warwick, 1967). The researchers decided very early that not enough was known about patterns of change to construct closed-response items. Therefore, they began with a series of open questions which allowed great freedom for probing by the interviewers, who were also the study directors. The following are some examples of the questions used:

> **1** One of the things we are quite interested in is the significant experiences students have while they are at Bennington. Some students feel that their college years are a period of great change in themselves, while others feel that they have hardly changed at all. Where do you think you would come with regard to these two extremes? In other words, how much and in what ways do you think you have changed since coming to Bennington?
>
> **2** Do you think that having gone to Bennington will make any difference in your later life? (In what way?)

As expected, these interviews produced some very spontaneous, rich, and diverse comments on change. They were then subjected to a careful content analysis, with every effort made to retain the words and flavor of the original. In the following year, the results were used to develop a structured questionnaire on change, and to add greater structure to the probes used in the second wave of interviews. The self-administered questionnaire which emerged is reproduced in part in Figure 6-1.

Another possibility lies in the use of the *funnel sequence* of topics. Here the interview opens with very broad, unstructured questions and then proceeds to more specific points in the same area. Kahn and Cannell illustrate this approach with the following series of questions.

> **1** How do you think this country is getting along in its relations with other countries?
>
> **2** How do you think we are doing in our relations with Russia?
>
> **3** Do you think we ought to be dealing with Russia differently from the way we are now?
>
> **4** (If yes) What should we be doing differently?
>
> **5** Some people say we should get tougher with Russia, and others think we are too tough as it is. How do you feel about it [1957, pp. 158-160]?

The great merit of the funnel sequence is that it allows for free response in the earlier questions, and yet arrives at a specific point to be commented upon by all respondents. It allows the investigator to discover the various frames of reference used by respondents in defining "relations with other countries," and at the same time to obtain data on the frequency of responses associated with the statements in the last question.

Kahn and Cannell also point out certain circumstances in which it may be better to use an *inverted funnel sequence.* In this case, the interview begins with fairly specific questions and then moves to broader issues or to questions about the strength of intentions, the intensity of attitudes, or the respondent's level of information. This sequence is recommended especially under two conditions: when the respondent is likely to know little or have a

BENNINGTON CHANGE QUESTIONNAIRE

Now we would like to know the extent to which you have changed in the areas just covered. In answering the following questions mention only changes that have occurred <u>since coming to college</u>. Please try to be insightful in answering these questions. Not every area should be one of <u>very much change</u>.

	Very much less this way now	Less now	No change in college	More now	Very much more this way now

Tolerant of behaviors that may violate my standards

: ___ : ___ : ___ : ___ : ___ : ___ : ___ :

Question or doubt the beliefs and values I brought to college

: ___ : ___ : ___ : ___ : ___ : ___ : ___ :

Similar to my parents in values and beliefs

: ___ : ___ : ___ : ___ : ___ : ___ : ___ :

Absorbed in studies and academic work

: ___ : ___ : ___ : ___ : ___ : ___ : ___ :

Bothered by ideas very different from my own

: ___ : ___ : ___ : ___ : ___ : ___ : ___ :

Individualistic—try to be myself

: ___ : ___ : ___ : ___ : ___ : ___ : ___ :

Committed to religious beliefs

: ___ : ___ : ___ : ___ : ___ : ___ : ___ :

Absorbed in social life and dating

: ___ : ___ : ___ : ___ : ___ : ___ : ___ :

Willing to support my convictions with action

: ___ : ___ : ___ : ___ : ___ : ___ : ___ :

Questions things about myself

: ___ : ___ : ___ : ___ : ___ : ___ : ___ :

Self-centered—not concerned with the problems and needs of others

: ___ : ___ : ___ : ___ : ___ : ___ : ___ :

Committed to learning for its own sake

: ___ : ___ : ___ : ___ : ___ : ___ : ___ :

Express myself—what I think and feel

: ___ : ___ : ___ : ___ : ___ : ___ : ___ :

Figure 6-1

poorly formulated opinion on the broadest topic in the sequence, as might happen when the subject is "interpersonal attraction" or "tariffs"; and when the person has such strong feelings on the general topic that his subsequent comments would be biased by them. This might happen, for example, if the first question taps his overall attitude toward a minority group and he is then asked to comment on prominent members of this group. The desire to be consistent with one's first assessment is likely to affect the later judgments.

An inventive combination of open and closed responses is seen in the random probe technique developed by Howard Schuman (1966) in his research on Pakistan. This technique is designed for use in surveys which rely mainly on closed questions, but where information is also needed on the respondent's interpretation of the question itself or the response options. Schuman describes his approach as essentially an extension of the traditional interviewer probe.

> The technique is direct and simple: each interviewer is required to carry out follow-up probes for a set of closed questions *randomly* selected from the interview schedule for *each* of his respondents. The probe does not replace the regular closed question in any way, but follows immediately after the respondent's choice of an alternative. Using nondirective phrases, the interviewer simply asks the respondent to "explain a little" of what he had in mind in making his choice. The recorded comments (or occasionally lack of comments) are used by the investigator to compare the intended purpose of the question and chosen alternative with its meaning as perceived and acted on by the respondent [p. 219].

The results of the random probe may be used either for a quantitative evaluation of the fit between the purpose of the question and the interpretation given by the respondents, or for a qualitative assessment of the data. Schuman developed the following evaluation code for quantitative ratings.

Code	Interpretation	Points
A	Explanation is quite clear and leads to accurate prediction of closed choice.	1
B	Explanation of marginal clarity and leads to accurate prediction of closed choice.	2
C	Explanation very unclear; cannot make any predictions about closed choice.	4
D	(a) Explanation seems clear, but leads to wrong prediction of closed choice. (b) Respondent was unable to give any explanation of his closed choice ("don't know"). (c) Respondent in course of explanation shifted his choice away from original.	5
(R)	(Explanation is simply literal repetition of closed choice; cannot judge respondent's understanding of question.) [p. 220]	(Omit)

When the random probe is applied systematically to each item in the questionnaire, the ratings based on this code can serve as a useful index of the validity of the questions. A qualitative analysis of the responses can provide further information about validity and help the investigator to interpret his findings. For example, in Schuman's study the following question was originally intended as a measure of religious versus secular attitudes:

> Some people say that the more things a man possesses—like new clothes, furniture, and conveniences—the happier he is. Others say that whatever material things a man may possess, his happiness depends on something else beyond these. What is your opinion? [Page 222.]

The results of the random probe showed that the Bengali phrase of "something else beyond these" was not limited to a religious interpretation. The respondents also included such conditions as having children, a good wife, and few enemies. Here the question was understood in most cases, but its meaning differed from that hypothesized in the study.

In short, the choice between open and closed responses does not have to be on an either-or basis. By allowing sufficient time for exploration and pretesting, and by using the techniques suggested here or some of their own, the investigators can often work out a solution that combines qualitative depth with quantitative data.

GUIDELINES IN QUESTION-WRITING

The aims of question-wording are the same as those of questionnaire design in general: to obtain complete and accurate information that is relevant to the purposes of the study, to maintain the cooperation and good will of the respondent, and to show respect for his dignity and privacy. Little is gained if one question produces highly detailed information but alienates the respondent for the rest of the interview. Similarly, there is no point in designing questions which are lively and entertaining, but which miss essential information. The guidelines which follow are aimed at striking a balance between these goals.

1 *Are the words simple, direct, and familiar to all respondents?* Two extremes of vocabulary should be avoided: technical jargon or concepts familiar only to those with specialized training, and slang or colloquialisms or "folksy" expressions which talk down to the respondent. A common pitfall for social scientists is to incorporate some of their theoretical or trade terms directly into the questionnaire. Respondents may not be able to identify their *marital status,* but they can say whether they are married, single, divorced, separated, or widowed. The adroit questionnaire designer will choose words which are understood by people at all educational levels and yet do not sound patronizing. At the same time he must be on the alert for regional differences in word usage, such as "elastic" versus "rubber band" or "pop" versus "soda" and "tonic," in different parts of the United States.

A subtle form of "talking down" may occur when the question mentions a reasonably common term, and then defines it, as in this case: "At the present time are you generally in favor of or opposed to the ABM, that is, the antiballistic missile system?" If the word is to be defined it is better to present the definition first, then the summary term (Payne, 1951).

2 *Is the question as clear and specific as possible?* A cardinal sin in question-writing is items that are too general, too complex, or otherwise ambiguous. To avoid these difficulties the investigator himself must have a clear understanding of the issues he is studying. A common source of ambiguity is an inadequate frame of reference for interpreting the context of the question. If a question asked, "What type of city is this?" any of the following dimensions could be applicable: size, shape, degree of comfort, cost of living, racial tensions, efficiency of government, or opportunities for entertainment. Also confusing are *indefinite words* drawn from everyday conversation. Some, such as *often, occasionally,* and *usually,* lack an appropriate time referent. For one person *often* may mean once a day, for another once a year. Other terms, such as *here* and *there,* are ambiguous in their spatial referent. If we ask a man in an apartment building, "How many people live here?" he should ask if we mean "here" in this apartment, this building, or perhaps this neighborhood. Other offenders include *many, any, fairly, near,* and *much.* As a general rule, questions asking *how much* should specify the appropriate units for the response, such as pounds, feet, percentages or acres, unless these are obvious from the context.

Just as the question designer can bend too far backward in simplifying vocabulary, so can he add confusion by trying to be too clear. One way is through *needless elaboration,* as in this fictitious example: "How well do you get along with your wife, that is, the woman you married and regularly live with?" Also dangerous is the *single illustration.* If we ask, "Do you respect political figures like Richard Nixon?" we will probably learn more about Nixon than political figures. Similarly, while some ideas might be expressed more precisely by *double negatives,* these should normally be avoided.

3 *Are any items "double-barreled"?* It is often tempting to save time and space with questions covering two or more issues at once, such as these:

Do you plan to leave your job and look for another one during the coming year?

Do you prefer cars that are big and powerful, or small and economical?

In comparison with your parents, do you agree that you have or have had the same kind of education and job aspirations?

The last question is actually quadruple-barreled, as well as leading and ambiguous. The basic difficulty with multiple-dimension questions is that one can never be certain, without further exploration, of the part that is being answered. A "yes" answer to the first example may mean that the respondent is going to leave his job and look for another, leave his job but not look for another, or look for another but not leave his job.

4 *Are the questions leading or loaded?* A leading question pushes the respondent in the direction of a certain answer, either by implication or outright suggestion. The proverbial "How often do you beat your wife?" approach works by implication—it tricks the respondent into admitting that he, in fact, does the deed. A less subtle version simply adds "don't you agree" or "wouldn't you say" to a desired statement. The loaded question biases the answers in even more subtle ways, especially through emotionally charged words, stereotypes, prestige suggestions, and the like.

In some works on survey research, the reader is advised categorically to eliminate all leading questions. This viewpoint has come under attack in recent years, especially after an impassioned defense of leading by Alfred Kinsey and his associates in their study of sexual behavior. Kinsey's argument was that if you ask people *whether* they engage in certain disapproved sexual practices they will lie and say "no." If you ask them *when, where, how often,* or lead them in other ways, the truth is more likely to emerge. Several recent works on interviewing have also argued that under certain conditions leading questions can increase rather than jeopardize validity (cf. Richardson, Dohrenwend, and Klein, 1965; Gorden, 1969). Gorden, for example, claims that leading may aid validity in the following situation:

> **a** Respondent has accurate information clearly in mind.
>
> **b** However, there is a tendency to withhold it, either because reporting the correct answer would violate the etiquette requirements of the situation, or because the correct answer is potentially ego threatening, since it admits a violation of public ideals of some type.
>
> **c** The respondent either accepts these ideals as valid or assumes that the interviewer does so.
>
> **d** Under the above three conditions the leading question is useful if it leads in a direction contrary to the public ideals [1969, p. 215].

Our own view is that leading questions or Kinsey-like interrogation tactics may be of value in certain types of interviewing, but they should normally be avoided in the household survey for reasons of ethics as well as public relations. Sophisticated respondents may see through such methods and inform the public that the survey is trying to prove a point rather than collect objective data. Others may rightly charge that leading questions show little respect for privacy, and treat respondents as suspects rather than citizens. In short, when the respondent goes to the interviewer, as in the Kinsey research, there may be some justification for leading; when the interviewer comes to the respondent there seems to be little.

Loading may take place in several ways. One is through *partial mention of the alternatives,* as in the following examples:

Are you familiar with any of the candidates for the School Board election next week, such as Walker and Jones?
How do you generally spend your free time, watching television, or what?

To avoid this type of bias, it is advisable to mention all of the likely options or none. When the list is long and an interviewer is present, literate respondents can be shown a card containing the options. In the self-administered questionnaire, the list can simply be included as part of the question.

A second form of loading results from the use of *emotionally charged words.* Each language contains words or phrases which arouse such strong and immediate feelings, either positive or negative, that they overshadow the specific content of the question. In the United States serious bias can be introduced by associating an opinion or position with *reds, fascists, socialists, big business, labor, black leaders,* or other major sources of cleavage in the nation. In many less developed countries the same effect can be produced by tying a position to *imperialism* and *colonialism.* Experimental studies have produced differences of 60 to 70 percent by linking the same alternative to different stereotypes.[7] A few examples will illustrate the dangers of using "red flag" associations.

1 There has been a great deal of discussion lately about having the federal government take over the costs of welfare. Which of the following statements comes closest to your own opinion?
   a It is up to the federal government to take care of people who don't work.
   b People who don't work already receive enough welfare—the federal government should not provide any more.
2 Do you agree with radical black leaders that more members of their race should be hired by the building trade unions?
3 The right of businessmen to make a fair profit has always been part of the American free enterprise system. Do you think the government should put higher taxes on their profits, or that present taxes are high enough?

Phrases such as "people who don't work," "radical," "fair profit," "American," and "free enterprise" hardly contribute to an objective frame of reference in answering. Of course, in some cases an issue is so completely bound up with a supercharged symbol that the two are hard to separate, as in

[7] See Payne (1951, Chap. 11) for examples.

"black power" or "recognition of Red China." Closely related to biasing through charged words is that produced by *linkage to the status quo*. The chances that one alternative will be endorsed are often increased when it begins with "according to the law," "as it stands now," or other phrases which play upon the widespread tendency to accept things as they are in the social order.

One of the more subtle forms of loading involves *appeals or threats to self-esteem*. The question designer must be continually on the alert for options which either flatter the respondent's self-image or injure his pride. Rates of endorsement are likely to rise if alternatives contain words such as *honest, fair, responsible, reasonable, experienced* or other terms bound up with self-esteem in that society. Similarly, a question on occupations will usually produce more "executives" if it asks the respondent to check one of ten broad categories than if it asks him specifically what he does. On the other hand, the investigator may be issuing an open invitation to deception if he asks the respondent to admit certain socially disapproved facts with no provisions for "saving face." Most male adults in the United States would find it embarrassing to say "no" when asked "Do you work?" but they might be more willing if the question were changed to read: "Do you have a job at present?"

The chances of endorsement of an alternative further depend on the *context* within which it appears in the questionnaire. A classic study of bias by Rugg and Cantril (1942) points up the importance of the balance among the various positions as well as their individual phrasing. In a controlled experiment, they found that the percentage of respondents endorsing isolationist and interventionist statements varied with the balance between these statements. When the list contained two interventionist and three isolationist alternatives, 35 percent of the respondents chose the former and 24 percent the latter. When the balance was shifted to four interventionist statements versus one isolationist, the interventionist sentiment rose from 35 percent to 47 percent, while isolationist supporters dropped from 24 to 7 percent. Except for experimental surveys aimed at testing the effects of context, the distribution of the alternatives should normally be equal. The effects of context are also seen when the items appearing early in the questionnaire create a mood or "set" which colors later responses. If the first ten questions touch areas in which a national government has experienced significant setbacks and these are followed by an item requesting a global assessment of the government's performance, the results are likely to be more negative than if the last question had come first.

Finally, opinion questions may be unintentionally slanted in a conservative direction through *personalization*. Experimental research suggests that an item worded "Do you think it is desirable to legalize gambling in this state?" will draw fewer positive responses than another reading "Is it desirable to legalize gambling in this state?" (cf. Blankenship, 1946). The critical question is whether the study aims at measuring the general direction of public sentiment toward an issue or at determining the characteristics of the respondents associated with their position on this issue. In the first case, the impersonal form is usually preferable, while in the second, where the emphasis is upon what the respondent *himself* thinks, the personalized version may be in order. If the investigator is in doubt about the likely effects of personalization, he would be well advised to experiment with a "split ballot" during the pretest.

  5 *Is the question applicable to all the respondents?* Few things are more irritating in the survey interview than to be asked "How old is your wife?" when you are single, or "How often do you spank your children?" when you have none. The following are typical of the inapplicable or irrelevant items commonly appearing in survey questionnaires:

  What is your present occupation?
  Where did you live before you moved here?
  For whom did you vote in the last presidential election?
  Do you generally approve or disapprove of the policies of the military government in Peru?

The first question assumes that the respondent actually has an occupation, the second that he has lived somewhere else, the third that he voted in the last election, and the fourth that he has heard of the government in Peru *and* is familiar with its policies—all very questionable assumptions.

  Inapplicable questions are not only irritating and confusing to the respondent, but also potentially misleading. The individual who is not working may list an occupation either to save embarrassment for himself or simply to oblige the interviewer ("He wants me to give an answer, so why not?"). The danger of "false positives" of this type is greatest when an admission of the truth might injure the respondent's self-esteem. People seem especially reluctant to admit that they have no opinion after they have been asked to supply one. The most effective solution to this problem is the device variously known as a *filter, skip pattern,* or *pivot question.* Its elements are simple; one part of the question determines whether or not a person qualifies (e.g., "Did you vote in the last election?"), and the remainder obtains information from those who do qualify. In highly complex studies, the use of filters can become quite complicated, requiring months of pretesting before they are finally clear. The effort invested is worthwhile, however, for a well-designed set of filters and skips can save confusion in the field and facilitate coding and analysis later. It may also eliminate the need for several sets of questionnaires to be used with different types of respondents. An example of a relatively complex filter-skip arrangement is contained in the specimen questionnaire at the end of this chapter.

  6 *Will the answers be influenced by response styles?* A response style is a tendency to choose a certain response category regardless of an item's content. Thus in a series of opinion questions using an *agree-disagree* format, some individuals may consistently check "agree," even when the content of the same item is reversed. Though psychologists have published literally hundreds of studies on response styles or response "sets" in the past two decades, it is still difficult to generalize about the influence of response styles on survey questions (see Rorer, 1965). Nevertheless, we can point to some likely pitfalls and offer a few suggestions that merit consideration in any event.

The response style of *acquiescence* has received the greatest attention from psychologists, and may affect many survey items. It is seen when respondents choose a disproportionate number of positive answers, especially "yes," "agree," and "true." There is some evidence, though far from conclusive, that certain individuals are chronic "yea-sayers" in answering opinion questions, especially those in which the answer categories are limited to single words such as *agree-disagree.* To avoid the potential effects of acquiescence, the response options should be given specific content. For example, rather than ask, "Do you agree or disagree that the cost of living has gone up in the last year?" the researcher could use the following wording:

> In your opinion have prices gone up, gone down, or stayed about the same the past year, or don't you know?

| UP | DOWN | SAME | DON'T KNOW |

This format puts pressure on the respondent to consider each option, and reduces the chances of a series of reflexive positive or negative answers.

A second response style seen with survey questions is *social desirability,* the tendency to choose those response options most favorable to self-esteem or most in accord with social norms at the expense of expressing one's own position. The general issues at stake were raised in the earlier discussion of loading. At this point we will simply underscore the importance of balancing the social approval or self-esteem value of the alternatives presented to the respondent, and of designing questions which allow for a differentiation between personal views and social norms. Studies of population, for example, are often concerned with respondents' notions of ideal family size. An effective way of separating personal views from social expectations in this area is to include items which explicitly draw this distinction, such as:

[SOCIAL NORMS] What would you say is the ideal number of children for a family in Pacifica?

[PERSONAL PREFERENCE] How about your own case? Suppose that you could choose just the number of children that you wanted to have. How many children would you want to have?

In opinion surveys it is sometimes advisable to take specific steps to ensure balance in the social desirability of the alternatives presented. Three steps are usually involved. First, the response options should be fully stated and assigned specific content, rather than expressed in an "agree-disagree" or "yes-no" format. This is partly to prevent acquiescence, partly to facilitate comparison on social desirability. Second, raters or judges should be asked to evaluate the social desirability of each option, for example, by using a five-point rating scale. Third, the response categories included in the final version of the item should be arranged so that their social desirability values are relatively close. Often this will mean rewriting certain options so that they become more plausible. For instance, consider the difference in the social approval values of the following items:

> Which of the following statements comes closest to your own feelings about spending time with your children:
> [ ] It's very important for a father to spend time playing with his children.
> [ ] It's not at all important for a father to spend time playing with his children.
>   OR
> [ ] A man ought to spend a lot of time playing with his children.
> [ ] A man should spend some time playing with his children, but not too much time.

Pretests on a national study of shift work showed that very few men were willing to admit that it is not important for them to spend time with their children, while more were willing to endorse a milder version along the lines of the last alternative (Mott et al., 1965). It should be emphasized that in the example cited and in many opinion studies, changing the social desirability of the alternatives does not simply mean asking the same question in different ways, but asking rather different questions.

A third response style consists of *ordinal or position biases* in answering multiple choice questions or using rating scales. Payne notes, for instance, that respondents will incline toward the middle in a list of numbers, toward the extremes in a list of ideas, and toward the second alternative in a list of two ideas (1951, Chap 8). Similarly, some individuals will consistently lean to the left, right, or center when using horizontal rating scales such as the following:

Very good	good	fair	poor	very poor

One solution to this problem of ordinal bias is to develop two forms of the questionnaire in which the order of the alternatives is reversed, or several forms allowing each alternative to appear in a given position (such as third from the top) with equal frequency. However, while the use of two forms is often manageable, the use of three or more may not be worth the added confusion for the interviewers and coders. Position bias may also be reduced by assigning more content to each of the rating categories. For example, the category of "very good" might be changed to read "shows initiative and originality in all of his work."

7 *Can the item be shortened with no loss of meaning?* One of the two final tests that should be performed on each item consists of paring away all excess verbiage. Survey questions should be as short as possible, in most cases no more than twenty words.

8 *Does the question read well?* Having satisfied all of the previous criteria, the investigator must still ask if the item will be properly understood by the interviewers and respondents. Several aspects of format and the order of ideas should be considered.[8] First, the key idea in the question should come last. To prevent the respondent from "jumping the gun," all conditional clauses, qualifications, and other less important material should precede the key idea. For this reason, the second item listed below is preferable to the first:

> In what city would you look for another house (apartment) if you were forced to leave the one you're in now?
>
> If you were forced to leave the house (apartment) you're in now, in what city would you look for another?

Second, commas, colons, dashes, and other punctuation marks should not be used if they cause a break in the flow of ideas. Punctuation must be geared to the needs of data collection rather than the rules of grammar. A comma placed in the middle of a sentence may lead the interviewer to pause and the respondent to think that this is the end of the question. Third, critical words should be underscored to ensure uniform emphasis by the interviewers or uniform interpretation by the reader. Finally, every word in the question and in the alternatives should normally be spelled out. Abbreviations may save space in the form, but can also generate confusion for the interviewers, especially when lighting is bad or interpersonal relations are tense.

QUESTION SEQUENCE

The order of the items in a survey questionnaire should take account of the expectations, logic, and limitations of both the respondent and the interviewer. The basic tasks are to arouse the respondent's interest, overcome his suspicions or doubts ("Is this another salesman in interviewer's clothing?"), facilitate recall, and motivate the person to collaborate throughout the interview. From the standpoint of the interviewer, the questions should flow in a clear and orderly sequence, with precise instructions on how to move ahead without having to look back to a previous section or ahead to another page.

---

[8] The following points were suggested by Payne (1951, Chap. 12).

Ease of administration is doubly important when the interview is conducted in a dark hallway, under a streetlight, in a farmer's field, in the presence of hostile kibitzers, or under other trying conditions. The aspects of question sequence most likely to affect the success of the interview include the opening questions, the flow of the items, and the location of sensitive questions.

The Opening Questions

The first questions in the interview should be pleasant, interesting, and easy. They should not only excite the respondent's interest but also build up confidence in his or her ability to answer (Kahn and Cannell, 1957). One of the points often overlooked is that respondents as well as interviewers may have doubts about their capacity to carry out the task expected of them. The opening questions should thus be written with an eye to both motivation and confidence building. Such seemingly appropriate topics as housing characteristics, who lives in the area, and length of residence in the area may do relatively little for either goal. Questions on age, sex, marital status, and other personal topics can do positive harm, mainly because respondents find them either prying or dull. Some of the better openers are those which are more conversational and encourage the respondent to express himself in positive terms, as in these examples:

We are interested in how people are getting along financially these days. Would you say that you and your family are better or worse off financially than you were a year ago, or about the same?

OR

Let's talk about schools. What would you say are the main differences between schools nowadays compared to what they were like when you went to school?

The Flow of the Items

The sequence of the questions in the body of the interview should be tuned to the logic of the respondent, and should aid him in providing the most accurate information possible. It is especially important that he see the relationship between a given question and the overall purposes of the study. If the survey is ostensibly concerned with housing and schools and the interviewer suddenly introduces a battery of items on sexual behavior, the individual may become understandably suspicious. The most effective means of maintaining a sense of legitimacy is through the apt use of transitional questions or explanations. These can be inserted to show how a new topic relates to what has been discussed previously and to the purposes of the study, or they can help to shift the respondent's attention from one "legitimate" topic to another. Transitional questions and comments can often be very brief, as in the following cases.

> We've been talking so far about general living conditions in this city. Now I'd like to ask you a few questions about your family. (Continues with question on date of marriage).

> To get an accurate financial picture of people all over the country, we need to know the income of all the families we interview (followed by questions on income).

> These are some questions that have been asked of people all over the country. We want to know how Boston people answer them.

The order of the questions can also be used to aid the individual's memory or to introduce him gradually to unpleasant or embarrassing topics. When the aim of the item is to obtain information on a sequence of events in the past it is often helpful to begin by asking about the events immediately preceding those of interest to the study. As Gorden notes, "this type of warm-up question gives the respondent a chance to get in a reminiscent mood which stimulates recall of a particular period of time" (1969, p. 258). In the household survey the "chronological lead-in" can be used with great advantage in questions about migration and reasons for migration, residential or job history, and the respondent's perceptions of earlier attitudes and values.

The Location of Sensitive Questions

What should be the policy on placing questions which deal with unpleasant or embarrassing topics and yet are vital to the study? We have already suggested that these items should not come at the beginning. Some have tried to solve this problem by using the "hit-and-run" method at the end: save all the sensitive items for the conclusion, ask as many as you can, and leave if there is any trouble. This approach is based not only on dubious ethics but also on a poor understanding of the psychology of the interview. After an hour or more of questioning, the respondent's interest in the study is often lagging and may drop even further if he is suddenly confronted with a barrage of prickly items. Some may decide to end the interview, while others may simply respond by gliding across the issues with very general and bland answers. In any case, this tactic will leave the respondent with negative attitudes toward the study and a bad taste for survey research.

Three general guidelines can be suggested for dealing with sensitive items. First, they should be introduced only at a point where the respondent is likely to have developed trust and confidence in the interviewer *and* the study. This point may come after two questions or twenty—only pretesting or prior experience with the subject matter can suggest when. Second, a sensitive item should be located in the section of the questionnaire where it is most meaningful in the context of other questions; it should not be left dangling in foreign territory. Third, a touchy issue should be introduced gradually by "warm-up" material that is less threatening. Studies on sensitive attitudes might begin with a few questions about the respondent's experience or behavior. For example, questions on job satisfaction and morale may be better received if preceded by factual items on the employee's length of service, the specific type of work done, and similar topics. Of course, here as elsewhere a great deal depends on the skill of the interviewer. If he shows confidence in the study and in his own abilities, and communicates this to the respondent, the problem of sensitive questions will be greatly reduced.

PHYSICAL LAYOUT OF THE QUESTIONNAIRE

Before the draft questionnaire is typed in final form or printed, specific attention should be given to its physical organization and design. It should be attractive, convenient for the interviewer or respondent to use, and easy to identify, code, and store. Neglect of these considerations often leads to questionnaires that are too bulky or too flimsy, confusing to read, and difficult to work with during coding. A common reason for poor layout is the desire to fit all of the questions on a single page, even if the type has to be reduced to minuscule proportions and the items crammed together. In Latin America one often sees questionnaires which look more like road maps than interview schedules. Though such forms mean savings on paper and printing expenses, they are ultimately wasteful if they reduce the quality of the information obtained.

Identification

To avoid confusion in the field and thereafter, each form should contain one or more identifying numbers. These may be assigned in advance and marked on each form, or assigned by each interviewer, who then maintains an accurate record of his own numbers. Clear identification is especially important when the same forms will be used in continuing surveys, such as studies of manpower or vital statistics.

Size

The questionnaire should be large enough to allow sufficient space between the items, but not so large that it is difficult to handle in the field or to file and store later. Here it is important to take account of the conditions under which the form will be used. In most household surveys, it is nearly impossible to juggle an oversize questionnaire and simultaneously maintain adequate eye contact with the respondent as well as record the answers. Unless there is good reason to choose otherwise, letter-size paper should be used. Smaller forms are most susceptible to loss, while larger sizes are unwieldy.

Numbering

The items should be numbered consecutively throughout the form, with no omissions or repetitions. This may seem obvious enough, but one sometimes sees questionnaires in which each new set of items, such as those covering education, housing, and employment, begins afresh with number 1. Consecutive numbering facilitates administration by the interviewer, and also avoids confusion in coding (it is easier to say "code question 4" than "code the third item from the top on page 5").

Space

As a general rule, the layout should err on the side of more rather than less space between the items. Generous spacing makes the questionnaire more attractive and easier to administer. Space should also be left on the back of the form for the interviewer's notes and observations or remarks by the editors and coders. A partial exception to this generosity principle concerns the format for open-end questions. The problem here is that large spaces may tempt some compliant interviewers to seek unnecessarily long answers. The best policy in this case is to gauge the average length of the responses fairly well in advance, and then allow sufficient space, but not much more.

The Paper

The quality of the paper used in printing the questionnaire can become rather important in some studies. When the forms will be exposed to rain or snow—a good possibility in household surveys—or when they will be handled by many different people, heavy paper or even card stock should be used. The paper should be tested to ensure that it will take ink and that the markings from ball-point pens do not penetrate or smear. Moreover, under these conditions it is advisable to print the questions on only one side of the page or, if both sides must be printed, to use opaque paper. If the questionnaire will be administered in group sessions, or if the forms must be prepared in duplicate or triplicate, the paper can be lighter. Similarly, the color of the paper can be varied either to make the questionnaire more readable and attractive, or to distinguish the various forms from each other.

Type Faces

Interviewers find it easier to administer a questionnaire when two kinds of content are readily distinguishable in the forms: (1) questions to be read verbatim to the respondent, and (2) instructions to the interviewer or code categories used in classifying the answers. The most effective means of preventing confusion between these elements (e.g., having the interviewer read code categories not intended for the respondent) is by separating them with different sizes or shades of type. One simple method is to print all verbatim questions in regular type, and instructions or code categories in capitals, as in the example below.

> 5   (INTERVIEWER: CHECK ONE)
>   [ ] R HAS BEEN IN CITY, TOWN, OR PART OF TORONTO <u>LESS THAN 5 YEARS</u> (ASK Q. 6)
>   [ ] R HAS BEEN IN CITY, TOWN, OR PART OF TORONTO <u>5 YEARS OR LONGER</u> (SKIP TO Q. 8)
> 6   Could you tell me where you were living five years ago—were you in the city of Toronto, nearby suburbs, or somewhere else?
>   [ ] CITY OF TORONTO
>   [ ] NEARBY SUBURBS
>   [ ] SOMEWHERE ELSE (SPECIFY): _____(SKIP TO Q. 8)

This example also illustrates the use of underlining to emphasize the most salient parts of the interviewer's instructions. When the questionnaire will be printed rather than typed, there is even greater freedom to use type faces in improving clarity. In the above example, three different sizes or shades of type could be inserted: one for the verbatim question, another for the instructions, and a third for the code categories. To avoid confusion, however, it is essential that the various type faces be used consistently throughout the questionnaire.

Illustrations and Symbols

Mechanical devices such as arrows, boxes, asterisks, and other symbols can often be used with good effect in guiding an interviewer or respondent through a complicated questionnaire. Arrows and boxes are particularly helpful when the sequence of items calls for many skips and filters. It may also be worthwhile to use a professional illustrator to increase the appeal of a long, self-administered form. This was done in a national study of shift work conducted by the Survey Research Center, from which the example in Figure 6-2 is taken. The respondents seemed to appreciate the extra attention.

ILLUSTRATIVE QUESTION TYPES

Questionnaire design leaves ample room for the study directors to exercise their imagination. The self-administered form typically allows greater freedom on this score, but even the interview schedule can use a variety of question types. There is not space in this volume to cover the intricacies of scaling and attitude measurement, but we can show some of the more common types of items used in sample surveys. Our aim in the following section is to offer examples rather than discuss the relative advantages of each approach.

Checklists

This is more or less a "cafeteria" question in which the respondent is shown a variety of possible answers, and asked to check those that apply. For example:

> Which of the following kinds of health problems do you have in your family? (Check as many as apply)
> [   ] No serious problems.
> [   ] One or more of the members is an invalid requiring a lot of care.
> [   ] One or more of the members is crippled or handicapped but doesn't require a lot of care.
> [   ] Very heavy hospital bills and doctor's fees.

Quantity and Intensity Scales

In this type of question, the respondent is asked to rate a concept, event, experience, or situation on a single dimension of quantity or intensity ranging from more to less. The item below comes from a study of managers in the U.S. Department of State.

> For what proportion of your activities as a manager do you *receive* clear and specific directions from your supervisors?
> Receive clear and specific directions:
> [   ] For almost all activities
> [   ] For most of my activities
> [   ] For about half
> [   ] For few of my activities
> [   ] For almost none of my activities

Probably the most common example of an intensity question is seen in the "Likert scale." Here each respondent is presented with an attitude statement, and then asked to place himself on a continuum running from "strongly agree" to "strongly disagree," as in the following item:

A man should run the family and make the big decisions around the house. (Check one)

[　] Strongly agree
[　] Agree
[　] Uncertain
[　] Disagree
[　] Strongly disagree

Frequency Scales

These are similar to the quantity scale, except that the dimension rated shifts from "how much" to "how often." This illustration is part of a suggested job motivation index developed at the Survey Research Center (Patchen et al., 1965).

On most days on your job, how often does time seem to drag for you?

(1)____About half the day or more
(2)____About one-third of the day
(3)____About one-quarter of the day
(4)____About one-eighth of the day
(5)____Time never seems to drag

Strictly speaking, the response options and the stem of this item do not match. The question asks "how often," while the responses speak of a proportion of the day. This may be a case, however, where the purposes of the study are well served by a question which does not meet rigid standards of logic.

Story Identification

One means of dealing with the response style of social desirability is to present the respondent with two reasonably plausible stories or vignettes, and then ask him to indicate his own position with respect to these. After a great deal of experimentation with other approaches, the technique shown in Fig. 6-2 was adopted in the shift work study to measure the relative importance of various family roles as well as several dimensions of self-esteem. Another way of using stories is to present a single account of a situation or set of events and then ask the respondent to indicate his reactions. The more indirect the stories become, however, the greater the risk of invading the respondent's privacy without his informed consent.

1.

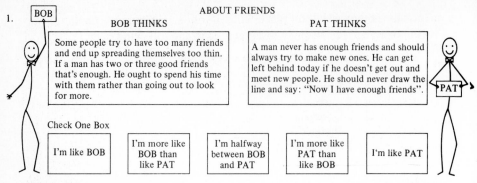

Figure 6-2

Sociometric Questions

The sociometric technique is used to obtain information on the pattern of positive and negative feelings among group members for each other. It operates by having each member of the group make a limited number of nominations indicating the others whom they most like, admire, dislike, and so on. The questions below are taken from a study of community norms at Bennington College (Newcomb et al., 1967).

Community Representative

Suppose there was to be an important gathering of representative students from every type of American college during the coming winter. Each of the colleges selected is to be represented by three students who are to be chosen by their fellow students—not for ability to speak in public nor for any other special ability, but merely as worthy representatives of their institutions. . . . It is fair to assume that Bennington College will be judged, to a greater or less extent, by the students who represent it. List three students whom you consider most worthy to serve as such representatives.

Admiration

Please enter below the names of at least two students (and no more than five) whom you particularly admire, no matter for what reason.

Semantic Differential Method

This method makes use of a series of seven-point rating scales, in the format shown below, with each end point anchored by an adjective or phrase (Osgood, Suci, and Tannenbaum, 1957). Extensive work with these scales has uncovered three recurring factors: an Evaluative Factor, covering such dimensions as good-bad, pleasant-unpleasant, and positive-negative; a Potency Factor, represented by dimensions such as strong-weak, hard-soft, and heavy-light; and an Activity Factor, with such scales as fast-slow, active-passive, and excitable-calm. Osgood and his associates refer to these as three directions in semantic space.

In a study of value change at Bennington College, Newcomb and his colleagues (Newcomb et al., 1967) used the general format of the semantic differential items, but changed the end points of the scales to reflect the specific concerns of the survey. The students were then asked to rate such concerns as My Mother, My Bennington Friends, My Non-Bennington Friends, and I Am, as in this example:

MOST OF MY FRIENDS AT BENNINGTON ARE:

interested in national and international affairs	___:___:___:___:___:___:___	: not interested in national and international affairs
conventional in dress or appearance	___:___:___:___:___:___:___	: unconventional in dress or appearance
conservative	___:___:___:___:___:___:___	: not conservative

Ranking Questions

In ranking questions, the respondent is asked to arrange a series of options in order according to personal preference or some other standard. The item below is part of a suggested index measuring an employee's identification with the work organization (adapted from Patchen et al., 1965):

> If someone asked you to describe yourself, and you could tell only one thing about yourself, which of the following answers would you be *most* likely to give? (Put "1" in the space next to that item).
> [   ] I come from (home state)
> [   ] I work for (employer)
> [   ] I am a (my occupation or type of work)
> [   ] I am a (my church membership or preference)
> [   ] I am a graduate of (my school)
>
> If you could give *two* answers, which of the items above would you choose second? (Put "2" next to that item.)
>
> If you could choose *three* answers, which of the items would you choose third? (Put "3" next to that item.)

Other ranking questions simply list five or six options and ask the respondent to give the order of their preference. When the number of possibilities exceeds five or six, it becomes increasingly difficult for the respondent to make the necessary comparisons.

Objective Information Questions

This is not a specific question type, but rather an organized strategy for seeking information about objective variables. In some surveys it is important to investigate the fertility history of a woman, the migration or employment history of a couple, family income or expenditures, and similar complex phenomena. How should the questionnaire be constructed to facilitate recall and accurate reporting? We will offer a few general suggestions and some illustrations.

A good rule to remember in designing questions about aggregated topics such as income and expenditures or about items of personal history is that the respondent has probably not thought about these questions at the level of detail required by the survey. In other words, we cannot expect the respondent to give us instant information which will match the ultimate code categories of the study. The study directors must think very carefully about the level of knowledge that the typical respondent will bring to the interview, and check these expectations against the results of some exploratory interviews.

More specifically, there are several proven aids to recall and accuracy. First, in questions dealing with individual or group history, it is often helpful to work either backward or forward from a natural starting point. In attempting to reconstruct a history of their pregnancies, for example, many women would find it most convenient to begin with the first and work ahead to the last. With employment history, on the other hand, it may facilitate recall to begin with the present job and move progressively back to the first. The main point is to establish the most comfortable procedure for initiating a step-by-step review of the history in question. Second, when the questions aim to cover rather broad and aggregated variables such as family income, it is usually advisable to arrive at the total by asking about the component parts. Sometimes, of course, a gross estimate of family income, accurate within broad income categories, is sufficient for the purposes of the study, and can be obtained by a global question such as the following:

> Considering all sources of income and all salaries, what was your total family income in 1974—before deductions for taxes or anything? Would you look at this card and tell me in which group the family income falls? (See Figure 6-3.)

When the research objectives call for fairly precise data on income, on the other hand, it is better to isolate its principal components and then collect information on these. The following are among the principal components of income included in the Survey of Consumer Finances by the Michigan Survey Research Center: wages and salaries; bonuses and commissions; income

INCOME SHOW CARD

A.   Less than $1,500

B. $1,500-$2,999

C. $3,000-$4,999

D. $5,000-$7,499

E. $7,500-$9,999

F. $10,000-$14,999

G. $15,000-$24,999

H. $25,000 or more

Figure 6-3

from farming, or unincorporated business; rent, interest, dividends, trust funds, or royalties; social security, pensions, annuities; alimony; unemployment compensation or welfare; and help from relatives. The questionnaire for this study collected detailed information on the income of the family head as well as the other members (Katona, Lininger, Mueller, 1964). No attempt was made during the interview to calculate total family income, which was left for the subsequent editing stage.

Third, accuracy can be enhanced by specific probes and cross-checks on the information obtained. Figure 6-4 provides a good illustration of this technique. This is a questionnaire designed to obtain data on family medical and dental expenditures as part of the Health Interview Survey conducted by the U.S. Public Health Service (U.S. Department of Health, Education and Welfare, 1963). Each subsection of the form typically begins with a general question about health expenditures, such as "How much did your hospital bills come to for the (one time, two times, etc.) you were in the hospital this past year?" This was then followed by a more specific item aimed at detecting omissions, such as questions 1b and 3b.

Finally, when objective information is being collected it is sometimes useful to encourage the respondent to check his or her records, for example, income tax returns or savings bank deposit books for economic surveys and birth or baptismal certificates for demographic studies. It might even be appropriate to ask another household member about the date of a past event such as a move, a job change, a period of unemployment, or a specific purchase.

DIRECT INTERVIEW QUESTIONNAIRE

COSTS FOR MEDICAL AND DENTAL CARE FOR PAST 12 MONTHS	
Ask question 1 *only* for 2 persons who were in a hospital (nursing home, sanitarium) overnight or longer during the past 12 months. *Ask* questions 2 - 6 for *EVERYONE*.	*(Check one box):* (1) ☐ In hospital (_____ times) ☐ Not in hospital
1. (a) How much did your hospital bills come to for the (one time, two times, etc.) you were in the hospital this past year? In case you don't know the exact amount of the bills, give the best estimate you can.	☐ No bills *(Free care)* $
(b) Does this amount cover ALL of the hospital charges–for example, in addition to the cost of the room, does it include charges for the operating (or delivery) room, anesthesia, X-rays, tests, special treatments, etc? If not included originally in 1(a), correct amount in 1(a)	From bills or records: ☐ All charges included ☐ Some missed - Now Corrected Not from bills or records: ☐ All charges included ☐ Some missed - Now Corrected
(c) Was any part of the hospital bill paid for by insurance, whether paid directly to the hospital or paid to you or your family? If "Yes" to 1(c), ask: (d) Was the part paid by insurance included in the amount (in 1(a)) you gave me? If not included originally in 1(a), correct amount in 1(a)	☐ No part paid by insurance ☐ Part paid by insurance and included in 3(a) ☐ Part paid by insurance, not originally in 3(a) - Now Corrected
(e) Besides these hospital bills you have already told me about, how much did the bills for any special nurses at the hospital come to?	☐ No other bills for special nurses $
2. (a) During the past 12 months did you go to a hospital for any minor operations, emergency treatment, outpatient clinic services, X-rays, tests or any thing like that for which you did not stay overnight? If "Yes" to 2(a), ask: (b) How much altogether did these kinds of hospital bills come to?	☐ Yes ☐ No $
(c) Was any part of these hospital bills paid for by any insurance, whether paid directly to the hospital, the doctor or to you or your family? If "Yes" to 2(c), ask: (d) Was the part paid by insurance included in the amount (in 2(b)) you gave me? If not included originally in 2(b), correct amount in 2(b)	☐ No part paid by insurance ☐ Part paid by insurance and included in 2(b) ☐ Part paid by insurance, not originally in 2(b) - Now Corrected
3. (a) How much did all of your doctors' and osteopaths' bills come to for the past 12 months? This amount should include all doctors' bills for home and office visits as well as for clinics and hospitals.	☐ No doctors' bills $
(b) Does this amount cover all doctors' bills–for example, operations or treatments, checkups or examinations, X-rays, tests, etc? If not included, correct amount in 3(a)	From bills or records: ☐ All charges included ☐ Some missed - Now Corrected Not from bills or records: ☐ All charges included ☐ Some missed - Now Corrected
(c) Was any part of these doctors' bills paid for by insurance, whether paid directly to the doctor or to you or your family? If "Yes" to 3(c), ask: (d) Was the part paid by insurance included in the amount (in 3(a)) you gave me? If not included, correct amount in 3(a)	☐ No part paid by insurance ☐ Part paid by insurance and included in 1(a) ☐ Part paid by insurance, not originally in 1(a) Now Corrected
4. During the past 12 months, about how much did you spend (for yourself) for prescriptions, medicines, tonics, vitamins, pills and things like that?	☐ No expenses for medicines, etc. $
5. (a) How much did all of your dentists' bills come to for the past 12 months?	☐ No dental bills
(b) Does this amount cover all dental expenses–for example, all fillings, extractions, cleanings, X-rays, bridgework, dental plates, straightening of teeth, etc.? If not included, correct amount in 5(a)	From bills or records: ☐ All charges included ☐ Some missed.- Now Corrected Not from bills or records: ☐ All charges included ☐ Some missed - Now Corrected
6. We are interested in OTHER medical expenses you may have had during the past 12 months but we don't want to include any insurance premiums you may have paid. About how much did all OTHER medical expenses come to for you during the past 12 months– for example, things like eye glasses, hearing aids, braces, chiropractors' fees, home nursing care and the like, not counting those you have already told me about?	☐ No other medical expenses $

Figure 6-4

PRETESTS

One point that cannot be overemphasized is the absolute necessity of pre-testing the questionnaire and other critical parts of the survey. Pilot tests should be used to evaluate not only the questionnaire items but also the adequacy of the sampling instructions, the quality of the interviews, the effectiveness of the field organization (see Chapter 8), the likelihood of controversy arising from the survey, the rate of and reasons for refusals, the cost and length of the interview (including call-backs), and the overall appropriateness of the survey method to the problem at hand. Though the temptation is often great, study directors should not succumb to the common practice of "swivel-chair" pretests carried out in the research office. The staff should certainly agree among themselves that a questionnaire is clear before sending it to the field, but office discussions are no substitute for direct contact with the population under study. It is essential to have interviews by others, and their reports. The lack of adequate pretesting is one of the major sources of failures in otherwise well-conceptualized surveys.

The process of pilot testing a survey questionnaire usually involves several stages. The first may be a set of exploratory interviews preceding question-writing and aimed at clarifying the frames of reference used by the respondents. A draft questionnaire is then constructed and testing is conducted under "real-life" conditions with perhaps 20 to 50 interviews. The results of this pretest can be used to sharpen the focus of the items, structure certain responses, and perhaps eliminate some questions. It is especially urgent at this stage, if not before, to determine whether the questionnaire will operate equally well in the different social classes and culture groups of the population to be studied. Ethnic, regional, and linguistic differences or widespread variation in literacy and education may suggest the need for more careful revision or alternate forms in different languages. In some cases, it may be necessary to carry out a small-scale sample to learn whether there is sufficient variability in the understanding of certain critical items, or to gain a better sense of the problems of interviewing likely to arise later.

The following are some specific questions that might be raised about the questionnaire during the pretests:

Do the respondents understand what the survey is about? After a reasonable "warm-up" period do they feel comfortable answering the questions?

Is the wording of the items clear? Do the respondents draw the same meaning from them as the study directors intended? Are the answers obtained adequate for the purposes of the study? Is there sufficient detail? In the case of open-end questions, is there so much variety that the results will be difficult to analyze?

Are there regional differences in the interpretation of the questions? Are there local expressions that should be incorporated into the items to avoid ambiguity?

Which items or sections are most difficult for the respondent to answer? Which seem to produce irritation, embarrassment, or confusion? Are there any items that the respondent considers irrelevant or comical? Is the questionnaire too long?

Is the questionnaire convenient for the interviewer to administer? Which parts, if any, are confusing? Do the filters and skips work properly? Are the instructions clear? Are the transitions from question to question and section to section smooth or abrupt?

In short, the pilot tests should be the proving ground for all of the issues raised earlier in the chapter—open versus closed responses, question wording, sequence, physical layout, and convenience of administration. We repeat: pretests are *essential.*

QUESTIONNAIRE DESIGN IN CROSS-CULTURAL SURVEYS

The sample survey is increasingly used as a means of gathering data on the same problems or issues in a variety of cultural settings (cf. Warwick and Osherson, 1973, Chap. 1). The units of comparison in such studies may be different *countries,* as in a study of crime in the United States, Great Britain, France, and Germany; different *cultures,* as in a survey of political alienation in the English- and French-speaking sections of Canada; different *regions,* as in comparisons of economic development in Central America and the Andean nations of South America; and other units. Even in a relatively unified country such as the United States, a sample survey may be confronted with cross-cultural problems in seeking conceptual and linguistic equivalence across different minority groups, such as Spanish-surname Americans, Native Americans (Indians), and Japanese- or Chinese-Americans. The researcher cannot simply assume that questions which prove intelligible to a pretest sample of English speakers in Ann Arbor or Chicago will have the same meaning or will even be understood by other groups. Such problems are greatly complicated in multiethnic and multilinguistic societies such as Malaysia, the Philippines, and India.

One of the most demanding tasks in cross-cultural or cross-national studies is to arrive at concepts and measures which are equivalent in meaning and response across all of the cultural units studied. The difficulties involved in reaching this goal are well illustrated in the best-known comparative survey published to date, the five-nation study of the "culture of democracy" by Almond and Verba (1961). While this is rightly regarded as a pioneering and creative research venture, its efforts to attain equivalence met with only partial success. Perhaps the main reason is that the interview schedules were developed in America by Americans for export to the other countries studied (Great Britain, Germany, Mexico, and Italy). Social scientists from these countries were not intimately involved in selecting the concepts and writing the questions, and there was very little pretesting *in situ*. The study also failed to draw an adequate distinction between the *formal equivalence* produced by translation and *functional equivalence*, or comparability of meaning from setting to setting. The primary emphasis was upon making the questions identical in form with the original version developed in the United States. As a result, it is often difficult to judge whether the national differences reported in *The Civic Culture* stem from variations in political systems or simply from the different meanings attached to the questions in the five countries (as well as nonequivalent sampling, field procedures, and response rates).

The Almond and Verba study points up four interlocking aspects of equivalence in questionnaire design: (1) comparability in the salience and meaning of the concepts, (2) functional equivalence in operational definitions, (3) linguistic equivalence through translation, and (4) comparability in the responses. Here we can only comment briefly on each and suggest additional reading for those interested in further exploration of the issues.

Conceptual Equivalence

The most basic question of all in comparative research is whether the concepts studied have *any* meaning or the *same* meaning in the various cultures. Concepts may differ, first of all, in their *salience* to the culture as a whole and to specific groups within the society. Thus the concept of "looking for work," a key element of international studies on unemployment, may be outside the experience of peasants in many parts of the "Third World." The problem of irrelevance is especially serious in opinion surveys asking about "international issues," "the national legislature," "public affairs," and other topics presuming awareness of national political units. Questions covering concepts on which the respondents have essentially no opinion may produce answers, but the information is likely to be a better index of social desirability than of the concept intended.

Even when issues are salient to a culture—people know what the interviewer is talking about—they may still differ in *researchability*. A prime obstacle to survey research in some areas is the respondent's unwillingness to discuss sensitive topics, such as politics and religion, with a stranger. Questions about political party preference will meet with a very different reception in the United States and Peru. In Latin American countries the subject of partisan politics in general seems to arouse strong suspicion about the purposes of the study, while in the United States respondents will talk more freely. Similarly, "it is said to be extremely difficult to obtain religious information in Moslem Pakistan, but relatively easy to do this in Hindu India. In some African areas, as well as in other parts of the world, there is reluctance to talk about dead children and the number of people in a household—obstacles to demographic researchers. In the Middle East there is a reluctance to discuss ordinary household events, and Chinese businessmen in any country are reported to be especially secretive about any and all facets of their work and political lives" (Mitchell, 1965, p. 675). Another reason that a concept may not be researchable is that respondents are unable or unaccustomed to discuss it. This problem is illustrated in a comparative study of adolescents in the United States and Denmark.

> The socialization of adolescents is a problem faced in every society and we reasoned that all socializing agents—including teachers—would act to examine and rationalize their influence. Accordingly, we approached the Danish teachers for cooperation on the premise that they would acknowledge the desirability of being analytic about teaching and the characteristics of their students. The first hint that this premise was not accepted by Danish teachers occurred during the pretest phase of developing the research instruments to be used with the teachers. Despite their apparent willingness to cooperate, the Danish teachers claimed that they were unaccustomed to considering analytically matters of adolescent interactions and could express no judgments or opinions. Indeed, despite strenuous efforts, only 30 percent of the Danish teachers completed the research materials in the study proper, forcing us to abandon this component of the investigation [Lesser, 1967].

In addition to the overall salience and researchability of a concept in a single culture, it is also essential to estimate differences in these characteristics *across* the units studied. For example, the concept of "national government" might be salient to 99 percent of the respondents in the United States and Western Europe, 75 percent in Mexico, 65 percent in Peru, 45 percent in India, and 30 percent in Gabon.

Equivalence in Measurement

The second challenge in designing a questionnaire for cross-cultural research is to devise comparable indicators for concepts judged to be salient and researchable. The task, in other words, is to take a generic concept such as "intelligence" and seek out behaviors that reflect this attribute equally in the cultures under study. A critical lesson drawn from existing comparative surveys is that different indicators may be needed to tap the same concept. If we define intelligence as the ability to adapt effectively to the social system in which one lives, the indices of adaptation will be at least partly different in the United States and rural Africa. A United States intelligence test might contain an item such as: "What is the difference between a hammer and a hatchet?" Obviously, knowledge of the difference between these tools will bear little relationship to effective adaptation in a society which does not use them. The same general point applies to measures of political involvement, religiosity, or other concepts studied through the sample survey.

How, then, might a survey researcher go about measuring a concept known to differ in specific meaning from society to society? His first task is to find out through exploratory interviews exactly what the concept *does* mean in each cultural unit. He may find that the notion of *political activity* is salient in the United States and Europe, but that the patterning of *activities* is different for each country. Przeworski and Teune (1966–1967, p. 558) suggest, for example, that the following might be equivalent activities in the United States and Poland:

United States	Poland
Contribute money to parties or candidates	Fight for execution of economic plans
Place sticker on car	Attempt to influence economic decisions
Volunteer help in campaigns	Join a party
Testify at hearings	Participate in voluntary social works
Write letters to Congressmen in support of or against policies or programs	Develop ideological consciousness

These authors also suggest an empirical procedure for determining whether or not these sets of indicators are, in fact, equivalent. Another useful approach was adopted by Inkeles and his colleagues in their six-nation study of "individual modernity" (Smith and Inkeles, 1966). This, in effect, is a compromise between the use of identical items in all cultures and completely different indicators, as suggested above. Their method involves first the selection of a general theme judged to be salient in all of the cultures, such as qualifications for holding office, and then the construction of response options adapted to the peculiar conditions of each society, as in this item:

> What should most qualify a man to hold high office?
> 1. Coming from (right, distinguished or high) family background
> 2. Devotion to the old and (revered) time-honored ways
> 3. Being the most popular among the people
> 4. High education and special knowledge

In short, an investigator can never assume that a question which is an adequate measure of a concept in his own country can be transported intact to another. The experience of the Inkeles study suggests that in certain cases there may be a middle ground between total ethnocentrism and complete cultural uniqueness in survey questions. More often, however, the quest for equivalence in measurement will require painstaking conceptual analysis, extensive pretesting, and perhaps items which are ostensibly very different in content.[9]

Linguistic Equivalence

The problem of attaining linguistic equivalence through accurate translations of identical items has received far more attention than any other aspect of cross-cultural survey research. Here, too, simple, "common sense" solutions have been found inadequate. In earlier surveys, translation was carried out by having a bilingual read the questionnaire and then translate it into his native language, with perhaps some cross-checking by another bilingual. Some studies exercised even less control, allowing bilingual interviewers to carry out the translations during the interview itself. The first approach came under heavy attack as experts in linguistics pointed out that bilinguals in general may use their native language differently than monolinguals in that society, and that there are several types of bilinguals (Lambert, 1955). A more sound approach used in recent cross-cultural surveys makes use of the technique known as "back-translation." In this case the questionnaire is translated from language A to language B by one person, then from B to A by another, A to B by a third, and so on until discrepancies in meaning are removed. This procedure is illustrated in the comparative study of adolescents mentioned earlier.

1 The original questionnaires . . . were translated into Danish, then another translator independently translated this Danish version back into English.
2 Original and re-translated English versions were then compared and discrepancies clarified and corrected.
3 A second Danish version was then pretested in interviews with individual adolescents, with probes used to assess the meaning of the questions to them.
4 Based upon this pretest information, the Danish questionnaire was again revised and then back-translated, this time into English and once again into Danish.
5 Field interviewing in small groups then constituted the next trial phase.
6 A final back translation was performed [Lesser, 1967, p. 13].

It should be emphasized that while back-translation is a highly useful means of guaranteeing equivalence in the translation of words, it provides no assurance of comparability in meaning. Other techniques for achieving equiv-

[9] For a more detailed treatment of this question see Warwick and Osherson (1973).

alence in translation are discussed by Deutscher (1968) and Anderson (1967).

Equivalence of Response

A survey question may be a good measure of a concept from a theoretical standpoint and may be well translated and still elicit a nonequivalent pattern of response across the cultures studied. Several factors may lead to differences in response. First, respondents in the various cultures may show *differential loquacity* in answering both open and closed questions. Mitchell writes that "if one examines marginals[10] from studies conducted in Malaysia, one will notice that the Chinese, when compared with the Indians, have a much higher proportion of 'no-answers' to precoded questions and fewer answers to open-ended questions. One of the reasons for this is that the Chinese are quite reticent, whereas the Indians are loquacious. This creates problems in comparing the two groups; and, of course, if the Chinese, Indians, and Malaysians are treated as a single national sample, the Chinese will be underweighted and the Indians overweighted" (1965, p. 676). Nonequivalence may also result from *differential response styles* in the several cultures. Serious problems of interpretation will arise if respondents in country A show a much greater proneness to acquiescence or social desirability than respondents in country B. Such differences may stem from cultural norms or values, or from differences in the social class composition of the various national samples. Landsberger and Saavedra (1967) present evidence showing that in Chile acquiescence is more common in the least educated strata than in the others. If in one country 40 percent of the sample has little or no formal education, and if in another the figure is only 20 percent, appropriate controls will have to be introduced to ensure that apparent cultural differences in response are not really educational differences.

Thus, the problems of designing questionnaires for cross-cultural studies are great and the solutions often complex. However, many of these problems can be kept to manageable proportions if the investigators take to the field early in the study and allow ample time and resources for pretesting the research instruments. The day when a single questionnaire is designed in the United States or Europe and sent to "hired hands" in other countries is hopefully over. Sophisticated control procedures such as the random probe, questions on the intensity with which opinions are held, and consistency checks are helpful adjuncts to proper conceptualization and pretesting, but they cannot replace them. As social scientists become increasingly aware of the advantages as well as the pitfalls of comparative research, we may begin

---

[10] Marginals are column or row totals of coded responses for an entire group.

to see cross-national surveys which meet the standards of quality seen in the better single-country studies. We may also hope that the latter will pay greater attention to the cross-cultural problems arising in a single nation-state.

A SPECIMEN QUESTIONNAIRE

One of the dangers of discussing the intricacies of questionnaire design is that the whole may be lost for the parts. To restore proper balance we would now like to introduce a specimen questionnaire designed for our hypothetical national urban survey in Pacifica. The forms which appear on pages 172-181 are shorter than a version which would be recommended for the survey at hand, but they are long enough to illustrate some of the points in this chapter.

Objectives and Design

The basic objectives of this survey are description and explanation. The study aims to provide descriptive information on such questions as housing characteristics; expenditures for rent and mortgage payments; patterns of employment; the rate of unemployment; various social characteristics, including age, sex, education, family size, ideal family size, and awareness of family planning methods; and attitudes of interpersonal trust.[11] The survey also has several explanatory objectives. The study directors hope, for example, to construct a measure of social class from the data on occupations, and then investigate social class differences in attitudes, ideal family size, and other conditions. The research also hopes to show the relationship between levels of education and occupational status.

The study design, which is outlined in greater detail in Chapter 5, basically calls for a one-time cross-section survey of the major cities in Pacifica. Thus no attempt is being made to assess change in the country. Interviews will be held only with heads of families, male or female, in order to avoid additional complications in this example.

The Cover Sheet[12]

The first form included is a two-page *cover sheet* for office and interviewer use only. It is designed for administrative control and for selected data which the interviewer can obtain by observation. The cover sheet has several specific purposes:

—to help the interviewers locate the sample dwellings. Items 5 to 9a on the first page are filled out in the research office before the interviewer sets out for the neighborhood, usually on the basis of information gathered through sampling field work.

---

[11] A more complete discussion of these data-collection objectives can be found on pp. 23-24.
[12] See Figure 6-5 on pp. 172-173.

—to check the accuracy of the sample listings. Questions 10, 10a, and 11 help to determine whether the original information about the sample address is correct.

—to maintain a concise record of what happens at each selected address. Question 12 provides a summary of the time, data, and outcome of each visit, while Question 16 records the reasons for noninterviews.

—to have the interviewer provide ratings on important characteristics of the respondent, the household, and perhaps the neighborhood. Questions 13 to 15 accomplish this purpose. During the training sessions the interviewers will also be instructed to prepare a brief "thumbnail sketch" of each respondent. This consists of general impressions of the individual as well as specific observations on his or her cooperation during the interview and any other information which might be helpful in interpreting the responses from this interview. The sketch can be begun in the section labeled "Comments," and continued, if necessary, on a separate sheet which can be stapled to the cover sheet.

Cover sheets are normally separated from the questionnaire and are coded separately. The main reason for this practice is to protect both the anonymity of the respondents and the confidentiality of the information. Thus, during the coding stage the coders would not be able to relate a sample address to a specific questionnaire, for only the study directors could put the two together. Cover sheets are often stored with sampling materials because they contain basic information about the sample addresses. The questionnaires, on the other hand, are filed or stored where the whole research team can have easy access to them.

The Survey Questionnaire[13]

Before the actual questioning process begins, the interviewer will have introduced himself or herself and briefly explained the study to the respondent (see Chapter 7). The remarks before the first question are intended to provide a transition between the general explanation and the opening questions.

Questions 1 to 3 serve two purposes. First, they collect information on the respondent's perceptions of changes in his or her neighborhood, economic, and housing situation. This will be coded and analyzed to assess the attitudes of residents toward their own situation and their cities. Second, the open-end format as well as the specific content of the items provides an opportunity for self-expression and confidence building. These are all subjects that will be familiar to most respondents, and on which they are likely to have opinions. On the basis of extensive pretesting the study directors concluded that these questions also helped to build an interest in the interview which carried over to subsequent items.

Questions 4 to 10 are objective information questions designed to provide an accurate picture of housing in the major cities of Pacifica. Question

13 See Figure 6-6 on pp. 174-181.

5, which applies only to renters, illustrates the use of several cross-checking items to obtain accurate information on rent. The question assumes that some respondents will include the cost of heat, light, cooking, and furnishings in their statement on rent mainly because this is the way that rent is calculated by the owners, while others will not. The sequence in Question 5a should remove any ambiguity on this point.

Questions 11 to 14 are also objective information items covering the sex, age, marital status, educational status, and relationship to head of household of each person living in the dwelling unit. This information will be coded in different ways to produce an accurate picture of family structure (e.g., number of children under five) as well as the density of housing and other conditions. Questions 12 to 14 are again checks on the completeness and accuracy of the information supplied earlier.

Questions 15 to 28a deal with various aspects of employment, unemployment, and occupations. This entire set of items is included partly to illustrate the uses of a complex skip pattern aimed at asking the right questions of the right respondents. The preliminary classification in Question 15 is used to direct the interviewer to the appropriate box containing follow-up questions. The items also illustrate the ways in which a complex and poorly understood concept such as unemployment can be broken down into its component parts in a survey. Thus, instead of asking the individual if he or she was unemployed, the questionnaire asks if the person worked for pay during the previous week and, if not, if he or she did anything to look for work, and so on as in Questions 23 and 24a,b and 26 and 27a,b. In case the respondent could not answer some of the factual questions, such as the number of weeks worked for pay during the past year, he or she might be encouraged to check any available records, such as pay slips.

Question 29 is designed to provide an estimate of income in broad categories. The question is accompanied by a "show card" (Figure 6-3) which is presented to the respondent. The responses will be useful in classifying families and their behavior by income levels, even though the measure is imprecise and incomplete.

Question 30 asks the interviewer to classify the respondent, R, on marital status as a filter for Questions 31 to 35a. This technique, which requires a decision by the interviewer as to the marital status of respondents on Question 11a, is helpful in screening out those for whom Questions 31 to 35a are inapplicable, that is, all who are not married heads of households.

Questions 31 to 35a aim to measure knowledge and attitudes related to family size and family planning. These questions were introduced at the request of the Ministry of Health in Pacifica. The government is in the midst of reviewing its population policies, and would like to know specifically if there is an acceptance of and possible demand for family planning services.

The items on ideal family size (Questions 31 and 32) together with the data on actual size (Questions 11 to 14) will provide some indication of whether couples have more or fewer children than they would have preferred. If on the whole, ideal size appears to be smaller than actual size, and respondents are not opposed to family planning (Question 34), the policy-makers might conclude that there is a significant demand for family-planning services. Question 35 is a general question about sources of information on family planning.

The final section on "Attitudes and Opinions" represents a modest effort to develop a scale of interpersonal trust and confidence. Several authors, including Inkeles and his associates (cf. Smith and Inkeles, 1966) have constructed items on these topics as part of more general scales on "modernity." Questions 39 to 43 were developed for the specific purposes of this study. The study directors plan to see if the questions on trust as well as those on "modern" experiences such as travel and mass media exposure are sufficiently interrelated to form single scales. If so, they will be combined as indices of trust and modernity (see Chapter 11).

FURTHER READINGS

Kahn, R. L., and C. F. Cannell, *The Dynamics of Interviewing* (New York: Wiley, 1967).

Oppenheim, A. N., *Questionnaire Design and Attitude Measurement* (New York: Basic Books, 1966).

Payne, S. L., *The Art of Asking Questions* (Princeton: Princeton University Press, 1951).

Warwick, D. P., and S. Osherson, *Comparative Research Methods* (Englewood Cliffs, N.J.: Prentice-Hall, 1973).

<table>
<tr><td>

┌─────────────────────────┐
│   │
│   │
└─────────────────────────┘
(for office use only)

</td></tr>
</table>

NATIONAL URBAN SURVEY

COVER SHEET

Survey Research Unit
National Planning Service   [ ] ORIGINAL   [ ] ADDITIONAL

1. Name of Interviewer _____   2. Your Int. No. _____

3. Length of Interview _____ (Minutes)   4. Date Letter Sent _____

Sample Information: 5. PSU _____

6. Tract No. _____   7. Segment No. _____   8. Line No. _____

9. Address or description _____

_____

9a. Additional information _____

---

10. Does the address given in Item 9 above include more than one Dwelling Unit?

   [ ] NO (COMPLETE ITEM 12 AND SIDE 2 OF COVER SHEET)

   [ ] YES, 2-4 UNITS (CONTINUE WITH ITEM 10a)

   [ ] YES, 5 OR MORE UNITS (CONTACT OFFICE FOR INSTRUCTIONS)

10a. How many dwelling units are there at this address? _____

11. Fill out a separate Cover Sheet for each unlisted Dwelling Unit at this address

   a. Be sure to mark it (them) "Additional";

   b. Enter the Sample Information from the original Cover Sheet in Items 5-9;

   c. Add enough information in Item 9a to distinguish each dwelling unit from the others—be sure
      that your description will identify the dwelling unit for someone visiting for the first time.

12. CALL RECORD

a. Call Number	1	2	3	4	5	6	7
b. Time of Day (AM or PM)							
c. Month and Date							
d. Day of Week							
e. Result (abbreviate)							

NAH	No one at home at time of call	REF	Refusal
RA	Respondent absent at time of call	HV	House (DU) vacant
INT	Interview taken	OTHER	Address is not a dwelling unit—
APPT	Appointment made		commercial, destroyed, etc.

IF FINAL VISIT TO DU,
FILL IN NON-INTERVIEW FORM ON SIDE 2

Figure 6-5

COVER SHEET—SIDE 2

13. Type of structure:
    [ ] DETACHED SINGLE FAMILY HOUSE
    [ ] APARTMENT IN PARTLY COMMERCIAL STRUCTURE
    [ ] 2-4 FAMILY HOUSE OR ROW HOUSE
    [ ] APARTMENT HOUSE OF 5-9 UNITS
    [ ] APARTMENT HOUSE OF 10-19 UNITS
    [ ] APARTMENT HOUSE OF 20 OR MORE UNITS
    [ ] OTHER (describe) _____

14. Type of neighborhood (consider 100 yards from sample DU)
    [ ] COMMERCIAL, INCLUDING STORES, FACTORIES, WAREHOUSES, TERMINALS,
        ETC. (EXCLUDE NEIGHBORHOOD STORES AND SCREENED SHOPPING
        CENTERS)
    [ ] LARGE APARTMENT BUILDINGS; TOWN HOUSES OF 3 OR MORE FLOORS;
        PROFESSIONAL OFFICE PREDOMINANTLY
    [ ] SMALL (2-4) MULTIPLE FAMILY UNITS; ROW HOUSING; TOWN HOUSES UP TO
        2½ STORIES
    [ ] BUILT UP AREA OF SINGLE FAMILY HOUSES WITH LOT FRONTAGE ON STREET
        OF 200 FEET OR LESS; (DEFINE "BUILT UP" AS HALF OR MORE LOTS OCCUPIED)
    [ ] SPARSELY SETTLED: LESS THAN HALF THE LOTS OCCUPIED; LOTS OF OVER
        200 FOOT FRONTAGE
    [ ] ISOLATED: NO STRUCTURES WITHIN 100 YARDS
    [ ] OTHER: EXPLAIN _____

15. Access to public transportation —frequent or regularly scheduled transportation
    [ ] LESS THAN 5 MINUTE WALK
    [ ] 5 BUT LESS THAN 10 MINUTE WALK
    [ ] 10 BUT LESS THAN 20 MINUTE WALK
    [ ] 20 BUT LESS THAN 40 MINUTE WALK
    [ ] 40 OR MORE MINUTE WALK; NO PUBLIC TRANSPORTATION

NON-INTERVIEW FORM

16. REASON FOR NON-INTERVIEW
    [ ] NO SUCH ADDRESS; COULD NOT LOCATE
    [ ] ADDRESS IS WITHOUT STRUCTURE
    [ ] NON-RESIDENTIAL STRUCTURE
    [ ] HOUSE OR DU VACANT
    [ ] NAH; NO ONE ANSWERED
    [ ] RA; R UNAVAILABLE
    [ ] PARTIAL REFUSAL
    [ ] TOTAL REFUSAL
    [ ] INACCESSIBLE; ADMINISTRATIVE REASON

COMMENTS: _____

_____

_____

_____

_____

_____

_____

_____

Figure 6-5 (cont'd)

NATIONAL URBAN SURVEY

Survey Research Unit
National Planning Service

We're interested in how people are getting along these days — their neighborhood and housing, their work and family. This house (apartment) was selected by chance so that we can get a good overall picture of people's opinions in Pacifica.

1. First of all, is this neighborhood a better or worse place to live than a year ago, or is it about the same?

   /BETTER/   /WORSE/   /SAME/   /D.K./

   (IF <u>BETTER</u> OR <u>WORSE</u>)

   1a. How is that? _____

   _____

   1b. Anything else? _____

2. Would you say that you and your family are better or worse off financially than you were a year ago, or about the same?

   /BETTER/   /WORSE/   /SAME/   /D.K./

   (IF <u>BETTER</u> OR <u>WORSE</u>)

   2a. How is that? _____

   _____

   _____

3. How about your house (apartment, room), is it better or worse than it was a year ago, or about the same?

   /BETTER/   /WORSE/   /SAME/   /D.K./

   (IF <u>BETTER</u> OR <u>WORSE</u>)

   3a. How is it different? _____

   _____

   _____

Figure 6-6

4. Do you rent this house (apartment, room), are you paying for it on a mortgage or land contract, do you own it outright, or do you occupy it without having to pay anything?

/RENTS/   /MORTGAGE OR CONTRACT/   /OWNS OUTRIGHT/ /D.K./

/OCCUPIES WITHOUT PAYING/ (EXPLAIN) _____

(IF <u>RENTS</u>)

5. How much do you pay a month for it? $_____/mo.

5a. Does that include: HEAT?   /YES/ /NO / /D.K./

    FURNISHINGS?   /YES/ /NO / /D.K./

    LIGHT AND
    COOKING?   /YES/ /NO / /D.K./

5b. How much were you paying a year ago? $_____/Mo.

(IF <u>MORTGAGE</u> OR <u>OWNS OUTRIGHT</u>)

6. What is the value of this house (apartment), that is, at what price would you sell it now?

    (IF <u>MORTGAGE</u> OR <u>CONTRACT</u>) $ _____

    6a. How much do you still owe on the mortgage (contract)? $ _____

7. How long have you lived here in this home?

/Less than / / 1-2 / / 3-5 / / 6-10 / /11 or more / / Don't /
/ 1 year /   /years/ /years/ /years/ / years   /   /know/

8. Where do you expect to be living next year at this time, in this house (apartment), another in this city, or somewhere else?

/THIS HOUSE/APT. /   /ANOTHER IN THIS CITY/ /SOMEWHERE ELSE/   /D.K./

(IF <u>ANOTHER IN THIS CITY</u> OR <u>SOMEWHERE ELSE</u>)

8a. Why do you think you might move?_____

9. How many finished rooms do you have, not counting bathrooms or the kitchen?   _____

10. How many of these rooms are used for sleeping?   _____

Figure 6-6 (cont'd)

HOUSEHOLD LISTING   3

11. We don't want full names, but would you help me list all the people who live here in this house (apartment, room)? Let's begin with the Head of the Household—who would that be? (FILL IN LINE 01)

11a. LIST ALL OTHERS RESIDENT IN DU. INDICATE RELATIONSHIP TO HH HEAD BUT ADD FIRST NAMES IF NEEDED TO KEEP THEM SEPARATE.

Line No.	Relationship to HH Head	Sex	Age in yrs (mos) completed	Marital Status: single, married, div., etc.	Education		Residency status: resident now, visitor, temporar. away
					Attending school now—(IF VACATION) last school yr.?	Highest yr. completed or now attending	
01	Head	M F			Yes No D.K.		Res. Vis. Away
02		M F			Yes No D.K.		Res. Vis. Away
03		M F			Yes No D.K.		Res. Vis. Away
04		M F			Yes No D.K.		Res. Vis. Away
05		M F			Yes No D.K.		Res. Vis. Away
06		M F			Yes No D.K.		Res. Vis. Away
07		M F			Yes No D.K.		Res. Vis. Away
08		M F			Yes No D.K.		Res. Vis. Away
09		M F			Yes No D.K.		Res. Vis. Away
10		M F			Yes No D.K.		Res. Vis. Away
11		M F			Yes No D.K.		Res. Vis. Away
12		M F			Yes No D.K.		Res. Vis. Away

(USE SECOND SHEET AS NECESSARY TO LIST WHOLE HOUSEHOLD)

12. Does anyone else live here now, such as (READ WITH PAUSES): another family? roomers or boarders? domestic help? visitors or guests? other friends or relatives? any children we missed? (LIST ABOVE)

13. Are there any other persons who live here temporarily—who slept here most of last week? (LIST ABOVE AND EXPLAIN HERE OR IN MARGIN)

14. Are there persons temporarily absent from the household now who usually live here most of the year? (LIST ABOVE AND EXPLAIN SITUATION HERE OR IN THE MARGIN).

Figure 6-6 (cont'd)

EMPLOYMENT HISTORY

4

15. Now I'd like to ask you about your present job. Were you working last week, unemployed, laid off, retired, in school, or what?

( ) WORKING — — — — — —(GO TO Q.16)

( ) UNEMPLOYED

( ) LAID OFF } — —(SKIP TO Q.21)

( ) RETIRED

( ) STUDENT

( ) KEEPING HOUSE } (SKIP TO Q.26)

( ) OTHER (EXPLAIN)

(IF <u>WORKING</u>)

16. What kind of work were you doing; what exactly do you do on your job?

_____

17. What kind of business or organization is that with, or are you self-employed, or what?

_____

18. Did you have a second or other job last week?   /YES/ /NO/ /DK/

19. How many hours did you work on your job(s) last week?

____ PRINCIPAL JOB +____OTHER JOBS = ____ TOTAL LAST WEEK

20. Aside from paid leave or vacations, how many weeks, if any, were you out of work last year, from January through December?

_____ weeks

(SKIP TO Q. 29)

Figure 6-6 (cont'd)

(IF <u>UNEMPLOYED</u>, <u>LAID OFF</u>, OR <u>RETIRED</u>)   5

21. What kind of work did you do when you were working, that is, exactly what did you do on your job?

_____

22. What kind of business was that with, or were you self-employed or what?

_____

23. Did you do any work for pay last week? /YES/ /NO/ /DK/

24. Have you done anything to look for work in the past week? /YES/ /NO/ /DK/

(IF <u>NO</u>) 24a. If you thought that there would be work available, would you be looking for work? /YES/ /NO/ /DK/

24b. Are you able to work now? /YES/ /NO/ /DK/

25. During 1973, January to December, did you do any work for pay? /YES/ /NO/ /DK/

(IF <u>YES</u>) 25a. How many weeks did you work last year?

_____

WEEKS WORKED LAST YEAR

(SKIP TO Q. 29)

(IF <u>STUDENT</u>, <u>KEEPING HOUSE</u> OR <u>OTHER</u>)

26. Did you do any work for pay last week? /YES/ /NO/ /DK/
(IF <u>YES</u>) 26a. What kind of work did you do?

_____

26b. What kind of business or organization is that with, or were you self-employed or what?

_____

27. Have you done anything to look for work in the past week? /YES/ /NO/ /DK/

(IF <u>NO</u>) 27a. If you thought there would be any work available, would you be looking for work? /YES/ /NO/ /DK/

27b. Are you able to work now? /YES/ /NO/ /DK/

28. During 1973, January to December, did you do any work for pay? /YES/ /NO/ /DK/

(IF <u>YES</u>) 28a. How many weeks did you work last year?

_____

WEEKS WORKED LAST YEAR

(SKIP TO Q. 29)

Figure 6-6 (cont'd)

29. Considering all sources of income and all salaries, what was your <u>total family income</u> in 1973 — before deductions for taxes or anything? Would you look at this card and tell me in which group the family income falls? (SHOW CARD)

    /A/LESS THAN $1500/ /B/$1500-2999/ /C/$3000-4999/ /D/$5000-7499/

    /E/$7500-9999/ /F/$10,000-14,999/ /G/$15,000-24,999/ /H/$25,000 OR MORE/

30. INTERVIEWER: CHECK ONE

    ( ) R IS <u>SINGLE</u>, <u>WIDOWED</u>, <u>DIVORCED</u>, OR <u>SEPARATED</u> — (SKIP TO Q. 36)

    ( ) R IS <u>MARRIED</u>

31. Now we want to ask you about your family. For example, in your own case do you want to have any (more) children?

    [ ] WANTS (MORE) CHILDREN
    [ ] DOES NOT WANT (MORE) CHILDREN — — ⎫
    [ ] UNCERTAIN — — — — — — — — — — — — — ⎬ — (SKIP TO Q. 32)
                                             ⎭

    (IF <u>WANTS (MORE) CHILDREN</u>)

    31a. How many (more) children would you like to have?   _____ CHILDREN

    31b. Among these children, how many boys and how many girls would you want to have?

                                              _____ (more) BOYS

                                              _____ (more) GIRLS

                                         _____ NO PREFERENCE

32. Now suppose that you could start your married life all over again and choose to have just the number of children that you would want by the time you were 45 years old. How many children would you want to have?

    NUMBER _____ , OR OTHER RESPONSE, VERBATIM _____

    (IF ANSWER IS: <u>UP TO FATE</u>, <u>GOD</u>, <u>CHANCE</u>, ETC., ASK:)

    32a. Many people feel as you do but still have some idea of what they would want God (chance, fate, etc.) to send them. How about you? How many children would you want to have?

        NUMBER _____

33. When you first got married, did you want to have your first child as soon as possible, or after some delay?

    /AS SOON AS POSSIBLE/ /AFTER SOME DELAY/ /DIDN'T CARE/ /D.K./

Figure 6-6 (cont'd)

(ASK IF <u>R</u> IS <u>MARRIED</u>)   7

34. Many couples do something to delay or prevent a pregnancy, so that they can have just the number of children they want and have them when they want them. How do you feel about this? Would you say that you approve, disapprove, or feel uncertain about this?

/ APPROVE /   / DISAPPROVE /   / UNCERTAIN /   / D.K., NEVER HEARD OF THIS /

35. Have you ever heard anything about any specific method of family planning?

/ YES /   / NO / — (SKIP TO Q. 36)

(IF <u>YES</u>)

35a. Did you hear about this from:	
Friends or relatives?	/ YES / / NO /
A doctor, nurse, or hospital worker?	/ YES / / NO /
Family planning workers?	/ YES / / NO /
A pharmacy?	/ YES / / NO /
A teacher?	/ YES / / NO /
Newspapers, magazines?	/ YES / / NO /
Radio or television?	/ YES / / NO /

ATTITUDES AND OPINIONS

(ASK EVERYONE)

36. Now I'd like your opinions on some questions about life in general and in this city. First, some say that if you aren't careful, people will take advantage of you. Others say that there's nothing to worry about because most people will not take advantage of you. How do you feel about this?

/ PEOPLE TAKE ADVANTAGE / / SOME OF EACH, MIXED, ETC. / / MOST DO NOT TAKE ADVANTAGE /

/ OTHER, Specify / _____   / D.K., NOT SURE, DEPENDS /

37. How do you feel about other people in this city — do you think that most people here can be trusted, that you can't trust most people here, or what?

/ MOST CAN BE TRUSTED / / HALF AND HALF: SOME OF EACH / / MOST CANNOT BE TRUSTED /

/ OTHER, Specify / _____   / D.K., NOT SURE, DEPENDS /

38. Still thinking about the people you know in this city, do you feel that they will take the time to help others, or that they are too busy taking care of themselves?

/ WILL TAKE TIME TO HELP / / HALF AND HALF: SOME OF EACH / / TAKING CARE OF THEMSELVES /

/ OTHER, Specify / _____   / D.K., NOT SURE, DEPENDS /

Figure 6-6 (cont'd)

39. What is the farthest that you have ever traveled away from this city?

8

        [   ]   R HAS NEVER LEFT CITY OR IMMEDIATE ENVIRONS
        [   ]   A PLACE IN THE SAME PROVINCE
        [   ]   A PLACE OUTSIDE THE PROVINCE BUT STILL IN PACIFICA
        [   ]   A PLACE OUTSIDE THE COUNTRY

40. Suppose that you had enough money and time to go on a long trip, anywhere you wanted. Where would you like to go?

        [   ]   R DOES NOT WANT TO LEAVE CITY OR ENVIRONS
        [   ]   MENTIONS ONE OR MORE PLACES IN THE SAME PROVINCE
        [   ]   MENTIONS ONE OR MORE PLACES OUTSIDE PROVINCE BUT IN PACIFICA
        [   ]   MENTIONS TRAVEL OUTSIDE OF COUNTRY

41. Have you ever flown in an airplane?

/ YES /   / NO /   / D.K. /

(IF YES) ▼

> 41a. How many times have you travelled in a plane?
>
> / ONCE / / TWICE / / THREE TIMES / / FOUR TIMES / / FIVE OR MORE / / DK /

42. Do you ever listen to the radio or watch television?

/ YES /   / NO /

(IF YES) ▼

> 42a. About how often? (READ ALOUD TO R)
>
> [   ]   Every day
> [   ]   Two or three times a week
> [   ]   Four or five times a month
> [   ]   Once or twice a month
> [   ]   Once every two or three months
> [   ]   Less often
>
> [   ]   D.K.

43. Do you ever read newspapers or magazines?

/ YES /   / NO /

(IF YES) ▼

> 43a. About how often? (READ ALOUD TO R)
>
> [   ]   Every day
> [   ]   Two or three times a week
> [   ]   Four or five times a month
> [   ]   Once or twice a month
> [   ]   Once every two or three months
> [   ]   Less often
>
> [   ]   D.K.

Figure 6-6 (cont'd)

Chapter 7

The Survey Interview

The survey interview is a form of verbal interaction designed to obtain information satisfying the objectives of a particular study. While the purpose of the interview is seemingly one-sided—to collect data from the respondent—the interviewing process is decidedly interactive, involving constant communication between the interviewer and respondent, and sometimes other parties as well. As a form of social interaction, the interview must operate within certain broadly defined social and ethical boundaries. Discussions which intrude too far into the life of the respondent, or treat him or her in a depersonalized manner, not only raise serious ethical questions, but may activate defenses frustrating the primary goal of data gathering. It is important, therefore, to understand the social norms and personal sensitivities at work in any interviewing situation. The aim of this chapter is to present a conceptual framework for understanding the processes of communication in the survey interview, and to offer various practical suggestions for conducting the interview.

AN INTERACTIVE APPROACH TO THE INTERVIEW

The success of the survey interview depends on the quantity and quality of the information exchanged between the interviewer and the respondent. This flow of information depends on four interacting parties and conditions: the interviewer, the respondent, the topic, and the immediate interviewing situation. The interviewer must communicate the question, motivate the respondent to cooperate, probe for additional information when necessary, and record the information thus obtained. The communication process may become distorted or biased at any of these points. Similarly, the respondent contributes his part by accepting the role of information-provider and giving answers which are complete, precise, true, and relevant. He, too, may disrupt the flow of information by refusing to take the questioning process seriously, or by introducing specific distortions into his responses. The study topic may affect the communication process either by stimulating the interest of the respondent and interviewer, or by arousing anxiety, hostility, fear, or lesser negative reactions in the participants. The interviewing situation includes such conditions as the time and place of the conversation, the presence of "third parties," and the general climate of opinion toward the survey in the surrounding community. The influence of the situation on communication will depend in part on the topic and in part on the respective characteristics of the interviewer and the respondent. The inhibiting effects will normally be greatest when the interviewer and respondent come from different and contending racial or ethnic groups, when the study touches on the conflict itself, and when the interview takes place before other community members. The mutual influence of these conditions is outlined in Figure 7-1.

An interactive approach to the interview suggests the importance of an intimate and advance understanding of the specific field situations to be faced in a given survey, and counsels against any simple approach to interviewer selection and training. Just as there is no standard respondent and no uniform interview situation, so there is no ideal interviewer in the abstract. Professional research organizations employ individuals who have demonstrated their competence in a wide variety of field situations, but even their talents are not universally applicable.

Every interviewer, no matter what his range of experience, is more at ease with some types of respondents, topics, and field situations than others. The critical task in planning the interview, therefore, is to devise a strategy which will produce an optimal fit between the talents of the field staff and the exigencies of data collection. In practice, this means close attention to the background characteristics and expectations of the respondents, the physical situations where the interview will take place, the hackles likely to

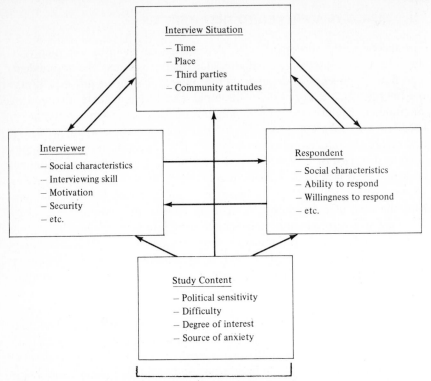

Figure 7-1 Factors influencing communication in the interview.

be raised by the survey questions, and the degree of competence that can realistically be expected from the interviewers. Energies spent in this direction are likely to be more productive than the vain search for the personality characteristics of the ideal interviewer.

The "systems approach" suggested here may show, for example, that the demands presented by a combination of sensitive questions and difficult interviewing conditions exceed the skills available among the interviewers. In this case, it may be better to shorten, simplify, and desensitize the questionnaire rather than run the risk of public animosity and questionable data. Similarly, if there are sharp racial or ethnic cleavages in the communities under study, it would be advisable to introduce racial and ethnic considerations in interview assignments. In short, there is no substitute for a detailed understanding of the actual conditions which will be faced in the field.

The Motivation to Participate

The goal of open, accurate, and relevant communication requires a motivational commitment to participation on the part of both the respondent and the interviewer. For the respondent, this means not only an initial decision to take part in the study, but a sustained willingness to share available information throughout the interview. In the case of the interviewer, the initial decision to participate can often be taken for granted, though commitment may be greatly weakened if there is actual or perceived danger. In the United States such fears have become increasingly common as interviewing in central city areas has grown more precarious. Whatever the initial commitment, the interviewer's motivation to work at a high level cannot be taken for granted.

For both parties the net commitment to participation can be viewed as the resultant of positive and negative forces. For either one, the balance of forces may shift at any time during the interview. It will depend on such conditions as the sensititivy or difficulty of the question, environmental distractions, fatigue, fear, and the response of the other. The respondent's balance may be tipped only slightly in favor of the interview at the beginning, but increase as the topic becomes more familiar and the discussion itself produces certain satisfactions. Conversely, the interviewer's reactions may be very positive at the beginning, when the challenge of winning over the respondent is great, but decrease as the process moves into more routine territory. Careful planning for field work requires consideration of both the positive and negative forces likely to arise for respondents and interviewers. We will now examine the individual motives affecting participation, and then turn to the impact of the interactions between the respondent and the interviewer.

The Respondent

Positive Forces   What are the various kinds of satisfaction which might be tapped to encourage a high level of participation by the respondent? The answer is, of course, culture-specific: the motives operative in the United States or Canada may be quite different from those in rural India. Nevertheless, we can suggest several broad categories which seem to be applicable to many cultures.

First, a strong motive for participation is the *desire for self-expression.* People often derive satisfaction from providing information or expressing their opinion on a subject which interests them. Such satisfaction may be enhanced by the belief that the information may have some effect on the policy of a nation, community, or employer. On rare occasions, a survey interviewer has been greeted by remarks such as: "I've been waiting for Gallup and you people to come around for twenty years—where have you been?"

A second positive force is the *desire for interpersonal response.* This is the satisfaction found in sharing information about important events in one's life with a sympathetic listener. One study conducted by the Michigan Survey Research Center suggests that this may be one of the most critical factors in respondent participation. Mail questionnaires were sent to respondents in national surveys to assess their reactions to the interview experience. "A high proportion of the respondents reacted in terms of feelings about the interviewer and the process of the interview rather than to the survey as such. It appears . . . that what impressed itself on the respondents was not the subject of the interview or the questions that were asked, but the relationships they established with the interviewers" (Kahn and Cannell, 1957, p. 47). Lest the significance of the interviewer be overemphasized, however, it should be noted that a later study by the same organization failed to find a direct relationship between the respondent's global reactions to the interviewer and the accuracy of reports on health information (Cannell, Fowler, and Marquis, 1968). Also, while identification with the interviewer can be a positive force, it may be counterproductive if extended too far. The respondent who "tells too much" at the beginning of an interview may later regret the indiscretion and become more guarded.

A third positive motive is the sheer *intellectual challenge* of some interviews. Respondents will sometimes agree to participate in order to satisfy their curiosity about opinion polls and surveys. Others may welcome the contact as a means of reducing loneliness or boredom. Intellectual satisfactions may be particularly salient for certain kinds of respondents, such as professors, writers, or busy members of a political elite. Such individuals may reject a highly structured questionnaire, but rise to the challenge of a dialogue on topics of interest. Lerner (1956) reports that a critical factor in winning over French intellectuals was the interviewer's ability to engage in intelligent debate.

Fourth, the survey interview may lead the respondent to *insights* which are helpful and rewarding. There is some evidence, for example, that interviews following on personal or social disasters, such as suicides among family members or tornadoes, help the respondent to make sense of highly disorienting experiences and to develop a greater sense of meaning in life (Gorden, 1969, pp. 92-93).

Fifth, the interview may stir *feelings of altruism*. The respondent may be motivated by a desire to help the interviewer with his or her job, or to aid a graduate student with a dissertation. In many studies, where the content is at least neutral, this may be the most important motive to participate. In some cases, altruism may take the form of identification with the sponsoring institution or with the content of the research, as in studies of cancer or mental health. Kinsey and his associates (1948) drew heavily on appeals to altruism in conducting their research on human sexual behavior, citing its relevance to marriage, child training, and other social concerns.

Sixth, the interview may provide an opportunity for *emotional catharsis.* This positive force is related to the desire for interpersonal response and to insight, but involves primarily the release of tension through the expression of feelings.

A seventh and critical source of satisfaction, and one which springs directly from the interview itself, is *gratification from the successful perform-ance of the respondent's role.* The talented interviewer will try to assist the respondent in accepting the role of information-provider, and provide him with the rewards of approval and esteem for a job well done. Much to the surprise of neophyte interviewers who see the respondent as a bundle of resistances, many individuals want to be good respondents and are pleased when they have reached a satisfactory level of performance. To be successful in this task of role-education, the interviewer must convince the respondent that *he* is vital to the study, that he has a unique contribution to make, and that what he has to say is interesting. Usually this process will be carried out more by the interviewer's own attitudes and behavior, including encourag-ing remarks, nods, and smiles, than by careful oral presentations. Needless to say, the message communicated must be sincere to be effective, and in principle there is no reason why it should not be, for all of the points sug-gested about the respondent's contribution are true. We shall see later, how-ever, that the communication process can become badly distorted by feelings and perceptions arising in the interaction between the interviewer and re-spondent. Finally, in some cases, the respondent may be motivated by *ex-trinsic rewards,* such as payment for the interview or fulfillment of academic obligations. The use of such incentives presents special and complicated problems that go beyond the present discussion.

Negative Forces   The optimism emerging from this review of positive forces must be balanced against the factors working against respondents' participation. Foremost among these is *fear,* whether of a potential assailant at the door, an ambiguous research process, or the uses of the data. In many cities at the present time, considerable resistance arises from the reluctance of the residents to open their doors to strangers. Interviewers in the less developed countries, moreover, often find that respondents have no category in their perceptual universe for "research" and "personal interviews." For this reason, there is often strong suspicion of any organized data-gathering activity, including the most thoroughly legitimated national census. The whole notion of engaging in a question and answer process with a stranger may be very foreign, producing uncertainty and confusion among the in-tended respondents. Even where survey techniques are more familiar, as in North America and Western Europe, respondents may still be wary of the interview process, or about the effects of revealing sensitive information. They may suspect the interviewer of being a tax collector, a disguised police agent or representative of a credit bureau, or some other imposter. Those with a major stake in the social order may also worry that the disclosure of confidential information about themselves may bring a loss of esteem in the community, or harm to the larger group of which they are members, such as an ethnic minority. Even when the interviewer assures them of confidentiality, they may doubt his or her credibility.

Another force working against participation is a *perceived invasion of privacy*. For many citizens a crucial aspect of personal freedom lies in the right not to discuss or disclose certain beliefs, attitudes, or behavior. Often the desire for privacy is tied to one's freedom to maintain some degree of geographic seclusion. In the United States some critics have argued that the growing use (and misuse) of surveys and polls is a threat to such seclusion (cf. Miller, 1971; Warwick, 1973).

Third, the motivation to participate may be reduced by *hostility toward the interviewer*. Immediately sensed differences in personality or background may tip the balance of forces in a negative direction. The same effect may be produced by *hostility toward the sponsor*. For some individuals, the fact that a survey is sponsored by a university organization or a government agency brings forth feelings of loyalty, while in others it has a completely negative effect. For this reason it is always essential to know in advance how the sponsor is regarded in a given area.

A final negative force lies in *threatening subject matter*. This point refers less to the uses which will be made of sensitive information than to the intrinsic dissatisfactions associated with discussions of the question topics. Some topics, such as partisan politics in Latin America, may be intrinsically interesting to respondents, but off limits for the sample survey because of fears about the uses to be made of the data. Others, such as death or debts, may evoke few fears of misused data, but generate considerable anxiety because of their personal sensitivity. The conflict between the forces working for and against participation in the respondent is illustrated in Figure 7-2.

The Interviewer   The interviewer is probably the major influence on the motivation of the respondent and on the quality of the responses received. Though research on this question is sparse, there is some evidence that the

The Respondent's Motivation to Participate in the Interview

Positive forces		Negative forces
Self-expression	→ ←	Fear: survey, interviewer, uses of data
Interpersonal response	→ ←	Hostility toward interviewer
Intellectual challenge, curiosity	→ ←	Perceived invasion of privacy
Insight	→ ←	Threatening subject matter
Altruism	→ ←	Hostility toward sponsor
Performance of respondent role	→ ←	Costs in time and energy
Extrinsic rewards	→	

Figure 7-2

face-to-face exchanges during the interview are more highly related to the accuracy of the information than any other conditions, including the demographic characteristics (age, sex, etc.) of the interviewer and respondent (Cannell, Fowler, and Marquis, 1968). If the interviewer is interested in the study, enthusiastic about his or her work, and likes the respondent, these feelings will usually be communicated to the respondent with positive effects on the latter's participation. On the other hand, if the interviewer is languid, lackadaisical, or bored, he or she may hurry the respondent to the detriment of completeness and accuracy, accept a response of "don't know" where a bit of probing would have produced a positive answer, or show carelessness in recording the answers given.

Positive Forces Many of the satisfactions provided by the survey interview are identical for the interviewer and the respondent. These include the desire for interpersonal response, intellectual curiosity, and identification with broad social concerns. Still, there are several differences in the incentives available to the two participants, stemming mainly from the fact that survey research is a profession (or at least a job) for the interviewer, and a temporary avocation for the respondent. This situation results in several sources of satisfaction that are specific to the role of interviewer.

Perhaps the most important is a sense of professional identification. The notion of a profession implies that an individual has acquired a specialized body of knowledge or skills, and has demonstrated his proficiency to some certifying group. Thus, one of the prime forces encouraging serious interviewers toward a high level of participation is a desire to demonstrate and improve their skills at interviewing. It is significant that problems of falsified or "curbstone" interviews are much less frequent when the staff consists of professional interviewers than when it is made up of temporary recruits. In the former case, identification with the organization as well as the profession serves to counteract the temptation toward faking. Closely related to professional identification is the intrinsic satisfaction accompanying the mastery of interviewing techniques, quite apart from the recognition accorded such mastery by others. Further, some interviewers may view their work not only as a skill and a professional activity, but as a means of career advancement. The career ladder may be confined to the interviewing profession itself, or extended to related fields, such as research administration. Also, in survey interviewing, as in most occupations, important satisfactions are derived from peer support or group morale. For some interviewers, one of the delights at the end of a long day is the opportunity to exchange stories and insights with others on the team. Back and Stycos, in their report on field work in the Jamaica Fertility Investigation, underscore the importance of peer relations in interviewing teams.

We have noted that the field work was accomplished by interviewing teams— groups of four to five workers who would travel together to the area and complete the canvassing and interviewing in a week or so. Because of the remoteness of most rural areas, teams had to reside in the field during these periods. Since we felt that these primary group conditions could be both a source of friction and a source of strength the make-up of the teams was determined at the conclusion of training largely by means of sociometric choice. Moreover, each team was in charge of a Group Captain also appointed largely on the basis of sociometric choice. Although friction did occasionally occur, we believe that the team organization was a decided asset if not an absolute necessity. It is hard to conceive of a solitary interviewer surviving under the field conditions in the rural areas without group support [1959, p. 23].

Interviewers working in large urban areas rarely develop the close-knit relations described here, but the general point raised about group support remains valid.

The interviewer's motivation to perform well in his work will also be higher if he understands and identifies with the purposes of the survey. It is, thus, important for the study directors to take time to explain why the research is being carried out, what they hope to learn from it, how the various sets of questions fit into these goals, and how the performance of the interviewers can affect the quality of the data. It is not advisable, however, to lay out the major theoretical hypotheses guiding the research. Studies of experimenter effects (Rosenthal, 1966) suggest that expectations by researchers of how the findings should appear may increase the likelihood that they do appear. Specifically, interviewers who know that Group A is hypothesized to be different from Group B in certain ways, may unwittingly influence the responses obtained in the direction of the hypothesis. Once the field work has begun, the study directors and supervisors can further contribute to the interviewers' motivation by providing feedback on how the study is progressing and, in other ways, treating them as "insiders."

Negative Forces Among the dissatisfactions leading to a reduction of the interviewer's motivation are those stemming from absence of the conditions just noted—professional identification, the mastery of basic interviewing techniques, opportunities for career advancement, and group support. Another negative force which assumes greater prominence as a survey progresses is boredom with the interview. At the beginning, the topic may be new and the respondents' observations intriguing. Later, the questionnaire may become routine and the interactions dull. The degree of boredom will vary with the topic of the study and the degree of flexibility enjoyed by the interviewer. Questionnaires consisting exclusively of closed-ended items will usually lead more rapidly to boredom than those which allow for free responses and require considerable probing.

Fatigue, fear, difficult travel conditions, inconvenient hours, and frustration in locating respondents are other common inhibitors of the interviewer's motivation. In the United States, interviewers may be thoroughly frightened by assignments in areas with high crime rates, even when accompanied by another staff member. Fear of robbery or assault will mount if the study requires extensive work at night in such areas. Similar problems are seen in other countries. In rural Peru, for example, some areas are so inaccessible that an interview may require several hours of precarious travel by horseback—something of a challenge to a citydweller. Trips in other areas might require a police escort because of the high concentration of contraband operations. For some individuals, interviews in dangerous areas can be thrilling, but for many they reduce motivation. Other negative forces include dissatisfaction with the field supervisor, pay, or related employment conditions (see Chapter 8).

The Interviewing Situation   The balance between positive and negative motivation for both the respondent and the interviewer will further depend on the peculiar circumstances of the interviewing situation. Though we sometimes speak of human personality as if it were fixed and stable, individuals show quite different "personalities" according to the situation in which they find themselves. A white interviewer conducting a survey in Harlem may appear to be a rather different person than in a predominantly white section of New York. Thus if we wish to predict how an interviewer and a respondent will behave in a particular set of circumstances, we must try to understand the forces at work in that situation.

The situational forces bearing most strongly on the survey interview are the place and time at which it occurs, the presence of "third parties," the attitudes toward the study in the surrounding community, and the sequence of the interviews. The place at which the interview is carried out may be psychologically significant because of the associations or memories evoked in either party, especially the respondent. Employee surveys carried out in the factory manager's office may elicit different attitudes than comparable questions asked in a union hall. Another important consideration in choosing a location is freedom from noise, distractions and interruptions, all of which affect the motivation to participate. In some situations, however, complete privacy may violate cultural norms and cast suspicion on the entire process. This is especially true with male-female interactions.

The timing of the interview will also affect the willingness of both parties to participate, and sometimes the respondent's ability to provide the desired information. Visits to households immediately before dinner are likely to meet with little success, and will often be resented. Moreover, in studies focusing on a specific sequence of experiences or events, the timing of the interview may influence the validity of the data collected. Gorden writes: "The most significant effect of timing is upon the respondent's ability to remember, accurately and completely, avoiding chronological and inferential confusion. To avoid simple forgetting, the interview should take place as soon after the relevant events and experiences as possible. Much psychological research indicates the distorting effects of fading memories" (1969, p. 159).

The presence of "third parties" or sometimes entire groups of spectators poses special problems for the validity of survey data. This is a very common situation in the less developed countries—in some areas almost all interviews at least begin in the presence of others—and not uncommon in other countries (Mitchell, 1965, p. 679). The major problem with this practice is that the presence of outsiders places the respondent under pressure to distort his answers in the direction of community norms or what will make the observers think well of him. This situation can have major repercussions when the first interview is carried out with a community leader or headman whose opinion then becomes a semiofficial standard for subsequent interviews. The effects of outside observers need not always be negative, however. As Mitchell points out, "In studies seeking information rather than attitudes, third parties may help keep the respondent honest and also help him to remember the requested information. In other instances especially with women and younger people, respondents may refuse to be interviewed unless a third person is present" (1965, p. 679). Some practical techniques for dealing with this problem will be suggested later in the chapter.

Community attitudes toward the survey may also affect the motivation to participate, as well as other facets of the interview. When the research is carried out in hostile territory, both the interviewer and the respondent are likely to be more guarded in their communications than if conditions were more favorable. Similarly, when the community itself is highly polarized, for example, along ethnic lines, the interviewer will have to be doubly careful to avoid identification with one or the other of the contending parties, unless such identification is part of his research strategy. This point is closely related to decisions about the sequence in which interviews will be conducted. An obvious situational factor affecting the interviewer's work is the degree of legitimacy he or she has acquired from previous contacts. This consideration is especially important when the respondents selected are in close communication with each other, as in single organizations such as a factory, or in small communities. Here it is often advisable to begin with the highest-status individuals in the sample, and also with those known to be most willing to participate (Gorden, 1969). Early success, especially with opinion leaders in the area, helps to legitimize the study as well as the interviewer.

Interviewer-Respondent Interaction

The survey interview is a short-term human relationship in which each participant will seek cues about the emotional reactions, attitudes, social status, and expectations of the other. As Cannell, Fowler, and Marquis observe, the household interview

> . . . may be so out of the ordinary stream of daily events that respondents really have no cognitive "set" which they bring to the situation. The situation is so new that it is difficult to generalize their associated feelings, attitudes, and expectations. Therefore . . . the respondent must look to the interviewer or some other source for cues as to her expected behavior [1968, p. 35].

The interviewer may have very similar reactions because of the wide differences among respondents. On the basis of an intensive study of the interviewing process, the above authors conclude: "It may be this cue-searching process which accounts for the very strong tendency of interviewer and respondent to behave at the same level of activity in the interview" [p. 35].

The specific cues will vary with the society and culture as well as with the immediate interviewing situation, and also with the stage in the interview itself. At the beginning, each will try in some fashion to locate the other in the social structure and his or her own perceptual framework. For reasons already indicated, many respondents will find themselves without an appropriate frame of reference, particularly if the culture contains no slot corresponding to the role of interviewer. The most common way of classifying the other at the beginning is to rely on such time-honored social characteristics as age, sex, and social class. The respondent will typically pay careful attention to such overt indicators of position as the interviewer's dress, manners, accent, skin color, bearing, age, and sex. Some, of course, may short-circuit this process and immediately classify the visitor as "professional interviewer" or "census taker." The interviewer may also unconsciously try to place the respondent, although he or she will have considerable advance information from the sample description, and the general character of the neighborhood. After the participants have tentatively completed this preliminary sorting, the search will normally shift from overt characteristics to more subtle qualities such as signs of approval, disapproval, pleasure, encouragement, and hostility. The research by Cannell, Fowler, and Marquis (1968) also suggests the importance of the activity level of the participants. If either the interviewer or the respondent is highly expressive, laughs, or jokes, the other may follow suit. Given the desire for interpersonal response, both will be on the watch for signs of how well they are meeting the overt and subtle expectations of the other (cf. Goffman, 1969).

First Impressions Psychologically, a great deal happens in the first part of the interview, even before the conversation begins. The tentative judgments made at this stage may be reversed later, but only if there is contrary evidence. To understand both the operation and implications of this process of impression formation, we must examine the structure of the society and community in question, especially the major lines of cleavage. Such characteristics as age, sex, and social class have no automatic influence on the outcome of the interview. Their influence, as well as that of other social characteristics, depends entirely on the emotional connotations which they evoke, and on their relationship to the issues under investigation in the survey.

We would suggest three questions related to impression formation that might be raised in planning for field work. First, what are the most significant bases of stratification and the most prominent lines of cleavage in the area under study? In other words, what are the social-structural and cultural factors most likely to bear upon the process of interpersonal communication? Age and sex will figure prominently in almost every society, though their importance will also differ greatly from one setting to another. Standard demographic characteristics such as these offer a useful starting point, but they are insufficient for many situations. In some societies, such as, perhaps, the border regions between India and Pakistan, religious differences (for example, Hindu versus Moslem) may well be more important than class. Similarly, even within the same nation, regional origins, reflected especially in accents and manners, may be critical for some studies. Quite probably a local interviewer will have more success in rural South Carolina than one perceived as a Northerner, especially if the survey touches on regionally sensitive issues. The use of university students as interviewers may also create obstacles to communication among those who bear one type of resentment or another against this group as a whole. Furthermore, in cross-national research, the nationality of the interviewer will assume great importance, and not always in predictable directions. A study of political elites in Europe and the United States (Hunt, Crane, and Wahlke, 1964) revealed that American nationality was a definite asset in gaining access to legislators in Austria and France.

Second, what are the external indicators of these social-structural and cultural factors? Given the perceptual vacuum at the opening of the interview, both parties are likely to focus on such immediately observable characteristics as dress, accent, racial or ethnic traits, posture, and gestures, as well as the year and make of the other's automobile, the quality of housing and furniture, and the social status suggested by other possessions. These visible characteristics are significant in the interview only insofar as they evoke emotional and perceptual reactions. For this reason, each interview situation must be analyzed in detail to uncover the most salient points of divisiveness.

Third, what is the likely relationship between social cleavages, observable personal characteristics, and the subject matter of the study? In general, the more closely the topics of the interview touch the tender nerves of the social system, the more important will be the social differences between the interviewer and the respondent. For example, the use of white interviewers in black neighborhoods in the United States will probably have a smaller impact on a study of consumer reactions to brand names than on one dealing explicitly with racial attitudes.

There is now a sizable body of evidence dealing with the impact of social characteristics on survey interviews in the United States. The studies concerned with race suggest that, in general, interviewers from the same racial backgrounds as the respondents will have greater success than others in collecting valid data on racially sensitive and, perhaps, even other topics. A carefully controlled investigation by Hyman and his associates (1954) showed that black interviewers were better able than whites to obtain information from black respondents on their resentment over discrimination. Other research (Athey, Coleman, Reitman and Tang, 1960) indicates that white interviewers working with white respondents obtained more expressions of hostility against nonwhites than did black or Oriental interviewers. Williams (1964), however, adds an important qualification to these conclusions. His results show that the interviewer's race is likely to produce bias only when there is a large status gap between him and the respondent, and when the questions are threatening. Other research by the same author (Williams, 1968) suggests that the effects of race will further depend on the objectivity of the interviewer's attitudes and the degree of "rapport" during the interview. This finding is in general agreement with the approach taken in this chapter, which argues that first impressions based on race and similar characteristics are important, but that subsequent interactions will also affect the quality of the information collected.

Age, sex, and class differences may also affect communication during the interview itself. The studies by Hyman et al. (1954) suggest that the communication of sensitive, personal, and especially sex-related information is easier when both parties are of the same sex. Thus female interviewers might be more successful than males in discussing contraception and abortion with female respondents, while males might hold the edge in a study of working conditions in a steel plant. The interviewer's age may also create barriers between him and the respondent, though again much will depend on the specific topics and the situation. Interviewers of high school age will probably have little success in gathering information from a cross section of a large city, but they may do better than their elders in a study of teenage drug use carried out at high schools.

The effects of social class are very similar to those of race. Katz (1942) discovered that working-class interviewers elicited more radical opinions on issues such as labor unions than did their middle-class counterparts. Lenski and Leggett (1960) also found that middle-class interviewers may create pressures toward deference or acquiescence among working-class respondents. Cannell et al. (1968), on the other hand, found no relationship between the social characteristics of interviewers and respondents and the accuracy of reporting health information. One reason might be that the content of the questions—illnesses, impairments, accidents, injuries, use of medical and dental facilities, and so forth—is less susceptible to distortion from social influences than questions on politics, sexual behavior, or drug use.[1] By contrast, the effects of status, race, ethnicity, or class may be greater in some of the less developed countries than in the United States. The main reason is that in these countries many respondents have no conception of the role of "survey interviewer," and thus turn to traditional social characteristics in defining the interview situation (cf. Mitchell, 1965; Berreman, 1973).

The previous discussion should not be taken as a firm recommendation that interviewers and respondents regularly be matched on social characteristics. Rather, the available evidence suggests that in selecting interviewers and making assignments, it would be very wise to consider the probable effects of social characteristics, interacting with the subject matter of the study, on the validity of the data. In some cases, matching may well be in order, while in others it would be totally unjustified.

Interpersonal Exchanges   Fortunately, first impressions are not necessarily irrevocable. What happens during the interaction itself will further affect the quality of the information and may completely override the initial attitudes formed on both sides. The critical task facing the interviewer, at this stage, is to train or "socialize" the respondent into the role of information provider, while simultaneously maintaining an attitude of objectivity and a high level of motivation in his own work. More is involved in this process than just establishing warm relations or "rapport," though this is crucial. The interviewer must also communicate to the respondent that the quality and accuracy of his responses are highly important. An emphasis on rapport alone may lead to a situation in which all are at ease and in good spirits and both parties report enjoying the interview, but the data collected are shallow and incomplete. The respondent should not be misled into thinking that the visit is a social occasion in which information gathering is secondary. It is both more ethical and more effective to make it clear from the beginning that *what* is said is essential to the task at hand. Thus the inter-

---

[1] For an overview of the literature on response invalidity and bias see Phillips (1971, pp. 12–49). The author also offers a number of useful suggestions for increasing our understanding of the sources of bias, including greater attention to the processes of data collection and increased reliance on direct observation. He does not appear to recognize, however, that the correctives he proposes are also open to bias. In social research there is no easy or direct route to objectivity, whether through introspection, observation, or interviewing.

viewer is forced to walk a tightrope between critical objectivity toward what is being said and an attitude of warmth and acceptance toward the person who says it. There is no necessary incompatibility between these tasks, and no need for insincerity. Many forms of social interaction involve similar tensions. What, then, are some of the forces which work for and against the joint goals of objectivity and rapport during the actual interactions?

Positive Forces   The interviewer commands a number of interpersonal rewards which can be used to stimulate a high level of participation and a proper role definition on the part of the respondent. Basically, these involve communicating positive feelings toward the study, the interview, the respondent as a person, and the contribution that he can make. Much of this communication occurs at the nonverbal level through facial expressions, tone of voice, mannerisms, and the like. Among the most critical attitudes to be communicated is a sense of self-assurance and ease about the task of interviewing. The respondent will be more likely to accept his own role, and also more flattered, if the interviewer makes it clear that he knows what *he* is doing. If he is apologetic about the study, fumbles with the questionnaire, overexplains the purpose of the questions, and generally seems frightened by the whole experience, the respondent will understandably be confused about what is expected of him and reluctant to proceed. Building upon a solid foundation of self-confidence, the interviewer may draw upon other rewards to encourage the respondent to accept the proper role. Such encouragement may take the form of sounds and phrases showing that the interviewer is paying close attention to what is being said and finds it interesting. Vocalizations such as "uh-huh," "I see," "I understand," as well as nods of the head or other motions communicating the same message, tell the person that he is moving in the expected direction. When used with skill, short periods of silence may have the same effect, especially when the interviewer's posture and expression indicate approval of the preceding remarks and an expectation of more like them. One must be cautious of overusing these signs of encouragement, however, especially the stronger varieties such as "that's right," "good," and "I agree." With these the interviewer runs the risk of biasing the responses in the direction rewarded (cf. Hildum and Brown, 1956). And, as noted earlier, research on "experimenter effects" in social research (Rosenthal, 1966) suggests that the interviewer may unconsciously use such rewards to elicit responses which fit *his* expectations. A study of the 1948 presidential election (Wyatt and Campbell, 1950) showed, for example, that interviewers who expected respondents to have discussed the campaign with others obtained answers in line with this expectation. These findings argue for the use of more neutral encouragements, such as "uh-huh," rather than "good."

In any survey interview there is always the danger that rapport-building efforts may be *too* successful. This happens when the respondent becomes overly dependent on the rewards provided by the interaction situation, and biases his responses in a direction which will produce more of the same. If the respondent comes to have a strong attraction to the interviewer, either for personal reasons or because of what the interviewer represents (the middle class, education, the government), he may be tempted to slant his responses toward what he thinks the interviewer wishes to hear.

Negative Forces Counteracting these several rewards are various interpersonal inhibitors of rapport and accurate communication. Here we shall consider only those under the control of the interviewer, but it is important to recognize that the respondent is also the source of parallel negative forces. One of the most common deterrents to the respondent's participation is a feeling of uncertainty about what is expected of him in the interview situation. This may stem from a lack of self-assurance or an overly apologetic attitude on the part of the interviewer, or from broader cultural factors such as sheer unfamiliarity with any type of social research, including surveys. Respondents will also be resentful if they feel that they are being "used" as a pipeline for information with no respect or consideration for themselves as persons. This feeling is most likely to arise when the interviewer emphasizes objectivity to the detriment of rapport. The interviewer may further inhibit communication by forgetting important pieces of information mentioned earlier in the conversation, such as whether or not a couple has children; by failing to express interest or failing to probe for further details on points which the respondent considers crucial to the study; by constantly interrupting the individual as he tries to speak; by communicating boredom with the questioning process; and by immersing himself so completely in note-taking that he rarely looks up from the questionnaire. Respondents are normally quite astute in assessing the interviewer's true feelings toward the interaction, basing their judgment both on what he says and does.

At this point, we should introduce a note of caution about the cross-cultural applicability of the previous remarks. The bias of our suggestions, and of those in most works on interviewing, is toward an egalitarian style in the interview. We feel that the bulk of the evidence gathered in the United States supports our assumptions in this regard. We would also have to admit, however, that these assumptions may not hold up for interviewing in other cultures. Some researchers have suggested, in fact, that the egalitarian approach may well have to be modified in situations where the interviewer and respondent come from different social classes, and there are distinctive patterns of interaction between these classes. Mitchell writes:

> Research agencies in many countries recruit their interviewers from among college students, which means that they come from middle- and upper-class

backgrounds, and are themselves educated people. This type of interviewer creates a communication problem, since there are certain traditional ways in which members of different classes interact. For example, custom demands that lower-class persons use polite forms of address, and not express themselves freely to members of the upper classes [1965, p. 683].

Similarly, Back and Stycos (1959) report that interviewers in their study of fertility in Jamaica occasionally made use of quite authoritarian high-pressure tactics, seemingly with success. The use of these tactics may have been dictated in part by the reception sometimes given the interviewing teams.

> Respondents or others in the community would put on a show to test the extent to which an interviewer could "take it." In some instances they would seem to engage in leg-pulling for sheer sport. Perhaps it was gratifying to frighten, humiliate, or anger someone of higher status. But perhaps they were unwilling to carry it through when confronted by someone of still higher status, or someone willing to call the bluff or even "pull rank" [p. 10].

The authors conclude:

> Whether such techniques are more harmful or valuable is difficult to judge. It can only be said that they were cited almost invariably as instances of *successful* techniques by interviewers who had had the dangers of biasing respondents drummed into them during training. Conceivably, they are effective among types of respondents found particularly in the lower class where an authoritarian approach from a person of higher status is expected. One must at least consider the possibility that such an approach seems more sincere than an equalitarian one which might give the impression of condescension or even of fawning as a means to an end.

The fact of the matter is that, aside from this report and two or three others, we know very little about the dynamics of interviewing in other cultures. Given this state of the art, we can only encourage survey researchers in any culture to pay careful attention to the issues raised by status discrepancies, and to carry out some informal experimentation during the pretests.

Interaction and Bias

Thus far we have considered the impact of the initial predispositions, background characteristics, behavior, and attitudes of the interviewer and respondent, as well as the interview situation itself, on each other's motivation to participate. We turn now to the specific kinds of distortion and bias introduced by poor participation and faulty communication. These errors may arise from the behavior of either or both parties.

Bias from the Interviewer   The interviewer may adopt a variety of attitudes toward the respondent. He can see him as a person who will be interesting to interview, one whom he would like to impress with his education or professional skill, a rather "low-class" individual who will probably not understand what the interview is about, or a disagreeable character who will make the whole experience unpleasant. These attitudes in themselves are of little consequence if they do not reduce the quality of the data, but research as well as experience suggest that they often will. The following are among the sources of bias resulting from improper attitudes and motivation on the part of the interviewer.[2]

1 *Errors in asking the questions.* On the basis of his early interactions, the interviewer will often arrive at tentative conclusions about the respondent's ability to answer the questions and about the type of answers which "fit" his situation. Acting on these assumptions, the interviewer may introduce bias by shortening the question or simplifying the language, by using a higher level of vocabulary to impress high-status respondents, or by suggesting answers, either through changes in the wording of the question or through indirect means such as vocal inflection. For example, in interviewing a person with little formal education, he may shift from a question worded "Have any of your children attended college?" to "I don't imagine any of your children have gone to college, have they?" He may further bias the responses by assuming that one of five opinion statements included in a certain item is inappropriate to the respondent's situation, and thus read only four. There are, of course, times when the interviewer must improvise in the field, but here we refer to those situations in which improvisation is not called for by the objective circumstances.

2 *Errors in probing.* Among the interviewer's most critical responsibilities during the questioning process is that of probing to obtain complete, accurate, and relevant answers to all items. Here, too, the expectations and stereotypes built up during the course of the interaction, as well as his feelings toward the respondent as a person, may lead to biased data. The most common error is insufficent probing. This often results from the interviewer's belief that the person has little to say about a given subject because of limited education, intelligence, or his position in society. Then, too, if the interviewer has strong expectations about what the respondent *should* say, or strong feelings about what he would like him to say, he may cease probing when the "right" answer has been mentioned. Even more serious errors may be introduced when probing is used positively to lead the respondent in the direction of a certain answer. When one feels he has come to know the respondent quite well, there is often a very real temptation to begin questions with phrases such as: "I suppose you would agree that . . ." The proper uses of probing will be discussed later in the chapter.

3 *Errors in recording answers.* Attitudes and expectations may also affect the accuracy with which the respondent's answers are recorded in the

---

[2] This discussion of interviewer errors draws on Kahn and Cannell (1957, pp. 189-193).

questionnaire. The interviewer may be tempted to improve the grammar or level of language used, omitting profanities or grammatical errors. Or he may simply not "hear" certain statements which are inconsistent with his own attitudes and expectations, or with what the respondent has said earlier in the interview. The respondent, however, is not the only source of pressures to modify the material recorded. The interviewer may feel that what the respondent has said is so incoherent that he will be criticized by his supervisor or the coding staff if it is presented in its original form. To avoid this situation, he may edit out some of the inconsistencies, make the sentences more complete than they actually were, and otherwise improve on the final product.

    4 *Errors in motivating the respondent.* The success of the survey interview rests heavily on the ability of the interviewer to maintain a high level of relevant motivation in the respondent. At times, the interviewer may fail to carry out this responsibility, perhaps because he dislikes the respondent or finds him dull or threatening, or because he would like to terminate the interview as soon as possible. This failure may take at least three forms. The first consists of *undermotivating* the respondent, for example, by failing to arouse interest in the study or by ignoring his needs for encouragement and information during the interview. Under these conditions, the respondent may continue with the interview, perhaps out of politeness, but provide a level of information which is quite shallow. A second error is that of *inappropriate motivation.* This would be seen when the interviewer encourages the respondent to become too dependent on his approval, or, at the other extreme, when he tries to induce participation by appealing to his official status in the government or university. A third error may result from the interviewer's *insensitivity to the situation of the respondent.* Kahn and Cannell write: "Basically, the interviewer must sense the respondent's needs, and adapt his efforts to motivate accordingly. He must 'pick up the respondent where he is.' Respondents will differ in resistance, in their interest in explanations and purposes, and in their need for encouragement and support" (1957, p. 192). Insensitivity is seen when the interviewer fails to relate what he does to the psychological situation of the respondent at that moment.

    Bias from the Respondent The respondent's attitudes, expectations, and feelings can also be a major source of bias in the survey interview. There are numerous reasons why he may be unwilling to "open up" with complete and accurate answers, including uncertainty about the purposes of the study, fear of the reactions of others in the community or organization, the feeling that the interviewer is so different in background that he will not understand his point of view, and fear that the interviewer may disapprove of responses indicating violations of some social norm. Bias and error will be reflected in responses which are incomplete, imprecise, inaccurate, false, or irrelevant. The following are several of the more common forms of respondent bias.

1 *"Courtesy bias."* This form of distortion was first suggested in an article on survey research in Southeast Asia (Jones, 1963). According to the author, courtesy is a pervasive value in this region, and one with important implications for social research. Its central elements include maintaining a pleasant and agreeable atmosphere, avoiding open disagreement with a person of higher status, saying what is pleasing and avoiding discussions which would affront or cause hurt to others, or which the other would not like to hear. While this source of bias seems especially applicable to Southeast Asia, it is found in other regions as well, including the United States. In a general sense, courtesy bias can be viewed as the tendency to limit one's answers to topics which are pleasant and cause little discomfort or embarrassment to the interviewer. At times it may be reflected in answers which move a painful interview to a speedy conclusion.

2 *Ingratiation bias.* This is similar to the courtesy bias, but more complex in its dynamics. It occurs when the respondent distorts his answers in directions intended to win the approval, attention, or favor of the interviewer (cf. Back and Gergen, 1963). To do so the respondent must develop hypotheses about what the interviewer wants to hear, and then aim his answers at fulfilling these expectations. This type of bias is most likely to arise when the interviewer becomes too friendly with the respondent and places excessive emphasis on the rewards available in the immediate interviewing situation. It may affect not only the direction a given response takes, but also its length. This point becomes important when the length or "number of reasons" is taken as an indication of interest in a topic, as sometimes happens with open-ended questions. The ingratiation bias has sometimes been caricatured in stories about native respondents providing interviewers with lurid but inaccurate details of village life because "that's what they wanted to hear."

3 *"Sucker bias."* This arises when respondents make a deliberate effort to mislead or deceive the interviewer, not so much to hide the truth as to trick, outwit, embarrass, or play a hoax on an "outsider." This situation is described in the Keesings' (1956) study of elite leadership in Samoa, but is not unknown in other populations, including university students.

4 *Social desirability bias.* This occurs when respondents distort their answers to conform to the prevailing norms and values in their own community or the larger society. It differs from both courtesy bias and ingratiation bias in that the main point of reference is the norms of the society rather than the expectations or feelings of the interviewer, although the two are intertwined. As a result of social desirability, the respondent may be reluctant to mention behaviors considered immoral, such as violations of sexual codes, or the use of alcoholic beverages. This list might be extended to include other biases, such as the apparent tendency of Japanese respondents to underevaluate their rank and accomplishments (the "humility bias"), or of many respondents to restrict the direct expression of emotions during the survey interview (affectivity bias).

PRACTICAL GUIDELINES FOR INTERVIEWING

Because no two surveys are alike, it is impossible to set down ironclad rules for conducting the survey interview. A study which makes use of highly experienced and carefully trained interviewers will allow for greater flexibility in procedures than another with a totally inexperienced field staff. Also, much depends on the aims of the study and the character of the questionnaire. A survey whose goal is mainly to explore issues and isolate problems for future study will require less attention to the specific manner of asking questions than another seeking precise quantitative information. In every case, it is essential to relate procedures followed in the field to the goals of the study, the capacities of the interviewing staff, the attitudes of the respondents, and other factors considered earlier in this chapter.

With this caveat in mind, we will now offer various practical suggestions for carrying out the survey interview. These should not be regarded as rules to be followed in every situation, but rather as guidelines applicable to many household surveys, especially in North America. To lend a note of concreteness to the discussion, we will introduce relevant materials from the *Interviewer's Manual* used by the Michigan Survey Research Center.

Preliminary Preparations

Preparing the Community   It is usually helpful to make a few advance contacts in the community or organization before the interviews. Particularly important in the United States are "legitimizing" contacts with the local Chief of Police, Better Business Bureau, and Chamber of Commerce. The field staff should not attempt to obtain their approval for the study, but rather make clear that theirs is a legitimate research organization. This step has become increasingly important as growing numbers of cities have passed ordinances exercising control over house-to-house solicitation, including survey research. One report (Arnold, 1964) indicates that at least 250 communities in 34 states in the United States have provisions for some type of control of surveys, ranging from registration with law-enforcement officials to complete prohibition.

In some cases, especially in cities with numerous interviews scheduled, it is worthwhile to submit a story about the survey to a local newspaper and then have the interviewers carry a clipping in a plastic cover. If respondents see that the study has received coverage from a respected newspaper, they will be less likely to reject the interview as a form of sales solicitation or political canvassing. Interviewers from the Survey Research Center have also found it helpful to carry reproductions of articles about the Center from *Time, Newsweek,* or other magazines.

Field work carried out in rural areas of the less developed countries may demand much more elaborate preparations. In certain cases, it may even be necessary for the interviewers to spend a month or more as residents before attempting any interviews at all. Wilson (1959) suggests that it may be prudent to carry out token interviews with a few leaders who do not fall into the sample, both out of deference to their position and as an example to others in the community. This procedure would be unethical, however, if the individuals were led to believe that their responses counted in the survey results when, in fact, they were to be discarded. It might be more sound to select a special sample of community leaders and to use their responses, perhaps treating them as a separate unit of analysis. Short of living in the village or interviewing community leaders, it is important at least to contact the key individuals, explain the survey to them, and secure their support.

Advance preparations can, however, be carried too far. Large-scale publicity campaigns will often arouse both interest and opposition. The dangers from overexposure are particularly great in potentially controversial studies, or in areas where the survey is relatively unknown. Often it is best to keep publicity to a minimum and to rely mainly on personal contacts.

Preparing the Respondent   The essential point here is the same as in the case of the community: some advance preparation may be helpful, while too much is often counterproductive. Some research organizations, including the Survey Research Center, find it useful to send a letter to the respondent mentioning that an interviewer will be calling at his home within a day or so. Experience suggests that these letters are most effective when they are brief and general, and when the gap between the arrival of the letter and the arrival of the interviewer is very short. There are at least three problems with long introductory letters. First, in many cases they will simply not be read. Second, it is difficult to choose a level of language appropriate to all respondents. If the level is too high, the less educated group may be offended, while if it is too low, it will be considered insulting by the better educated respondents. Third, if the respondent receives too much information about the study, he may either decide that he does not like it or lose the sense of curiosity which serves as one of the strongest sources of motivation to participate. The typical letter used by the Survey Research Center now says little more than "You have been selected and our interviewer will be calling you," with a few general comments about the study and an assurance of confidentiality. Given the ambiguous results produced by letters of introduction, we would suggest that study directors carry out informal experiments testing their effects.

Introducing the Study

The Interviewer's Appearance   We have already noted that from a psychological standpoint a great deal happens during the first thirty seconds after the interviewer and respondent see each other. The most immediate and obvious clue to the interviewer's background is his general appearance, especially his dress. It is important for the interviewer to ask himself each day how his appearance will be interpreted in the particular areas in which he will be working. Usually he should dress in a simple, neat, and inconspicuous manner. While on the job, he should also avoid wearing pins, lapel buttons, rings, or other forms of identification which associate him with a particular social group or cause, such as a fraternal order, political party, or antiwar movement. These signs may call attention to themselves and thereby either distract the respondent's attention from the study or lead him to slant his comments in a direction related to the perceived identification, or both. In planning for the study, policies should also be set regarding other aspects of personal appearance which may evoke strong reactions in respondents, such as hair styles, bare feet, and hemlines. Though these are difficult questions to raise with interviewers, especially those with no prior experience, it is much better to do so at the beginning than to attempt to sort out the effects of appearance after the data are collected.

Timing   As part of planning the day's work, the interviewer should make a specific assessment of the time of day which will be most convenient for the respondents in that area. Male heads of households, for example, will rarely be at home during the day, while they may be much more accessible in the early evening and on weekends. Housewives may be at home in the morning, but resent interruptions of their work schedule during those hours. In such cases, the interviewer may meet with greater success in the early afternoon before the children return from school. Similarly, when the community contains a high proportion of night and shift workers, it is best to schedule interviews at a time which will not interfere with their sleep schedule, such as immediately after lunch. Other considerations which might enter into planning for the initial contact include the timing of popular television programs, preparation for church services, holidays, and fears of burglary or other crimes.

Opening remarks

Gaining entry   The interviewer should appear relaxed and confident as he approaches the respondent's doorstep; he should be neither impatient nor overly casual. He should be serious, but not grim. Also, to avoid being mistaken for a salesman or a tax assessor, he should carry his interviewing materials in a folder rather than a briefcase. The larger survey research centers in the United States and Canada typically provide each interviewer with a plastic folder bearing the official seal of the organization. When he arrives at the sample household, he should knock or ring briefly, and allow ample time for the occupants to answer. The main challenge when someone appears is to present enough information to obtain an invitation into the house, where the study can be explained more conveniently. The SRC *Interviewer's Manual* offers several suggestions on this point.

> *Doorstep introductions should be brief.* . . . The doorstep is not a very convenient place to carry on a conversation and to establish a friendly relationship. For this reason, the doorstep introduction usually should be brief, just sufficient to get you inside the house. Once inside, you are in a better position to convince the person of the value of his cooperation. It is easier for the respondent to say "no, thank you" on the doorstep than it is in his living room.
>
> At the doorstep the interviewer should not ask questions to gain permission for the interview but should suggest the course of action which she desires. For instance, instead of asking, "May I come in?"—to which a respondent could easily say "No"—say, "I would like to come in and talk with you about this." Other examples of questions to avoid are, "Are you busy now?" ("Yes, I am.") and "Could I take this interview now?" ("No, not now.") and "Should I come back?" ("Yes, come back later.") Questions which permit negative responses can lead the respondent into refusing to be interviewed.
>
> The interviewer should assume the respondent is *not* too busy, and should approach her meeting with the respondent as though the interview were going to take place right then, at the time of contact. By all means, make arrangements to return at a more convenient time *if the respondent suggests this,* but accept this situation at the respondent's instigation. Suggest it yourself *only* as a last resort when you want to leave the door open for another try at a time when the respondent might be more willing to be interviewed.
>
> The first few contacts with a respondent should be made by a personal visit, and not by telephone. This is simply because it is so much easier for a respondent to say "no" and hang up the phone than to say "no" when you are standing in front of him. The exceptions to this are telephone surveys and addresses where numerous attempts to reach a respondent have been unsuccessful.

As soon as possible after meeting the respondent, the interviewer should introduce himself by name, say where he is from, and identify his research organization. It is helpful to support these statements by showing an official identification card. By mentioning his name and showing his credentials, the interviewer helps to reduce suspicions about the study.

Explaining the Study   By this time the respondent will be curious about the purpose of the research and about why he was chosen. The interviewer should now provide a brief and clear explanation which would handle these questions honestly, but without raising new objections to participation. The statement on the purposes of the study should be fairly general, including enough information to stimulate interest and legitimate further questioning, but not so much as to overwhelm the respondent. New interviewers sometimes feel obligated to answer every possible objection to participation, and in the process raise more difficulties than they solve. It is usually most effective to begin the questioning process very soon after arriving, and to handle questions and objections only as they are raised by the respondent. It is also vital, however, to make the opening statement sufficiently broad to cover the specific areas of questions to be covered later. The respondent may rightly become suspicious if he is told, "This is a study of people's attitudes," and then is asked, without further explanation, to provide detailed information on income and birth control practices.

Many respondents will wish to know why they were included in the study and what will be done with the information gathered. The *Interviewer's Manual* suggests this approach to the first question:

> It is important that the respondent understand that he is part of a "cross section" survey, and that he was chosen quite impersonally only because he happens to be a particular person at a particular address. You may say something like this: "You see, in trying to find out what people in the country think, we don't talk with everyone, but we try to talk to men and women of different ages in all walks of life. We start by selecting certain counties or cities from all over the country. (Interviewer can show map from back of thank you card.) In each of these areas the Center selects smaller areas, such as blocks, and finally a selection of specific addresses is made. Then when the interviews from all these addresses are combined, we have a cross section of the people."

The interviewer should emphasize that the person's answers are confidential; neither he nor his address will be identified when the results are published. In most surveys, it is inaccurate to say that the answers are anonymous, since the household is identified by a specific address. Moreover, when the study is carried out in a single organization or community, the respondent may ask, and has a right to be told, if the results will be presented in such a way that he can be indirectly identified as a member of a group, such as a work group in a factory.

Very often it will help in explaining the study to show a newspaper clipping about this particular survey or previous studies carried out by the research organization. Similarly, the interviewer may wish to carry a few copies of published reports of other studies to answer questions raised by more sophisticated respondents.

Handling Field Problems   Each interviewer should be prepared to deal with several difficulties commonly arising in the first stages of the interview. These include:

1   *"I'm too busy."* In this case, the interviewer must first decide if the respondent is really too busy, or if his comment reflects suspicion or a lack of interest in the study. If circumstances (guests, housecleaning, children) suggest that the interview should be postponed, the interviewer should try to set a definite time when he will return, preferably within the following twenty-four hours.

2   *Respondent is away.* If the person selected is absent but other family members or roommates are at home, the interviewer should give a brief explanation of the study to one individual and try to determine when the respondent will return. It may also be necessary to explain why one of the other residents cannot be interviewed in his place. To avoid confusion and maintain comparability, the study directors should set a definite policy on the maximum number of call-backs allowed.

3   *"I'm not interested."* When comments of this sort are made, the interviewer should maintain his composure and try to determine the specific objection to the survey. If the person has no specific objections, it may be possible to win his cooperation by asking the first question. Usually it is advisable to avoid protracted debate about the study unless the individual presses a particular point. If he is concerned about the uses of the findings, for example, the interviewer should deal directly with his questions, emphasizing the confidentiality of the results, and perhaps showing him a specimen report illustrating the statistical nature of the analysis.

4   *"Who's behind this?"* Generally it is best to say that the study is sponsored by the research organization, such as the Survey Research Center, rather than to mention the specific details of funding. Statements such as, "This study is being carried out by the Survey Research Center with funds from the federal government" may lead to extended debate about the government, the advisability of spending money on this project, etc. If, of course, the respondent knows something about how survey research is funded and asks a specific question about sponsorship, he should be given an honest answer. But here, as elsewhere, there is no point in raising unnecessary objections.

5   *"Do I have to do this?"* Some respondents may confuse a survey with the national census, which is legally obligatory. This question can be answered by comments such as "There is no legal obligation for you to take part in this survey, but we do need information from persons like you if our results are to give an accurate picture of conditions in the whole country."

6 *"What good is this?"* Interviewers should think about this question well in advance of the first contact. It is both unethical and practically inadvisable to suggest that the survey will lead to specific benefits, such as better jobs, more parks, or less crime in the neighborhood. The most accurate answers are those indicating the general usefulness of surveys in uncovering the major problems facing people in the area, planning for the future, or measuring public opinion on important issues.

7 *Refusals.* In almost every survey, some respondents will say that they do not wish to be interviewed. The interviewer must exercise considerable judgment in these situations, for the same words may conceal rather different feelings toward the study. For some the refusal is really a veiled request for additional reassurance about the legitimacy of the study or the sincerity of the interviewer. In survey research as in romance, a "no" should not always be accepted at face value. But there will also be others who are adamant in their refusals, and whose wishes should be respected. Under these circumstances, the interviewer should leave politely, with no extended debate or "parting shots." A graceful exit may facilitate a second visit by another staff member. Sometimes, even with a firm refusal, it is possible to lay down the questionnaire and say: "All right, I won't try to interview you, but could you tell me just two things for the record." Two questions carefully chosen in advance may be very helpful in assessing the degree of bias in the results by comparing respondents with refusals. Also, the apparent reasons for the refusal should be noted as soon as possible in the space provided on the questionnaire.

Dealing with "Third Parties" We have already suggested that the presence of outsiders may lead to bias or distortion. This problem is especially serious when the study includes questions related to controversial community norms or sensitive aspects of personal behavior. Several avenues are open to the interviewer for dealing with "third parties."

1 The simplest and most effective approach is often to explain to the outsiders that their presence may bias the results. The interviewer can point out that people often do not express their real feelings when others are listening. Back and Stycos (1959) report that this technique of "role-educating" the outsider was used with considerable success in their study of fertility in Jamaica. Some interviewers in this study used a doctor-patient analogy, others flattered the outsiders by emphasizing their power or influence over the respondent, while one underscored the tendency of others present to answer questions themselves rather than leaving them for the respondent.

2 After arousing the respondent's interest in the study, the interviewer can encourage him to find a place where they can talk alone. This can either mean finding another location for the interview, such as a nearby room, a porch, or the backyard, or having the respondent ask the outsiders to leave.

This approach will be most effective with children and others lower in status than the respondent, but it may succeed with other adults as well. In making such requests, however, one must be sensitive to cultural norms regarding "excessive" privacy, especially in male-female contacts. In rural areas of Asia, Africa, and Latin America the purposes of the study will often best be served by moving to an area which affords some privacy, but which is still somewhat visible.

3   The interviewer can point out that it may save time to conduct the interview in private. If the respondent objects that he will probably need help in answering some of the questions, the interviewer can agree to call the others when necessary. Once the interview begins, the perceived need for help may quickly disappear.

4   When the questionnaire begins with fairly broad items which are not likely to be sensitive to the influence of outsiders, the interviewer might allow these individuals to remain for a few minutes to satisfy their curiosity, and then ask them to leave. This approach may backfire, however, if the outsiders become so interested in the previews that they wish to remain for the full feature. Back and Stycos (1959) suggest several other methods for gaining privacy, such as having the interviewer "freeze out" the outsider through pregnant silence and pointed glances, offering a modest bribe to children, and arranging to have family members go off on errands for the duration of the interview.

Any attempt to deal with the problem of "third parties" will be greatly aided by advance knowledge of the cultural norms and practices found in the interviewing sites. With such knowledge the entire issue can be discussed intelligently during the training sessions, and policies can be set on the techniques judged appropriate for the study at hand. While considerable discretion must always be left to the interviewer, it is important to define the range of acceptable techniques in order to assure comparability from one setting to the next. We would argue, for example, that the use of payments to "third parties" is a particularly dubious policy, especially if there is any possibility that later surveys will be carried out in that area. Payments of any sort usually have the effect of setting a precedent which creates serious problems in future interviews.

Asking the Questions

The interviewer has two major responsibilities in asking questions: developing and maintaining "rapport" with the respondent, and following standard procedures in using the research instrument. Since both are essential, one should not be emphasized to the exclusion of the other. The purposes of the study will not be served if the interviewer changes or omits questions in an effort to improve his relationships with the respondent. By the same token, little is gained if the questions are asked in such a rigid and mechanical fashion that the respondent becomes emotionally detached from the interview. We would offer the following as general guidelines for achieving both rapport and standardization.

1 *Use the questionnaire carefully, but informally.* The interviewer should treat the questionnaire as a tool for data collection, rather than a master controlling all his actions in the interview. In order to achieve this level of informality, he must be thoroughly familiar with the purposes of the study and especially the wording and order of the items. Informality and a relaxed attitude toward the questionnaire and the questioning process need not imply carelessness or arbitrariness in asking the questions.

2 *Know the specific purpose of each question.* Both to satisfy the purposes of the research and to increase his own ease in using the questionnaire, the interviewer should be clear about what is considered an *adequate* response for each item. In questions about occupations, for example, it is important to know in advance how the information will be coded. For certain purposes an answer of "factory worker" may be sufficient, while for others it may be important to know if a man is a machine operator or a janitor in the factory. Instructions about the purposes of the questions and the level of information required are usually considered in the training sessions, but they should be carefully reviewed before the field work begins.

3 *Ask the questions exactly as they are written.* Experiments on question-wording (see Chapter 6) indicate that even minor changes in wording can alter the meaning of a question, thus reducing its comparability from one person to another. The success of the survey depends in large measure on standardized conditions in the interview, especially in the way that questions are asked. For this reason the interviewer should avoid omissions, improvisations, explanations, or abbreviations of the items. Ad hoc interpretations of "what they're looking for" are especially precarious, for they may easily have the effect of suggesting an answer to the respondent. If the person does not understand the question the first time, the interviewer should repeat it just as it is written. Sometimes silence may mean doubt or confusion rather than a misinterpretation of the item. In this case, the interviewer might make a neutral comment, such as: "We're just interested in what people think about this—there are no right or wrong answers." If all else fails it may be necessary to restate the question with slight changes, but this should happen very rarely, and the exact changes should be written in the questionnaire. Sometimes interviewers feel that they must have the freedom to modify the questions in order to appear natural and comfortable during the discussion. We would argue that the same effect can be produced, with fewer risks, by intimate acquaintance with the questionnaire and the occasional use of neutral phrases or comments.

4 *Follow the order indicated in the questionnaire.* In the well-designed survey questionnaire, the order of the questions receives careful attention,

and is chosen for specific reasons. These reasons may include stimulating the respondent's interest in the study, facilitating transitions from one topic to the next, aiding the respondent's memory of past events, standardizing the conditions for asking opinion questions, experimenting with the impact of varying question placement, and maintaining rapport as the interview moves on. Arbitrary changes in the order of asking the questions may not only reduce the comparability of the interviews, but also introduce serious bias into questions sensitive to sequence. Also, if the interviewer skips around the questionnaire, he may omit important questions or find, to his embarrassment, that he is lost.

5 *Ask every question.* Sometimes, in answering one question, the respondent will make comments which seem to answer a later question as well. Similarly, especially in the case of opinion questions covering more or less the same topic, he may say, "Put me down as 'yes' (or 'no') for all of them." The best policy to follow under these circumstances, even when it seems clear that the respondent has answered the later questions, is to repeat the question. It may help, however, to say: "You may have answered this question before, but I want to be sure to put down your own answer. So, to refresh my memory . . ." The reason for this policy is that prior responses often prove to be inaccurate when the later question is asked directly. The SRC manual offers sound counsel on this point:

> Write down the initial answer under the question when it occurs. Then ask the partially answered question when you get to it, but preface it with some remark which will show the respondent that you haven't forgotten what he said earlier and haven't rejected his earlier answer. Such a remark might be: "We're asking people on this survey about each one of these, and I'd just like to make sure how you feel about each one separately . . ." In those few cases where the question has been clearly answered, the interviewer might say, "You've told me something about this, but the next question asks . . . ?"

6 *Do not suggest answers.* It is often tempting to suggest a response which seems to fit the respondent, especially after he greets a lucid restatement of the question with a blank stare. The danger in suggesting answers is that the respondent will accept them, perhaps feeling that the interviewer's answer must be the right one, or that this is the fastest way of ending the interview.

7 *Provide transitions when needed.* A major challenge for the interviewer is to ensure that the questioning process flows smoothly from one item and one section of the questionnaire to the next. In some cases, the questionnaire itself will aid him in this task by providing standard transitional phrases. At other points he should feel free to improvise with neutral comments showing how one section connects with the next, or simply that the interview is shifting to a new topic. He should not, however, attempt to summarize groups of questions when his summary might create a "set" or affect the respondent's attitudes in other ways. In most cases it is sufficient to say, in effect, "We're through with questions on that topic. Let's move on to the next."

8 *Do not leave any question blank.* Even when every effort has been made to anticipate the situations which will make a question inapplicable or inappropriate, an item may not be relevant for certain respondents. In this case it is best to cross it out and add a note about why it is inappropriate. When items are left blank it is difficult for the coders and study directors to determine if they were omitted by oversight, or were judged to be inapplicable. Aside from questions covered by "skips," every item should have either an answer or an explanation of why it does not.

Obtaining an Adequate Response

Often during the course of the interview, the respondent will give answers which are incomplete, unclear, irrelevant, or otherwise inadequate to the purposes of the study. These situations provide the interviewer with one of the most critical tests of his skills—the ability to probe for answers. The term *probing* refers to a variety of techniques used to stimulate discussion and focus the flow of information without suggesting answers. The specific aim of the probe is to obtain information which satisfies the purposes of the question. To carry out this task successfully, the interviewer must be thoroughly familiar with the objectives of each question. Without such knowledge, he will have no yardstick for gauging the adequacy of the responses. Several kinds of neutral probes may be used in the survey interview.

1 *The silent probe.* A well-timed pause is perhaps the simplest and most neutral way of stimulating further discussion by the respondent. Inexperienced interviewers often feel uncomfortable with this technique, fearing that silence will be interpreted as incompetence or that it will embarrass the respondent. The difference between the "pregnant pause" and the "embarrassed silence" usually lies in the attitudes and behavior of the interviewer. The pause is likely to be most helpful when the interviewer waits confidently and expectantly for more information, perhaps encouraging the respondent with a nod of the head or his facial expression.

2 *Overt encouragement.* This form of probing draws on brief assertions of understanding and interest indicating that the interviewer accepts what has been said up to that point and would like to hear more. These include remarks such as "uh-huh," "I see," "yes," "hmmm," or "that's interesting," as well as nonverbal expressions such as a nod of the head. As suggested earlier, overt encouragement is often combined effectively with the silent probe.

3 *Elaboration.* This consists of neutral questions or comments used to obtain more complete or accurate responses. When elaboration probes are used they should always be recorded in the interview schedule, preferably in abbreviated form. The *Interviewer's Manual* suggests the following helpful phrases, together with "key word" abbreviations.

> "How do you mean?" (How mean)
> "Could you tell me more about your thinking on that?" (Tell more)
> "Will you tell me what you have in mind?" (What in mind)
> "I'm not sure I understand what you have in mind." (What in mind)
> "Why do you think that is so?" (Why)
> "Could you tell me why you feel that way?" (Why)
> "Which figure do you think comes closest?" (Which)
> "What do you think causes that?" (What causes)
> "Do you have any other reasons for feeling the way you do?" (Other)
> "Anything else?" (AE or else)
> "Repeat question." (RQ)

These examples make it clear that the probe chosen must be adapted to the specific form of inadequacy seen in the answer. The ability to match the probe to the problem increases with experience in interviewing and with knowledge of the specific objectives of each question.

4 *Clarification.* Here the interviewer not only asks for more information, but specifies the kind needed. Clarification probes are in order when the responses given appear inconsistent, contradictory, or ambiguous. In these cases, the interviewer might introduce questions such as the following:

> "I'm sorry, but I'm not clear about what you meant by that—could you tell me a little more?"
> "I'm not sure I understand. Did you say a few minutes ago that (repeat earlier and seemingly inconsistent response)?"
> "Could you tell me why you felt that way?"
> "About when did that happen?"

Great skill is required of the interviewer in probing to resolve inconsistencies, contradictions, or obviously inaccurate statements. The challenge is to elicit correct information without appearing to be carrying on a cross-examination. The challenge will be particularly great when, as sometimes happens, a respondent will cling tenaciously to seemingly inconsistent statements. In certain cases, especially when rapport will be seriously damaged for other questions, it may be necessary to forgo probing completely, or to return to the inconsistencies at the end of the interview.

5 *Repetition.* Sometimes the interviewer can probe by repeating what the respondent has just said as he records this information in the questionnaire. Respondents will often treat such "echo probes" as a request for additional information, and will respond accordingly. This approach can sometimes go awry, however, when the phrase repeated is relatively insignificant, or when a sophisticated respondent sees through the technique. It is very embarrassing for the interviewer to be greeted with the comment, "Yes, that's what I just said."

Probing is helpful only when it is *neutral*. By their very nature, interview probes are difficult to control and can easily lead to bias or distortion in the information obtained. The greatest danger of all lies in questions which implicitly suggest an answer or direct the respondent's attention to one alternative rather than others. For example:

Q. Are there any magazines that you read regularly?
A. Yes, a few.
Q. What are the names of these magazines?
A. I don't remember exactly right now. Let me think a minute.
Probe: Well, how about *Time* and *Newsweek*. Do you read either of these?

This is an illustration of a leading or directive probe which introduces basic changes in the content of the original question. In the example above, the most appropriate probe might well have been a few moments of silence, followed by a neutral question such as "Can you think of any at all?" The risk of leading probes can often be reduced by requiring the interviewers to record all of their remarks in the questionnaire.

Another aid to the accuracy of survey information, particularly when the study deals with objective questions such as income or fertility, is to encourage the respondent to check relevant and available records. One danger in this procedure, of course, is that the search for records may take so much time or become such a distraction that the interview itself is placed in jeopardy. Nevertheless, as noted in Chapter 6, it is sometimes helpful to ask the respondent to check financial records, birth certificates, receipts, or other records. However, the interviewer should not ask to see these records. With objective questions, other members of the family might also be asked to supply information, particularly when they are better informed on the topic than the respondent. But the hazard arising from outside consultation is that the experts consulted may then be reluctant to depart from the scene. One way to deal with this problem is to structure the questionnaire so that the items likely to require information from others come near the end, and follow those known to be more sensitive to group influences.

Recording the Responses

The success of the open-end interview question depends on a complete and accurate record of what the respondent said and how he said it. It is not enough for the interviewer simply to include enough information to satisfy the coding categories used in the study. He must also try to convey, through the respondent's own words, a full and nuanced picture of his knowledge, attitudes, feelings, and other personality characteristics. Also, it is usually impossible to know in advance what answers satisfy the minimal requirements of coding. Often, during the coding or later, a researcher will return to the verbatim accounts and attempt to measure other, more subtle qualities, such as the person's verbal fluency, his attitudes toward the survey and the interviewer, or his covert feelings toward the government. In general, the simplest rule for open questions is to *provide a verbatim record of all material relevant to the purposes of the study.* Occasionally the respondent will digress to discuss sports, the weather, or other topics which clearly fall outside the scope of the research. In these cases, it is not necessary to provide a verbatim record, though the interviewer should note that the digression has taken place.

While the goal of providing a complete verbatim record might seem straightforward enough, the task of transcribing responses is complicated by other demands made on the interviewer's attention, such as asking questions, probing, listening, and looking at the respondent. To facilitate the recording process, it is usually advisable to observe three rules.

1 *Record the responses immediately.* The best time to transcribe the respondent's exact words is during the interview itself. As time passes, the interviewer's memory will fade so that he may easily distort what the respondent has said or miss many of his characteristic expressions. The problem, of course, is that it is difficult to take exact notes while simultaneously trying to hold the respondent's attention. Some have attempted to solve this problem by using a tape recorder, but this practice poses special problems which we shall consider shortly. Usually the most practical course is to jot down key words and phrases as the respondent is talking, and then reconstruct the full conversation immediately after the interview. It is also important to begin writing as soon as the respondent begins to talk. In general, one interview should be completed before another is begun.

2 *Abbreviate words and sentences.* The following abbreviations, for example, are commonly used by interviewers at the Survey Research Center:

R	Respondent
I'er	Interviewer
I'w	Interviewing
DK	Don't Know
RQ	Repeated Question
Q're	Questionnaire

Sentences can also be abbreviated by omitting articles and prepositions, and by giving priority to key words.

3 *Include all probes.* To have a complete record of the interview, it is important to know not only what the respondent said, but how the interviewer's comments affected the flow of information. The interview, as noted earlier, is an interaction between two or more parties, and this fact should be reflected in the final record. Thus, comments or interruptions made by outsiders should be noted as well. To allow for proper identification and coding of the recorded material, the interviewer's probes, remarks, summaries, and instructions should always be placed in parentheses.

Two additional procedures will often facilitate the mechanics of recording. First, both for purposes of clarity and to permit easy erasures later, the interviewer should record the answers with a black lead pencil rather than a pen. This suggestion is especially important if the interviewer uses shorthand or relies on key words to reconstitute the full response. Answers recorded in ink are difficult to change and also may become illegible if the questionnaire is exposed to moisture. Second, it is always advisable to carry a portable writing surface in the event that a table is not available. The folders issued to interviewers by large survey organizations are designed with this specification in mind.

Using a Tape Recorder   On the surface, one simple solution to many of the problems of recording would be to have all interviewers use portable tape recorders. This technique would clearly have the advantage of leaving the interviewer free to concentrate on the questioning process, and would virtually eliminate the problem of selective recall and distortion in recording. Frequently, however, respondents answer differently when they know they are being taped, while secret taping is highly unethical.

The tape recorder is invaluable for certain types and phases of interviewing, but it remains impractical for the typical household survey. It seems most appropriate when the purpose of the study is to explore complex issues in considerable depth with a limited number of respondents. This situation might well arise in the initial stages of a large survey. Here the emphasis is on defining the problem and isolating the precise dimensions for further study. Tape-recorded interviews with a few dozen respondents may serve this purpose admirably well.

When the survey involves relatively brief interviews with a large number of respondents, on the other hand, there are several disadvantages to tape-recording the responses. First, this procedure requires that much more time—perhaps double or triple that seen in the usual interview—be devoted to transcribing the responses. When the tape recorder is not used, the interviewer can write while he is talking with the respondent, and then spend a brief amount of time later in editing this material. If the interview is recorded, he must replay the entire tape after he leaves and transcribe the material onto the questionnaire. Second, respondents may feel uncomfortable when they know that the conversation is being recorded. The empirical evidence regarding the effects of tape recording on the accuracy of survey data is unfortunately very limited. Most of the existing research (Bucher, Fritz, and Quarantelli, 1956; Belson, 1967) has turned up no measurable differences in the quality of information obtained in recorded and nonrecorded interviews, but these studies have been neither comprehensive nor tightly controlled. However, Belson speculates, on the basis of his research, that

> using the tape recorder in the interview introduces a degree of formality or officialness such that the less-educated respondents are bluffed into making a greater effort to get their statements right, whereas members of the middle and upper social sectors become characteristically more wary about what they go on record as saying [1967, p. 257].

If the tape recorder is used, the interviewer should be thoroughly familiar with its operation, and should keep it as inconspicuous as possible during the conversation. Respondents are easily distracted by microphones and revolving spools.

Concluding the Interview

Before leaving the household, the interviewer should thank the respondent and any others present for their time and cooperation. His departure should never be hasty or brusque, but neither should he prolong the conversation unnecessarily. In most cases, the best approach is to ask the respondent if he has any questions about the study, and if so, answer them. Also, several large research centers find it helpful when concluding the interview to hand the respondent a "thank you" card signed by the interviewer. A typical card might read:

Name of Center

We wish to thank you for your cooperation and your contribution to the success of the _____ study.
Your interviewer was: _____ (signed)
If you have any further questions or would like additional information we will be happy to answer your inquiries.
Requests should be addressed to:

Name and address of Center
Telephone number

Immediately after leaving, the interviewer should take a few minutes to complete a "thumbnail sketch" of the respondent and the interviewing situation. This might include information about the respondent's attitudes toward the interviewer and the study, unusual aspects of the interview, such as "third parties," apparent inconsistencies, interruptions, tense situations, or language problems, the interviewer's reactions to the respondent, and any other information which would be useful in coding the data or interpreting the individual's responses. While these descriptions are usually not coded explicitly, they are almost always read by the coders and often by the study directors as a means of understanding and evaluating the interview.

In sum, the survey interview must be set in the context of the entire study, including the subject matter under investigation, the social structure and culture of the society, the immediate interviewing situation, the demands posed by the questions, and the ultimate requirements for data. While both social-psychological theory and cumulative experience with survey research suggest a number of general guidelines for interviewing, these can never be applied in a mechanical fashion. Proper planning for interviewing requires both a clear sense of study objectives and an almost clinical understanding of the conditions under which the data will be gathered. If this understanding is limited at the beginning, it would be well worth the investment to conduct pretests aimed as much at evaluating the conditions of field work as the questionnaire. Finally, we must admit that, while there has been a great deal of useful experience with survey interviews, to date there has been relatively little controlled observation of the interview process. Other studies such as that conducted by Cannell et al. (1968) are badly needed, and would provide a more sound basis for practical advice about interviewing.

FURTHER READINGS

Back, K. W., and J. M. Stycos, *The Survey under Unusual Conditions: Methodological Facets of the Jamaica Human Fertility Investigation* (Ithaca, N.Y.: Society for Applied Anthropology, 1959), Monograph No. 1.

Converse, J. M., and H. Schuman, *Conversations at Random: Survey Research as Interviewers See It* (New York: Wiley, 1973).

Gorden, R. L., *Interviewing: Strategy, Techniques, and Tactics* (Homewood, Ill.: Dorsey Press, 1969).

Kahn, R. L., and C. F. Cannell, *The Dynamics of Interviewing* (New York: Wiley, 1967).

Survey Research Center, University of Michigan, *Interviewer's Manual* (Ann Arbor, Mich.: Institute for Social Research, 1970).

Chapter 8

Organization and Administration of Field Work

Successful field work requires not only a theoretical understanding of the interview and a command of practical techniques, but also a carefully organized system of field administration. In most surveys the quality of the data is no better than the quality of the field organization. Interviewers may perform well as individual *virtuosi* during the training sessions, but fail to satisfy the needs of the study because of inadequate supervision or low productivity. Similarly, a study which begins well may lose momentum after a few weeks as a result of poor morale among the interviewers, high attrition because of other demands on the staff's time, improper scheduling, or a lack of coordination between the field sites and the central research office. These examples suggest that time invested in planning for the administrative side of field work will pay high dividends later. The most important organizational and administrative problems include the recruitment and selection of the interviewers, interviewer training, field supervision, quality control, and the control of materials.

Recruiting and Selecting Interviewers

There is no ideal interviewer in the abstract and, even if there were, the present state of research technology makes it difficult to identify him or her in advance. We suggested earlier that the choice of interviewers should be geared to the distinctive objectives and problems of the survey in question. Apart from certain early warning signs which often identify the very poor interviewer, it is very hard to know when this match has been made. Usually, all that is possible is to determine in advance which characteristics, such as age, sex, education, race or ethnicity, and personal appearance, will make a difference, and then proceed accordingly. It would be helpful, however, if more investigators carried out some type of subsequent evaluation research to check on the accuracy of their original hypotheses, and shared their findings with other researchers. At the moment, we can make a few practical suggestions drawn from research and experience, but the task of choosing good interviewers remains largely a matter of personal judgment.

Recruitment   The problems posed by recruiting new interviewers will vary with the size and geographic coverage of the survey and the staff already available. In large survey organizations the task of recruitment is greatly eased by the fact that only a few additions are made at a time. This situation is markedly different from that facing the "private" researcher who must recruit an entire staff ranging from a dozen to two hundred interviewers. Since many of the readers of this volume are likely to fall into the second category, let us review some of the possible sources of qualified recruits.

Experience in several countries, including our own in the United States and Peru, suggests four useful sources for obtaining large numbers of applicants: employment offices, school systems, colleges and universities, and ads in local newspapers. Employment offices have the distinct advantage of possessing complete files on potential candidates so that decisions on their general suitability can be made almost immediately. However, except during periods of high unemployment, few individuals qualified for survey research will be on file at these agencies.

Contacts with school administrators, on the other hand, are often more fruitful. Research organizations in the United States and elsewhere have found that primary and secondary school teachers prove to be excellent candidates for temporary employment as interviewers, especially when they can work in areas where they are not well known. Not only does their training and teaching experience serve as a helpful preparation for dealing with the public, but their schedules tend to leave them time for interviewing.

Universities, of course, are a familiar source of applicants, and one that should be explored. It is precarious, however, to rely exclusively or even heavily on students in a cross-section survey. First, their relative youth may create problems in having the study taken seriously by older respondents. This age factor could, on the other hand, prove to be a major asset in a study concentrating on persons under thirty. Again, it is essential to relate the background of the interviewers to the respondents and other demands of the survey. Second, as a body, university students in many countries tend to evoke stronger feelings—both positive and negative—on the part of respondents than older interviewers, and may thus introduce greater bias into the results. Third, because of fluctuating pressures (vacations, examinations, political crises, etc.) students hired at large may be less reliable than other interviewers in carrying out assignments or continuing to the end of a study. We also found in Peru that some students from the humanities and liberal arts were reluctant to accept the "rigidities" of survey interviewing, such as asking the questions in the way that they were written. A few expressed the feeling that, since they were educated individuals, they should be allowed to administer the questionnaire as they saw fit in each situation. As noted earlier, the advantages of such flexibility are usually outweighed by the distortion introduced by excessive personalization. These observations should be qualified by noting that student interviewers have been used with considerable success in many surveys, both in North America and in other countries. Much depends on the circumstances of the study as well as the degree of the student's involvement. Some of the disadvantages are lessened, for example, when the student is part of the research team, and will be active in subsequent stages of the research, including analysis and publication.

The use of ads in local newspapers usually has mixed results. The typical ad will lead to numerous applications, but produce very few qualified candidates. In a survey carried out in Arequipa, Peru, for example, a series of ads resulted in applications from about 150 individuals showing wide differences in age, education, and occupations. Of these only about 100 appeared at the first training session, and 50 passed an initial screening test. Very few of the 50 actually completed the training program and were selected as interviewers. Experience with recruitment in other surveys confirms these impressions. In short, there seems to be no one best way of locating interviewing talent. Often the best interviewers seem to come via personal contacts with the study directors or other members of the field staff.

Selection Many attempts have been made to develop standardized procedures for selecting survey interviewers. Unfortunately, all seem much better at rejecting poor candidates than in identifying excellent prospects.

The following are several devices which are helpful in evaluating potential candidates, though none is reliable when used alone.

1 *Application form.* Having the candidate fill out a handwritten application form accomplishes two purposes: it gathers the necessary background information with a minimum of effort, and it provides a rough sample of the person's handwriting and ability to follow instructions. The resulting information may be used immediately for selection and then later for evaluation and research. The form need not be elaborate, but should include questions on the following topics: name, address, phone number, age, marital status, number and ages of children and other dependents, health, education, present employment (if any), hours of work, length of employment, prior experience in interviewing, number of hours available, times available and not available, reasons for applying, and references. Hauck and Steinkamp (1964) made creative use of such straightforward information in a careful study of factors associated with the interviewer's performance in the field. Their research produced few positive findings in the case of interviewers finally selected for the study, though it is quite possible that background characteristics were useful in *eliminating* unqualified candidates. They do report, however, that the completion rates for persons with college or postgraduate work were higher than those for interviewers with a high school education or less.

2 *Personal interview.* Though the evidence on the effectiveness of this procedure is sparse, many study directors prefer to form their own impressions of applicants through a personal interview. This approach is probably most valid and effective when carried out by a panel of three or more interviewers, each of whom completes independent ratings on the candidate. Among the characteristics which might be evaluated in this way are the individual's self-confidence, poise, clarity of expression, appearance, alertness, friendliness, and overall personal impact.

3 *Psychological tests.* In some situations, particularly in less developed countries, a simple intelligence test serves well as a preliminary screening device, especially when there are large numbers of applicants of dubious quality. In carrying out field work in Peru, the authors found that a short measure of verbal intelligence used by the local employment service seemed to work well for this purpose. Given the questionable validity of many tests, it is important to set the cutting points fairly low so that good candidates unfamiliar with objective tests are not eliminated. In the Peruvian case, the research team decided that any applicant who could not attain a verbal intelligence score of 60 would probably not be able to handle a survey questionnaire of considerably greater complexity than the test. Subsequent research showed, in fact, that all candidates scoring less than 90 failed in subsequent parts of the training program. There is very little evidence, on the other hand, that standard personality tests, such as the Minnesota Multiphasic Personality Inventory, show consistent relationships to interviewer performance. While such instruments might be used for experimental purposes, they should not be used for screening unless there is some evidence showing their validity in this context.

4 *References.* Letters of recommendation, including ratings from persons who know the candidate well, may be the best "objective" tool in selection. Hauck and Steinkamp (1964) found that evaluations by outside referees of self-confidence and of appearance, manner, and poise were significantly related to pick-up rates (the percentage of completed interviews) for interviewers. These ratings, in fact, proved to be the best single predictors of interviewers' performance ($r = .58$ and $.55$).

Thus far we have considered only the criteria of selection available before the training sessions begin. In most cases, the study directors will also wish to base their decisions on the candidate's performance in the training program.

Interviewer Training

Just as there is no standard survey or interviewer, so can there be no standard training program. The type and intensity of training will depend on the goals of the study, the number of interviewers, their previous experience, the availability of time, and the difficulty of the questionnaire. Thus, the total time needed for training may range from two hours to three weeks of intensive effort.

The model we have in mind in the following discussion is the fairly typical household survey in which the questionnaire contains a combination of closed- and open-ended items, the subject matter ranges from the simple to the complex, and the study makes use of several new interviewers. Many of our suggestions would have to be modified if, for example, just one or two interviewers were being added to an existing staff. In any event, it is helpful to distinguish between initial training and continuing training.

Initial Training   The goals of the initial training program in the type of survey described above usually include the following:

—communicating factual information about the objectives, uses, and sponsorship of the present study.

—developing an interest in and commitment to this research.

—arriving at an understanding of the role of interviewing in the research process and the significance of nondirective interviewing.

—developing basic skills in interviewing (e.g., through role playing) as well as practical suggestions for introductions, closure, and difficult situations arising in the field.

—familiarizing the interviewers with the questionnaire in general, and with the specific objectives of each question.

—reaching agreement on administrative procedures to be followed in the field.

—coming to an understanding on standard procedures for dealing with difficult situations.

—developing favorable attitudes and morale among the interviewers and supervisors.

—screening out candidates poorly suited for interviewing.

Several specific steps can be taken to attain these goals.

1 *Preparing background materials.* It is advisable to spend considerable time before the training sessions preparing written materials which outline the purposes of the study, the objectives of the questions, and the major responsibilities of the interviewer. At the Michigan Survey Research Center, it is customary at the beginning of the training program to distribute an interviewer's kit containing several types of materials:

—an instruction book covering the purposes and background of the study as well as the goals of the questionnaire.

—an *Interviewer's Manual* providing general guidelines for carrying out the survey interview as well as other forms of field work, such as listing for the sample.

—a series of sampling exercises.

—examples of the codes which will be used in classifying the responses from the study.

The advantage of this material is that the interviewer can study it on his own time, and also refer to appropriate sections when he encounters difficulties in the field.

2 *Group sessions.* Group discussions involving the interviewers, study directors, supervisors, and other members of the research staff are highly useful in communicating substantive knowledge about the study as well as in building morale. The format of these sessions can vary a great deal, but they will usually include some combination of lectures, open discussions, role playing, and practical experience with the questionnaire. The following is one example of the way in which this part of the training program might be organized.

Session 1 (3 hours)

A   Distribute written materials, including administrative forms.

B   Introductions—interviewers and staff.

C   General introduction to survey research: why and how it is done, principles of sampling, importance of following established procedures in sampling and interviewing.

D Explanation of the study: purposes, sponsorship, some uses of the findings. (Refer to *Interviewer's Guide.*)

E Basic concepts and definitions: critical terms used in the study, such as the labor force, unemployment, a household, and a family. (These should also be included in the *Interviewer's Guide.*)

F Guidelines in interviewing: preparing the community, preparing the respondent, introducing the study, confronting common field problems, dealing with "third parties," asking questions, probing, recording the answers, ending the interview. (Refer to *Interviewer's Manual* for further study.)

G Discussion of points raised.

H Assignments for second session: read sections of *Interviewer's Guide* describing the purposes of the questionnaire and the objectives of each item; familiarize yourself with the filters, skips, and general flow of the questionnaire.

Session 2 (3 hours)

A Practice questionnaire: have all participants fill out a blank questionnaire with answers describing someone they know well.

B Question-by-question discussion: a detailed review of the objectives of each item and the level of information needed to satisfy these objectives. This can be enlivened by drawing upon the responses recorded by the participants in their practice questionnaires.

C Practice coding: participants should be asked to code typical responses (bordering on the difficult) for the most critical open-ended items.

D Practice examination: this might consist of a series of multiple-choice or fill-in questions testing the interviewer's knowledge of the points raised in the written materials and earlier in the session. It would be understood that this examination will not be used for purposes of selection, but that a final examination similar to it will be an important consideration in selection.

Session 3 (3 hours)

A Review of theoretical and practical considerations in interviewing.

B Illustrative interviews from tape recordings preferably showing examples of proper as well as improper interviewing techniques. If available, video tape would be even better for the same purpose.

C Role playing: participants pair up to act as interviewer and respondent for the introduction and parts of the questionnaire. Their performance is then discussed by the group, and another pair takes over. This exercise is generally extremely useful, and can again be improved through the use of video tape.

Session 4 (3-4 hours)

A Question period, clarification of problems raised in earlier sessions.

B Discussion of administrative details: expense vouchers, mileage statements, method of payment, record keeping, etc.

C   Qualifying examination: this should be a comprehensive test covering the interviewer's knowledge of the questionnaire, field procedures, the psychology of the interview, and other relevant matters. While this examination may be administered on a "take-home" basis, it may be a more accurate measure of the interviewer's *immediate* knowledge—the kind required in the field—if it is given in the classroom.

It has been our experience in both Peru and the United States that written examinations serve not only as a useful means of selection, but also as a vehicle of education for the interviewer. Because of the time demands on the field staff, it is usually necessary to use mostly "objective" questions in the examination. These can be quite straightforward, often drawing upon appropriate sections of the *Interviewer's Guide.* For example:

Interviewer's instruction: A house, an apartment, or other groups of rooms, or a single room is regarded as a dwelling unit when it is occupied or intended for occupancy as separate living quarters, that is, when the occupants do not live and eat with any other persons in this structure, and there is either (1) direct access from the outside or through a common hall, or (2) a kitchen or cooking equipment for the exclusive use of the occupants of the unit.

Examination question: (Cover Sheet) The address to which you are assigned contains three families: a father and a mother (age about fifty-five), and two married sons, each of whom lives at that address with his wife. Each of the married sons has a separate bedroom shared with his wife, but the three couples eat together and use the same household facilities (refrigerator, stove, iron, etc.). How many dwelling units is this?

  a   One
  b   Two
  c   Three
  d   We cannot decide without further information.

Interviewer's instruction: Never suggest answers.

Examination question: When the respondent understands the question but finds it hard to give an answer that fits his situation, which is the best strategy:

  a   Help him out a little by pointing out the answer that seems to fit him best.
  b   Wait until he gives his own answer, and then, if it is not clear, use a neutral question or probe.
  c   Repeat the question to give him more time to think.
  d   Go on to the next question so that he will not be embarrassed.

If a candidate passes the final examination but has missed some items, it is instructive for him to review his errors in consultation with one of the staff members.

3   *Trial interviews.* After the interviewers have been instructed in the use of the questionnaire and in the techniques of interviewing, they should

be given an opportunity for direct experience in the field. Candidates who do well on the "academic" portion of the training program are not necessarily adept at live interviews. If possible, one of the trial interviews should be carried out in the presence of a field supervisor. If time permits, the supervisor might begin with a demonstration interview which provides the candidate with a helpful model for his own work. While trial interviews are valuable as a means of training, they suffer from two limitations as tools of final selection. First, some candidates become highly self-conscious when their first venture into the field is observed by someone whose task is not only to be of help but also to screen him. Second, unless the trial interview is assigned by the field staff, there are great variations in the difficulty of the situation chosen. Some candidates will deliberately go to a working-class neighborhood to seek out families representing the most complex contingencies in the questionnaires. Others will choose to stay in more familiar territory, perhaps choosing a middle-class household in their own neighborhood. It is generally recommended, therefore, that if the trial interview is used for purposes of selection, it should be made clear to the candidates that the staff recognizes the difficulties involved, and will not eliminate anyone solely on this basis.

Continuing Training The process of training interviewers should not end with the initial program, but should be maintained, at a lower level, on a continuing basis. Interviewers share the general human tendency to become careless as their work progresses, and must be aided by frequent checks and reminders as well as positive encouragement. It is possible to forget important definitions or policies covered in the training sessions, and to develop one's own interpretations for handling difficult cases.

One effective method of continuing training is to have the field supervisor carry out periodic reviews of a sample of completed questionnaires before they are sent to the research office. In this way, the supervisor can point out strong, as well as weak points, in the interviewer's work, and discuss problems while they are fresh. Another technique is to have the coders maintain a check sheet of problems appearing in their work, and then send this list to the interviewers while the study is still in the field. There is one difficulty with this practice, however—it may be demoralizing for the interviewers to receive only criticisms from the coders. New interviewers often view the coding staff with a certain amount of suspicion, fear, and resentment; the issuance of such checklists may reinforce these feelings. To restore a proper balance, it may be more useful to have the field supervisor return to the coding office several times during the study, discuss the problems reported by the coders, and then review a sample of coded interviews to determine strengths as well as weaknesses in the interviewer's performance.

In studies which continue through several waves, as in panel studies of economic behavior, it is helpful to reconvene the interviewers for periodic review and refresher classes. These sessions can be used to point out common errors noted by the editors and coders, and also to discuss problems experienced by the interviewers and supervisors. They may also contribute to morale by making the interviewers feel that others experience problems and satisfactions very similar to their own. The study directors might further help to reduce errors by issuing occasional memoranda citing specific difficulties seen in coding. The same goal can be accomplished by preparing "home study" materials covering changes in field manuals, supplementary instructions, or other policy statements. When these materials cover significant changes, or if there is evidence that interviewers have forgotten important parts of the original instructions, it is helpful to include a "take-home" examination to be completed by the interviewers and scored in the research office. In general, however, impersonal reminders such as memoranda and self-study materials are most readily accepted and produce the best effects when combined with some form of personal contact, especially with the supervisor and study directors.

Field Supervision

While proper training is essential for effective field work, it is not enough. The success of field operations also requires careful and active supervision as well as positive leadership. The general principles of organizational management and leadership apply as much to survey research as to large corporations, and should be given close attention. The most critical tasks in supervision include organizing work groups, handling assignments, establishing production quotas, reviewing completed work, serving as a liaison with the research office, and maintaining a high level of commitment to the study among the interviewers.

Work Groups   In large surveys it is useful to develop a squad system in which five to ten interviewers work with a single supervisor in roughly the same geographic area. One advantage of this system is that the supervisor can come to know the strengths and weaknesses of each member of his squad and deal with their problems on a personal basis. Also, if the group proves to be compatible, the members will often form a sense of solidarity and perhaps a healthy sense of rivalry with other squads. Moreover, if all are working in the same region and face approximately the same difficulties, it is difficult for the laggard to claim that his output is low because of his unique problems.

Work Assignments   Assignment procedures often work in two stages. First, the central research office assigns a certain number of sample addresses to the field supervisor. Then the supervisor distributes a few of these at a time to the interviewers in the work group. Hauck and Steinkamp offer useful advice on this point:

> An important function of proper supervision is to avoid overloading interviewers with assignments. It would seem best to give each interviewer only a few assignments at a time, with a deadline date for these assignments. After *all* of these assignments have been turned in, a new group of assignments may be sent out together with a new deadline date. In this manner, tight control can be maintained over the flow of interviews. Also, this procedure serves to counteract the tendency of many interviewers to put off contacting sample members whom they feel may be difficult to interview. It is not unlikely that at times this very habit of tardiness makes a sample member less cooperative, especially when an advance letter has been received many days or weeks earlier [1964, p. 87].

Moreover, it is generally a useful rule to have the interviewers return all completed assignments, including refusals and noncontacts, no later than one day following their completion. This policy will stimulate the interviewers to carry out their editing before the details are forgotten, and will also ensure a rapid flow of questionnaires to the coding staff.

Production Quotas The social psychology of organizations suggests that interviewers will increase their output if they have a clear concept of the amount of work expected from them. Within a week or so after the field work is underway, it should be possible to establish general norms regarding completions per day, with perhaps some modifications for specific work groups. It is also helpful to post production figures for the staff as a whole and for each of the work groups, with perhaps a footnote to the effect that some areas are more difficult than others.

Quality Control

Every survey should include programs of quality control aimed at preventing falsified interviews and reducing interviewer bias in the data. The problem of falsification is sometimes overplayed in criticisms of survey research, but it is a very real danger nonetheless. The experience of the Survey Research Center suggests that this problem is greater when interviewers are hired for a single study than when they join the permanent field staff. Various techniques have been used to detect falsification, though often their main value is in deterrence.

The most commonly suggested approach is a *reinterview of the sample address* by someone other than the original interviewer. The U.S. Bureau of the Census and some survey organizations abroad reinterview up to 10 percent of respondents (postevaluation surveys) to validate certain estimates, estimate errors, and establish a deterrent for cheating. The reinterview is useful for these purposes, but it requires great tact, and is not always foolproof. Even with the cooperation of the respondent, the repetition of attitudinal questions may provide little or no valid information on falsification, principally because of the possibility that the individual's attitudes have really changed in the interim. Thus it is often necessary to limit the reinterview to such basic questions as age, sex, and number of children, or the time and place of the first interview and the characteristics of the interviewer. Given this limitation, the clever interviewer bent on cheating might spend five minutes with the respondent to obtain the verifiable data, and then fill in the other questions at home or elsewhere. The properly designed revisit could still detect such falsification, for example, by asking about the length of the interview, but it remains cumbersome. Because of these difficulties and the stability of its staff, the Michigan Survey Research Center uses reinterviews only rarely, and mainly where there are positive suspicions about the information.

A simpler device is the *postcard follow-up* to individuals presumably interviewed in the study. The cards typically contain such questions as: "How long was the interview?"; "How did you like the interviewer?"; and "What day did the interviewer visit you?" But there are two problems with this technique. One is that if the respondent was not interviewed, he or she may feel that the card is a mistake and discard it. Under these conditions the results would show a decided bias in the direction of favorable reports. Second, an experiment cited by Hauck and Steinkamp (1964, p. 72) revealed that some persons who were *never* interviewed returned postcards indicating the length of the interview and their opinion of the interviewer.

A third means of quality control consists of a *telephone validation check.* Here the interviewer tells the respondent that the supervisor will be calling him about the interview, and asks the individual for his name and telephone number. After the interview has been completed, the central research office carries out a sampling of respondents and then repeats questions from the beginning, middle, and end of the questionnaire. While this procedure solves some of the problems raised by the other procedures, it does create considerable inconvenience for the respondent, and may lead to doubts about the confidentiality of the findings.

In some ways, the most effective and least bothersome method of quality control is to rely on the reports of the coders. This approach assumes that coding and field work occur simultaneously so that corrections can be made if necessary. Also, it is now possible to use the computer to determine whether a given interviewer shows more or less variability than the group as a whole in the responses obtained with selected questions. However, this procedure is limited by the fact that some interviewers deal with more homogeneous sample members than others, so that one would expect to find a fairly narrow range of variation on the control items. Thus, there is no magic answer to the perennial problem of quality control, but it is possible to avoid flagrant abuses and stimulate high performance by making it clear to the interviewing staff that various checks will be made on their work.

Administrative Control

The research process will be greatly facilitated and costly errors avoided if the central and regional research offices take careful steps to control the flow of work and materials. The following are several control procedures which are highly recommended for any large survey.

   1 *A sample control book:* a list of the total sample by address, with space allowed to record the end result of each sample element, such as: interview, including the date received and the name of the interviewer; refusal; could not locate; no one at home, etc.
   2 *Record by interviewer:* a list of all assignments made to each interviewer, together with the result of each (completed, not at home, refusal, etc.). This information is especially helpful in appraising the quantity and quality of the interviewer's work.
   3 *Record by interview number:* to avoid confusion in identification procedures, it is advisable to assign numbers sequentially to completed interviews as they reach the research office, and to maintain a list indicating this interview number, the name of the interviewer, and the date completed. It is also helpful for the interviewers to number their own interviews sequentially in the field. However, to minimize the chances of error, the interviewer's number should be superseded by a continuous series once the materials reach the central office.

These procedures can save numerous administrative headaches during the field work itself, and later during coding and the analysis of the data.

Payment of the Interviewers

For years there has been a debate among experts in survey research about the relative merits of paying interviewers on a "piecework" basis versus by the hour. The proponents of the first method argue that it provides a strong economic incentive to the field staff, allows for greater predictability in costs, leads to a more efficient use of time, and ultimately results in more work per hour. The implication is that the interviewer who is paid on an hourly basis will be tempted to prolong his work, be less efficient in locating respondents, and introduce other inefficiencies into field work. Those who favor hourly payment, on the other hand, claim that payment by the interview generates pressures to favor quantity over quality and leads to a neglect of the more subtle motivational dimensions of the interview. It is also commonly suggested that the rate of fabricated interviews is higher with "piecework." At present, the debate remains unresolved, with some organizations paying by the hour and others by the interview, and both claiming to have chosen the better method.

One bit of empirical evidence bearing on the question is found in the research carried out by Hauck and Steinkamp (1964). They experimented with both of the usual methods of payment and concluded that payment for work completed was more efficient, provided that the study made use of careful supervision. They write:

> Experience with this method of compensation [hourly] brought out various shortcomings. For one thing, high variability in the cost per interview was apparent from one interviewer to another. In part, this may have been due to padding of time sheets by some interviewers. More likely, however, was padding of a different sort; namely, stopping at a sample address if it were near the place the interviewer was going, even though he had not really expected to find the sample member at home.
>
> Another problem was that the system of compensation failed to provide an incentive for interviewers to complete assignments on time. Furthermore, the 5 cent increase per wave seemed to provide virtually no incentive for the interviewers to remain with the study. . . . Particularly pertinent in this respect were the comments of many interviewers that being paid on an hourly basis was degrading in view of their professional status in their regular jobs [p. 82].

Clearly, all of the evidence relevant to the compensation question is not yet in. While a strong case can be made for payment by the interview, directors of professional survey organizations, including the Michigan center, argue for payment by the hour. Given the risks of "piecework" for the quality of the survey data, our own preference would also be for the latter approach.

In conclusion, we would suggest that careful attention be devoted to these organizational and administrative questions during the planning stage of the research. The issues of selection, training, assignments, quality control, and the like are easy to overlook because of their relative lack of glamour. When field work is underway, however, they prove to be exceedingly important.

FURTHER READINGS

Back, K. W., and J. M. Stycos, *The Survey under Unusual Conditions: Methodological Facets of the Jamaica Human Fertility Investigation* (Ithaca, N.Y.: Society for Applied Anthropology, 1959), Monograph No. 1.

Hauck, M., and S. Steinkamp, *Survey Reliability and Interviewer Competence* (Urbana, Ill.: Bureau of Economic and Business Research, University of Illinois, 1964).

Chapter 9

Editing and Coding

Editing and coding are related processes designed to translate the information recorded in the questionnaires into a form suitable for statistical analysis. The task of data reduction becomes especially significant when the study makes use of open-end questions. The challenge in this case is to devise sets of categories which are faithful to the data, and yet permit the responses to be classified in meaningful ways. Editing is a preliminary step in which the responses are inspected, corrected, and sometimes precoded according to a fixed set of rules. Coding is a technical procedure for converting verbal information into numbers or other symbols which can be more easily counted and tabulated.

EDITING

The editing process typically consists of two sets of activities: checking of the questionnaires, and procedures to facilitate subsequent coding. The first aims not only to improve the quality of the single questionnaire but also to provide a means of improving interviewing and the quality of supervision in the field. The second set of procedures looks ahead to make the coding as efficient as possible by dealing with difficult or complex information.

Checking Procedures

Checking begins when the interviewer reviews each questionnaire in the field and continues with various other checks by the field supervisor and the office staff. The basic aim of checking is to eliminate incomplete or inconsistent responses as well as errors in the use of the questionnaire. The theory behind it is that interviewer efficiency and the quality of supervision are aided by the prompt detection of errors and rapid feedback to their source.

Interviewers should review their work after each completed interview, or, at the latest, by the end of the day. Particular attention should be given to the legibility of the information and to missing answers, inconsistencies, or a lack of uniformity. Despite the apparent simplicity of this task, interviewers should have explicit instructions about interpolating answers or resolving apparent inconsistencies. Often the best policy is to change nothing in the forms, but to be more alert to the occasions of error in subsequent interviews. Particular caution is necessary in supplying missing answers. Here the interviewers can easily be influenced by their attitudes toward the respondent, or their expectations about "appropriate" behavior for such a person.

In many studies, a second edit is carried out by the field supervisor upon receiving the questionnaires. While this review may be very detailed and complex, its purpose is normally to detect obvious errors which can be corrected in the field. Frequent or persistent errors by the same interviewer may reveal the need for additional training or even dismissal. Similar errors by different interviewers, on the other hand, may point to inadequacies in the questionnaire or in the instructions for its use. These problems would also call for more training.

Another quick edit may be done when the questionnaires reach the main research office. The purpose is again to judge whether the information is sufficiently complete to be coded. If not, the case may have to be considered a noninterview or, in rare cases, the questionnaire might be returned to the field. A critical advantage of these preliminary checks is that they point up deficiencies in the interviewing process and allow them to be corrected while the field work is still in progress.

Several specific questions should be considered during these checking operations. The first is *completeness*. Every questionnaire should be reviewed to determine whether each question has an answer and, if an item has been left blank, to try to find out if the omission occurred because the interviewer forgot to ask the question, because he failed to record the answer, or because the respondent refused to answer. Omissions can occasionally be corrected, but the interviewers should be reminded of the need for complete information in future interviews.

Second, editing should aim to clear up *logical inconsistencies* in the responses. Sometimes, for example, a respondent may be simultaneously listed as "male" and "wife." In this case, the editor would want to scan the questionnaire for other indications of sex and marital status. To avoid arbitrariness the study directors should develop a precise set of procedures for reconciling apparent inconsistencies such as the one cited. It is also important to distinguish between *definite* inconsistencies (a mother listed as dead in one question and working 40 hours a week in another) and *probable* inconsistencies (a twenty-year-old wife with five children). In the latter case, the editors are well advised to seek out information from other parts of the questionnaire which may clarify an ambiguous response. If they find none, the best policy, usually, is to leave the answers as obtained.

A third task is to ensure *comprehensibility* for the coders. Occasionally, a problem arises when a response that seems perfectly clear to the interviewer is less clear to others unfamiliar with the total situation. Another major source of difficulty is illegible handwriting in the questionnaires. In either case, it may be necessary to consult the original interviewer for clarification before the coding begins.

Procedures to Facilitate Coding

Editing to facilitate coding is normally of two types: (1) the precoding of difficult information, such as occupations; and (2) the preparation of complex computations or summaries to simplify the subsequent coding process. These kinds of editing are best carried out in the research office by a few specialists who can master the intricacies involved and demonstrate high consistency among themselves. With information on occupations, for example, it is often helpful to have the coding done by two or three specialists who become thoroughly versed in the material rather than by a large group of coders.

At times this editing process includes the preparation of special worksheets on each family or respondent in the study. The worksheets are particularly useful in organizing and summarizing complex data on such variables as income or debt. There are often sound methodological reasons to spread different questions on key variables throughout the questionnaire. A variable such as consumer indebtedness may require complementary questions on different sources of debt at various points in the questionnaire.

Questions about debts arising from major purchases (cars, refrigerators, etc.) will often fit best into a section dealing with the purchases themselves. At other points, questions could be included to cover other sources of debt. It would normally be a serious mistake to bring all of these items together in one sequence just to facilitate coding. At the same time, it is usually inadvisable to have the regular coders skip over several pages to summarize the information from several items. The best approach may be to have the summaries prepared on special worksheets by a few editors who specialize in this task.

Given the importance of clear and consistent procedures in preparing summaries of information, it is helpful to have special worksheets for complex major variables. A family income worksheet, for instance, might have separate columns for each potential income recipient (head of family, wife or husband of head, children, etc.), and separate rows for the various types of income: wages and salaries, interest and dividends, profits or income from self-employment, rent, and so forth. A clear and logically organized worksheet provides a guide for editors to consider each possible entry. When well prepared, it can also permit direct keypunching without additional coding.

If conventional data processing equipment or computers are available, the more routine parts of office edits can be replaced with machine checks. However, some degree of judgmental editing is typically necessary to place the data in proper form for machine processing. Not even the most sophisticated computer system can solve the problem of illegible handwriting. Nevertheless, in cases where the raw materials for editing are available on punch cards, tape, or discs (see Chapter 10), and the editing operation involves combinations of punched data, the error rate will probably be lower if the calculations are carried out by machine. Mechanical edits are particularly useful in identifying impossible or inconsistent responses, and errors in keypunching or coding—topics to be discussed in the following chapter.

The editing process can also be used to resolve problems arising from nonuniformity in administering the questionnaire or recording the answers. One common error stems from the respondents being asked inapplicable questions, as when a divorced person is asked a sequence pertaining only to married respondents. Another frequent problem is the interviewer's failure to obtain information in the units specified by the study. Difficulties arise in coding when income is recorded by the month rather than by the year, or hours of work by the week rather than by the day. To ensure uniformity, it is usually advisable to have the necessary adjustments made by a few editors following common rules rather than by the entire body of coders.

CODING

The essential task in coding is to have the data represented with numerical or other symbols permitting rapid and flexible storage, retrieval, and tabulation. Earlier we stated that the prime challenge is to devise categories which are both faithful to the data and meaningful for the analysis to follow. Several specific questions arise in coding: the principles of code construction, including the characteristics and limitations of the punch card system; the training of coders and production coding; and problems in coding. These are treated through the rest of this chapter. At the end of the chapter, we include an illustrative code for parts of the questionnaire presented in Chapter 6.

Code Construction

In developing an effective code, it is necessary to consider simultaneously the aims of the study, the information needed to satisfy these aims, the structure and content of the questionnaire, and the physical limitations of the punch card or other storage unit to which the data will be transferred. The purposes of the study should determine the general nature of the code categories, as well as the level of detail sought. Normally, the code should not try to be much more comprehensive than is necessary for the study itself, unless there is reason to believe that others may wish to use the data for secondary analysis.

The Punch Card   Although the technology of information storage has shown marked advances in recent years, the great bulk of coded survey data continues to be entered on punch cards. From there it may be transferred to magnetic tape or discs (see Chapter 10), but so long as the punch card is used in the first stage of storage, its characteristics must be taken into account in developing the code.

The punch card is organized to contain information recorded in the form of holes or punches. The most commonly used type is the Hollerith card, known more generally as the IBM card. This is shown in Figure 9-1.

In the Hollerith system, the card is divided into 80 columns, each of which can receive 12 possible punches. The various punches can be used to record digits, letters, or other symbols used in writing. Within a given column, it is possible to punch digits from 0 to 9, corresponding to the numbers printed on the card, or two nonnumerical punches variously identified as X and Y, + and −, or 11 and 12. These latter punches are entered in the top two rows of the card, where there is no printing. Columns 69–78 of Figure 9-1 show the punches corresponding to the digits 0 to 9 respectively, and

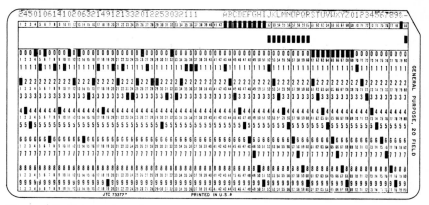

Figure 9-1 The punch card.

columns 79 and 80 show the + and − punches. Capital letters and other symbols can also be formed through combinations of the 12 punches just mentioned. For example, the letter A is represented by a double punch of + and 1, B by + and 2, and so on. Due to limitations of space, the Hollerith system does not permit the use of lowercase letters. Punches corresponding to the alphabet from A to Z are shown in columns 43-68 of Figure 9-1.

Although each column is independent of every other column, it is possible to use two or more to record information which is too complex for the single column. Income figures, for example, can be recorded in six digits across six columns, age in two digits, etc. The punches in the first 37 columns in Figure 9-1, for example, record the information shown in Table 9-1, following the code for Card 01 at the end of the chapter.

In addition to considering the implications of the punch card or other storage units, the study directors should also take explicit account of the limitations imposed by the data processing equipment or the computer programs which will be used. Particular attention should be paid to punches which are illegitimate and may cause cards to be rejected, and to the proper format for missing data. The time spent in looking ahead at this stage will often save time, money, and headaches during data processing and analysis.

Developing Code Categories The set of categories developed to code the data from a single question or set of questions should be exhaustive, mutually exclusive, and adapted to the aims of the study. That is, the code categories should cover all of the possible answers for any given question; they should not overlap, and should be explicitly oriented to the analysis objectives of the study. With questions on age, sex, and marital status it is relatively easy to develop a code which meets all of these requirements. Age,

Table 9-1

Column	Item	Response
1–3	Study Number	245
4–5	Card Number	01
6–9	Interview Number	0614
10	Cover Sheet	Original
11–12	Name of Interviewer	M. Valdez
13–14	Your Interview Number	06
15	Length of Interview	35 min
16–19	Sample Information	2149
20	Number of DU's	"No" to Q. 10
21	Number of Calls	Two
22	Final Result	Interview
23	Structure	Row house
24	Neighborhood	Row houses
25	Public Transportation	5-10 min
26	Noninterview	Interview
27	Sex	Male
28	Age	31
29	Marital Status	Married
30	Attending School	Not in school
31	Grade Completed	High school
32–33	Number in Household	Three
34	Number of Adults	Two
35	Number of Children	One
36	Age of Youngest Child	4 years
37	Spouse Present?	Spouse at home

for example, can simply be coded to the nearest year, with clear rules for rounding. Open-end questions pose much more of a challenge, for they can typically be coded in a variety of ways.

To illustrate the process of code construction, let us take an example from the specimen questionnaire at the end of Chapter 6. This is Question 2 and Question 2a, which read:

> **2** Would you say that you and your family are better or worse off financially than you were a year ago, or about the same?
> (IF BETTER OR WORSE) **2a** How is that?

Before attempting to build a detailed code, the study directors should review very carefully the objectives of the questions. The general objectives in the Pacifica survey were twofold: (1) to measure perceptions of family economic conditions and (2) to determine the prevalence of factors reported as the main influences on *changes* in such conditions. The study directors hope to use the resulting information to make comparisons among the sample cities on perceived economic well-being, and to relate such perceptions to more objective information, such as changes in rents. The study also had as a descriptive goal to report in fairly specific terms the factors thought to influence economic well-being. It was thus decided that the code would require more than a single column, but no more than two.

The process of developing a code for an open-end question of this type is usually a mixture of theory and empirical observation. That is, the study directors will often have some general notions about how the responses should be organized for purposes of analysis, but without sufficient specificity to construct an entire code a priori. The challenge in developing a comprehensive code, therefore, is to keep matching concepts against the empirical reality until most of the possibilities are covered.

In the Pacifica study, the economists on the study staff hypothesized that changes in the following conditions would make a difference in the perception of changes in one's financial situation: rate of pay, income from self-employment or property, availability of work and number of workers in the family, contributions from outside the family, prices, taxes, expenses, and asset and debt levels. These categories were suggested by economic theory and had been empirically validated in previous surveys of Pacifica.

The next step was to test the general categories against the responses obtained in Questions 2 and 2a. One of the study directors began reading each answer to these questions in about 50 completed questionnaires from different cities and economic levels. Both to provide a more concrete definition of each category and to illustrate the major types of response falling within it, he noted specific responses which, in his judgment, were covered by each term, such as "rate of pay." At the same time, he copied on cards the responses which did not seem to fit any of the categories mentioned, in order to find out whether certain responses occurred with sufficient frequency to warrant a separate category.

Out of this process came the following double-column code:

Table 9-2

(Questions 2 and 2a.) Factors affecting financial situation.
Reasons Why Financial Situation Is Better
- **10** *Better pay:* raise in wages or salary on present job, promotion, higher commissions, change to higher paying job.
- **11** *Higher income from self-employment or property:* higher business profits or farm income, higher dividends, royalties, or rents; more income from professional practice or trade; include "more money" reported by independent street merchants and artisans.
- **12** *More work, hence more income:* head (or wife) started working (again); more members of family working; higher income, why N.A. (not ascertained).

Table 9-2 (cont'd)

---

13 *Increased contributions from outside sources:* from government pension; re-
    lief; welfare; family allowance; or from private sources, such as relatives.
14 *Lower prices:* decrease in the cost of living.
15 *Tax cut.*
16 *Decreased expenses:* fewer people to be supported by family unit; children
    out of college, on own, etc.; spending less, not ascertained whether 14 or
    16.
17 *Better asset-debt position:* more savings; lower debt; business or farm worth
    more; has more business, farm, or personal assets.
18 *Other reasons why better off:* greater security (job more permanent, psycho-
    logical security, better opportunities) higher standard of living, have more
    things (new car, etc.), better outlook for future, more stable government
    will mean less inflation.

Reasons Why Financial Situation Is Worse

50 *Lower pay:* decrease in wages or salary on present job; change to lower
    paying job; lower commissions.
51 *Lower income from self-employment or property:* lower business profits or
    farm income; lower dividends, royalties, or rents; less income from profes-
    sional practice or trade; "less money" reported by independent street mer-
    chants and artisans.
52 *Less work, hence less income:* head unemployed, laid off, sick, retired, de-
    ceased, on strike, unsteady work; less overtime; fewer hours; fewer mem-
    bers of the family working; lower income, why N.A.
53 *Decreased contributions from outside family unit:* parents cannot help out
    any more; off welfare or relief; pension or annuity ran out.
54 *Higher prices, inflation:* increase in cost of living, money not worth as much
    as last year.
55 *Higher taxes.*
56 *Increased expenses:* more people to be supported by family unit; one or
    more children entered college or university; heavy medical or dental bills.
57 *Worse asset-debt position:* savings used up (wholly or partly); increased
    debt; less business, farm, or personal assets.
58 *Other reasons why family unit worse off:* less security (job less permanent,
    psychological insecurity); fewer opportunities; dark outlook for future;
    lower standard of living; poor government creates uncertainties.

Other

99 *N.A. (not ascertained):* situation reported to be better or worse, but no
    codable reason given.
00 No change.

Note that all the reasons for a better financial situation are included in codes under 50, while the reasons for a worse situation are assigned numbers 50 or over. This division has two advantages: (1) it allows the codes under 50 to be expanded should it become apparent that more categories are needed; and (2) it immediately tells the analyst that any number under five in the first of the two columns refers to a factor tending toward "better" conditions, while any number of five or over (short of 99) refers to a "worse" response.

The code illustrated above allows only one "reason" to be coded in a single pair of columns. Many respondents will give more than one reason to a question such as this, and they may even give offsetting reasons. Study directors who wish to capture this detail would have to assign additional pairs of columns to record additional reasons.

A common problem in designing a code extending over more than one column is the level of detail to be provided in the code categories. The previous illustration, for instance, could have been converted into a three-column code, along the following lines:

100-199   *Better pay:*
   101   Raise in wages on present job
   102   Raise in salary on present job
   103   Increased fringe benefits from employer
   104   Higher commissions
   105   Change to higher-paying job
                                        etc.

The directors of the Pacifica study decided that this level of detail was unnecessary, and that the time involved in coding the more precise information would result in a significant increase in coding costs. Moreover, given the sample size, it was evident that it would not be possible to carry out a useful analysis with the three-digit information, mainly because there would not be enough cases in each category to permit reliable cross tabulations. It is also possible that many respondents would not provide sufficiently detailed information to permit making the fine distinctions in such a code with the result that the number of "not ascertained" cases would be increased.

How does a researcher decide whether to include more or less detail? In general, two extremes should be avoided. The first is excessive detail—a common temptation in a first survey or one whose analytic objectives are not clear at the time of coding. In a typical sample survey with a sample size of 2000 or under, a three-column code will often be excessive unless there is a specific need for the more refined data. The second pitfall is a code which is too condensed. For example, a code for age with only two categories—under

40 and over 40—would normally be a mistake. Such a gross classification would make it impossible to study the precise relationships between age and key variables without recoding the data. With age it is almost always advisable to use two columns to code the exact figures (19, 45, 66, etc.). These can easily be collapsed for cross tabulations. Most research centers find it helpful to retain more rather than less detail, largely because it is easier to combine detailed codes than to break down those which are too highly condensed.

The Coding Unit[1] The coding unit is the specific set of information taken into account in coding. The unit most frequently used is the single question, but it is also possible to code the responses to several questions or even an entire interview. In the example from the Pacifica study, the unit was two items: Questions 2 and 2a. When the coding unit is a series of questions, it is called a *block* or *area code;* when it covers the entire questionnaire, it is called an *overall code* (Muehl, 1961). Block codes are often used to tap concepts which are too complex to be covered with a single question. For example, the coder might be asked to review five open-end questions about employment conditions and then rate the employee's attitude toward his supervisor. Occasionally, a study may use a *sieve code* to catch comments of a certain type which appear anywhere in the questionnaire. The coder could be instructed, for instance, to review the entire interview to determine whether or not the respondent has made any mention of a certain minority group and, if so, what attitude was expressed toward the group.

Coding Conventions For various reasons, it is advisable to develop coding conventions to handle comparable responses to different questions. Specifically, it is recommended that responses such as "Yes," "No," and "Don't know" be given the same code numbers throughout the study. This procedure has at least three advantages: it reduces clerical error by simplifying the task for the coders; it increases the speed of coding through standardization; and it facilitates the interpretation and analysis of data by obviating the need to check the code book to see, for example, what a "9" or "0" means with a five-point scale. Both the coders and the analysts can become confused if "Don't know" is coded as "4" in one question, "7" in the next, and "8" later. Coding conventions must sometimes be violated in the interests of other goals, such as the need to fit nine response options into a single column. However, in most cases, much is gained and nothing is lost by developing and following a set of conventions. The set shown in Table 9-3 is commonly used at the Michigan Survey Research Center and in several other centers.

---

[1] Much of this discussion of coding draws on the practices followed at the Survey Research Center of the University of Michigan, as summarized by Muehl (1961).

Table 9-3   Coding Conventions

Numerical category	Response category
1	"Yes" (or most positive response on a scale).
5	"No" (or least positive response).
7	"Other"—responses which are valid but which do not fall into any of the remaining code categories.
8	"Don't know" or other answers indicating that the respondent cannot give an answer, including memory failure or lack of an attitude or opinion on a given question.
9	"Not ascertained." This category would be used when: (1) there is no response to the question, (2) the response is irrelevant or otherwise inappropriate, (3) the coder cannot understand the response, and (4) the response is so ambiguous or broad that it could fit equally well into several response categories.
0	"Inapplicable" or "None." The "0" code is used when a question or set of questions is inapplicable or inappropriate for a certain type of person, such as questions about employment which do not apply to the unemployed. It is also used to show the absence of a particular item ("None") such as an automobile or television set.

Types of Codes   There are literally dozens of ways in which the information from a survey questionnaire can be coded, especially when it contains many open-end questions. However, most codes fall into certain patterns. The following, suggested by the *Manual for Coders* used at the Michigan Survey Research Center (Muehl, 1961), are among the most common.

1   *Factual or listing codes.* In this case, each separate item mentioned by the respondent is placed in a separate category. Consider the question, "What magazines do you read each month?" While the responses could be handled in many ways, the listing code would simply provide a separate column for each possible magazine or set of related magazines. It might read:

Column	
41	1. Mentions *Time* or *Newsweek* or both
	5. Does not mention above
42	1. Mentions *Reader's Digest*
	5. Does not mention above

While this approach has the advantage of simplicity, it requires a great amount of space on the card, especially when the number of possible options is great, as in the case of magazines.

2 *Field codes.* These are codes in which numbers representing amounts, age, or similar items are recorded more or less as they were given by the respondent. They are often used in questions dealing with income, debt, prices or values, taxes, profits, and other subjects in which the responses involve units of currency. The principal problem with field codes is that of rounding the figures to fit within the columns available. Here, as elsewhere, a uniform policy is needed to ensure consistency in the data. Dollar amounts, for example, are often rounded to the nearest dollar. Thus $5234.25 would become $5234, $9.95 would become $10, etc. The major difficulty in rounding concerns numbers, such as $$11.50, which would logically fit into two code categories, in this case $11 or $12. This problem can be handled by adopting a rounding convention, such as moving to the nearest odd number ($11). The examples below show how a given amount can be coded in different ways according to the amount of rounding involved.

Amount reported	Coded to nearest dollar	Coded to nearest $100
	(4-digit code)	(3-digit code)
$ 11.13	0011	000
$ 229.95	0230	002
$ 76.50	0077	001
	(rounded to odd number)	
$9000.00	9000	090

Sometimes a few responses, such as $11,000, will not fit within the four-digit code to the nearest dollar shown in the example above. This situation should be anticipated and codes should be built with enough digits, but in an emergency it can be handled by having the highest number in the code (**9999**) include all responses of that magnitude *or more.* If the analysis plan calls for the calculation of means or other figures based on the unclassified data, it is advisable to keep a record of the figures which exceed the code limits so that they can be entered by hand into the calculation.

3 *Bracket codes.* These are also used with numerical data, but they involve the assignment of a code category to a range of numbers. Bracket codes are often used to simplify the analysis of data on age, income, expenditures, and debt by grouping the information into a limited number of categories. For example, the information on rent derived from our specimen survey could be coded as follows:

Column	Monthly rent
56	1. $1–$99
	2. $100–$199
	3. $200–$299
	4. $300–$399
	5. $400 or more
	8. D.K.
	9. N.A.
	0. Inapplicable, does not pay rent.

As in the case of field codes, bracket codes require a consistent method of rounding. A policy may also have to be developed to handle answers given as a range of figures ("My rent is between $250 and $300"). One approach is to adopt the midpoint of the estimate ($275) as the figure to be coded. Under some circumstances, it may be more realistic to code either the upper limit ($300) or the lower limit ($250). Much depends on whether distortion is likely, and the direction it will take. People may exaggerate the number of their friends, for example, so that it would be advisable to code the lower limit, while they may understate the number of traffic tickets received in a given year, so that one might code the larger figure. Whatever policy is adopted should be applied consistently throughout the coding.

4 *Pattern codes.* These are designed to cover combinations of responses to a single question or to more than one question. In our hypothetical urban survey, the code below could be used to summarize analytically relevant combinations of responses to questions on home ownership and plans to move.

Column	Home ownership and plans to move
47	1. R owns home—plans to stay
	2. R owns home—plans to move
	3. R owns home—moving plans N.A. or D.K.
	4. R rents home—plans to stay
	5. R rents home—plans to move
	6. R rents home—moving plans N.A. or D.K.
	7. R neither owns home nor rents
	9. NA or DK whether owns home or rents

The pattern code is particularly useful when the researcher knows in advance which combinations will be of interest, and when there is limited space available on the punch card. However, this type of coding might best be handled by first coding the responses to the separate questions, and then having the combinations worked out by data processing equipment. Machine processing tends to be less expensive than hand coding, and normally results in fewer errors.

5 *Reasons codes.* As the name implies, this type of code attempts to classify the respondent's reasons for giving a certain answer or holding a certain opinion. The problems posed by reasons codes are among the most difficult in the entire coding operation. Some of these were suggested in the discussion of the double-column code developed to handle Questions 2 and 2a in the specimen questionnaire. Consider also the construction of a code for Questions 1, 1a, and 1b, which ask:

> . . . is this neighborhood a better or worse place to live than a year ago, or is it about the same? (IF BETTER OR WORSE) How is that? Anything else?

The responses to these questions pose at least five problems for coding. First, what types of categories should be used in developing the reasons code? Should they be developed empirically by scanning the range of answers actually obtained, or should they be based on a priori theoretical concepts? While the answer should clearly reflect the aims of the study, it is scarcely worthwhile to develop an elaborate, theoretically oriented code which bears little resemblance to the information at hand. As noted earlier, it is often possible to begin with theoretical concepts or categories, test them against the responses, modify them, test them further, and in this way achieve a balance between the a priori and the empirical reality. Second, what procedure should be followed to determine the number of separate reasons in a response? Answers to open-end, "why" questions often blend into each other. A single sentence may contain five reasons in one interview, while in another, one reason will extend across six sentences. Muehl (1961) suggests that close attention be paid to conjunctions such as "and" and "but," as well as pauses in the conversation. Third, how many reasons should be coded? Some codes are designed to accommodate only one or two, while others allow for five or six. Given the difficulties in analyzing information from this type of code, most researchers code no more than one or two reasons. Fourth, when not all the information can be coded, how should the coder decide which reason to include? One method is simply to code the first two mentioned. Another is to "flag" certain reasons as priority items, and to code them first. Finally, should there be consistency between the position on the scale of better or worse and the reasons to be coded, or should offsetting or inconsistent responses be treated equally? For example, a respondent may reply that he is better off now and then explain that he earns less because the opportunities to work overtime have diminished recently, but that his wife recently began a part-time job. The authors' preference is not to discriminate against inconsistent responses. As should be evident, reasons codes are a mixed blessing. They often provide vital information, but are sometimes difficult to code and even more difficult to analyze.

Card Design   Two other important decisions in code construction concern the way in which information from the study will be organized among cards, and the way it will be organized within a card. In most surveys, including the illustrative survey used in this book, the amount of coded data will require more than the 80 columns available on a single card. In fact, in most studies of any length, six or more cards may be needed per interview.

When studies obtain information about several different populations of interest to the analysts, one or more cards may be needed to record the information about each of them. One card should always be designed for the sample information. The basic purpose of this card will be to represent each and every address selected and what is known about it, plus the outcome of the attempt to interview at the address. For the most part, the information will have been recorded by the sampling staff and the interviewers on the front and reverse side of the Cover Sheet.

Another card can be designed for the population of all persons identified as residents at the selected addresses. Information about adults and children will for the most part be found on the Household Listing Sheet. This card will be useful for tabulations about all household residents and selected subgroups such as adults, school-age children, women in childbearing ages, and so forth. Several other cards are likely to be required to record the information from the body of the questionnaire where most of the information obtained from the respondent is found.

In designing the card layouts, two basic guidelines should be observed, one dealing with the users, the other with the coders. First, the cards should be organized around the major topics in the study. This will be relatively easy when the questionnaire itself is divided into sections corresponding to the main topics. The reason for this suggestion is that it is helpful in analyzing and interpreting the data to have natural clusters of items, as on education, income, and "modern attitudes," near each other in the code book. Moreover, if the data processing options are limited to conventional equipment (still a possibility in some areas), there are additional advantages to having all of the information about one topic on a single card. With conventional equipment (see Chapter 10) data processing is facilitated if tabulations can be prepared from one card, rather than from items scattered across seven or eight cards. With computers, on the other hand, there is much more flexibility in this respect.

Second, to speed coding and to reduce errors, cards should be organized to follow the sequence of items in the questionnaire. In the Specimen Code at the end of this chapter, all the information for Card 1 (Sample, Interview, Household Characteristics) is to be found either on the Cover Sheet or on the Household Listing section of the questionnaire. Similarly, Card 2 (Household Residents) uses only the Household Listing, while Cards 3 and 4 cover contiguous sets of items in the main questionnaire. This suggestion underscores the significance of coding considerations in questionnaire design. With an approximate idea of card layout, the researcher can group the items in ways that will facilitate coding. It is generally wasteful and confusing for the coders to have to skip back and forth in the questionnaire. Moreover, if they feel that the form was poorly designed, their own motivation to produce high quality coding may drop.

Once the organizing principle of the card is clear, a decision must be made on how to distribute the 80 columns available. Typically, up to ten columns of each card must be given over to identification. To avoid confusion during analysis, the same columns should be used for identification on each card. The following procedure will be followed in the Specimen Code:

Columns 1–3	Study Number (245)
Columns 4–5	Deck (Card) Number
Columns 6–9	Interview Number (respondent identification)

When data processing is limited to conventional equipment, it is advisable to assign no more than about 60 of the 80 columns to identification and coding of original data. The reason is that, with these machines, it is often necessary to transfer information from one card to another. If, for example, the analysis plan calls for a cross tabulation of employment (Card 4 in the Specimen Code) and housing characteristics (Card 3) it may be necessary to transfer the housing "controls" (the information needed in the tabulation) from Card 3 to Card 4. The process is much easier if about 20 columns are left free for subsequent transfers. While it may be tempting to "save money" by filling up all of the cards, in our experience this is a false economy, particularly since the cost of additional cards is small and the inconvenience of working with full cards great. When the study has access to modern computers, of course, there is no need to reserve columns for transfers.

Training the Coders

The coders play a key role in the transition from questionnaire responses to analyzable data. Much of what was said about the role of training in developing motivation among interviewers applies here as well. The coders must be given careful instruction not only in the general principles and techniques of the trade, but in the nature and objectives of the study. To maintain their motivation throughout what can become a difficult and tedious task, they must be made to feel that the study is theirs. If they begin to feel ignored by the study directors, or sense that they are regarded as data-processing robots, both the quantity and the quality of their work may fall off. In many studies, the research staff relies heavily on the coders to catch ambiguities or to note other points affecting the reliability and validity of the data. Poorly motivated coders may continue to meet the minimal norms of productivity, but pass over points which would require a positive effort on their part.

The training sessions might begin with one or more meetings between the coders and the study directors. The opening discussions could cover such topics as the background of the study, its subject matter and goals, the basic concepts and definitions guiding the research, and the general design of the questionnaire, including the meaning and purpose of the various sections as well as of specific questions. Note that this training format is very similar to that followed for the interviewing staff, though its aims are different.

Usually the most valuable part of the training program consists of practice sessions. These might begin by having all the participants carry out an independent coding of the same questionnaire, usually one prepared by the study directors and the coding supervisor from actual responses that will illustrate important and difficult coding decisions. The coders and the study directors can then review the results, question by question, and attempt to clarify the discrepancies which emerge. Training sessions of this type are valuable not only in establishing the general norms to be followed in coding, but also in arriving at a constant frame of reference for difficult items. Such discussions are useful for the study directors as well as the coders. When difficulties of interpretation arise, the study directors can immediately set to work to revise the code before production coding takes place.

Production Coding

Production coding refers to the coding of all completed questionnaires for a given study. Check coding, in turn, is the systematic recoding of information from a sample of the questionnaires.

A question which often arises during production coding concerns the amount of time that the study directors should spend with the coders (assuming that they are separate individuals). At the Michigan Survey Research Center, at least one study director is expected to be present during the entire process. Though this practice can be inconvenient and boring for the study director, his presence can be helpful in various ways. For one, he can answer questions about the meaning of various items as they arise. In this way he can discourage coders from making arbitrary decisions about ambiguous situations. Moreover, it is to his advantage to be available for discussions of proposed revisions in the code. In most studies the code will have to be amended (usually expanded) to take account of unforeseen possibilities. The study director should be a part of these decisions, for they may have important implications for the analysis and interpretation of the data.

It is also helpful if he takes part in the check coding. This experience provides important insights into the quality of the questions as well as the information they yield. Finally, his presence can help to bolster the morale of the coders and maintain a seriousness of purpose in their work.

Another problem during this stage involves revisions of the code. Even with the best of preparations, it is often necessary to add categories to cover unanticipated responses, or delete others which are never used. As a general rule, new categories should be added only when they are relevant to the analysis goals of the study, when the responses in question occur often enough to deserve separate categories, and when none of the existing categories can be expanded to accommodate the additional responses. Code changes should be as few as possible, for they require the coders to go back over all of the questionnaires completed to that point. Sometimes it is hard to know in advance if a given response will appear often enough to warrant a separate category. In this case, it is advisable to code the response as "other," but also to prepare a small card with the following information: interview number, column number, question number, the code number actually chosen, and the verbatim response. As the cards accumulate, the study directors will be in a better position to decide on a code change. If a new category is added, the recoding of earlier responses is relatively simple with the information available on the cards. These cards can also be useful in preparing a report on the survey, especially to illustrate the range of answers obtained with a given question.

Check Coding

As noted earlier, check coding refers to the systematic recoding of a certain proportion of the questionnaires as a form of quality control. Usually this operation is carried out by the coding supervisor, study director, or experienced coders who specialize in check coding. Its main purposes are to reduce errors and to determine the overall reliability or consistency of the data. Only by having two or more persons code the same questions is it possible to know whether the code is understood and used consistently. A rough estimate of reliability can be calculated by dividing the number of differences found between coders by the total number of responses covered in the check coding. If, for example, the check coders review 1000 code categories and find disagreements on 50, the consistency rate would be 95 percent. Check coding is also a helpful means of training, for it brings out misunderstandings or differences of interpretation suggesting the need for further discussion and clarification.

A critical decision in this process concerns the percentage of questionnaires to be recoded. This decision should take account of the difficulty of the code itself, the importance of certain kinds of information for the goals of the study, and the extent of differences found in preliminary check coding. We would recommend that the first few questionnaires be completely recoded, both for purposes of training and to uncover weaknesses in the code. Once the discrepancies reach an acceptable level (such as 2 percent or less) check coding can be reduced to 20 or 30 percent, and later to perhaps 10 percent. Even here it may be advisable to carry out a 100 percent check coding of certain items which lend themselves to coding error and which are central to the study. A similar policy could be recommended for questions which are likely to generate controversy when the findings are published.

Problems in Coding

Many of the difficulties which might otherwise occur in coding can be prevented by a careful system of editing. Even with such a system, three kinds of problems may still arise.

   1 *Two answers to the same question.* Occasionally, a respondent will give two responses where one is required or he will begin with one response and change it to another. Specific policies should be developed to handle such contingencies, which can be quite confusing for the coders. The following policies are used at the Survey Research Center.
   Responses beginning "I don't know, but." Some individuals will open a discussion by saying that they are unable or unqualified to answer, but then proceed to provide a perfectly acceptable response. In this case, phrases such as "I don't know" can often be regarded as a means of gaining time or general qualifiers on the respondent's expertise. Usually such introductions can be disregarded and the remainder of the response coded.
   Changes of opinion. What should be done if the respondent gives one complete answer and then changes it to another? Unless there is good reason to regard the first opinion as the more valid or important (as in a free association test), the second should be coded.
   Changes in frame of reference. Respondents sometimes misunderstand the question and begin with an answer based upon an inappropriate frame of reference. When a relevant answer is obtained, either spontaneously or through probing, this should be coded and the first disregarded.
   Two or more legitimate responses. If more than one response can be expected from a question, the code should provide explicit instructions on how to handle the various answers. A commonly used convention is to code the first response mentioned unless some other category is given priority.
   2 *Improper probes.* Problems occasionally arise in coding because the interviewer has made improper use of probes (see Chapter 7). The coder may notice, for example, that the interviewer used a probe which suggested an answer to the respondent. In such cases, the best policy is to have the

coder consult the coding supervisor or study director for advice. It may be necessary, for example, to determine if the probe used was, in fact, leading, or if the interviewer simply did not record it properly. In clear cases of inappropriate probing, the supervisor may decide to code only the information obtained before the probe was introduced.

   3 *Questions answered by the wrong respondent.* It may happen during an interview that one or more questions is answered by someone other than the proper respondent, such as his wife or another family member. A man may turn to his wife and say, "You can answer that better than I can," and then refuse to add anything to what she says. During coding, this situation might be handled as follows. The answers obtained from other respondents can be coded if the questions deal with facts rather than attitudes and opinions, and if the conscripted respondent is qualified to provide the information. If, for example, a father asks his son to describe the latter's job, it would be reasonable to assume that the son could do so at least as well as his father. However, when questions deal with matters on which there is likely to be considerable variation from one respondent to another, as in questions on attitudes, beliefs, expectations, and the like, the safest policy is to code only the information obtained from the properly designated respondent. If he or she does not answer, the item should be coded "Not ascertained" and the other information ignored, at least for purposes of coding.

   In conclusion, it bears repetition that the coding process should be guided, above all, by the objectives of the survey. It is very easy, especially in a first survey, to become mired in the technical details of designing and using a code. These details are important, especially when they touch the reliability and validity of the data. But it is also essential to view coding continually in the perspective of the questions which the study is trying to answer and the specific goals of analysis, as well as the data processing capabilities available to the research staff. Coding, in other words, should never be viewed as a disconnected process or as an end in itself.

An Illustrative Code

Following the pattern of earlier chapters, we will now present an illustrative code designed for the responses obtained in the specimen questionnaire (see (pp. 172–181). The aim of this example is both to show one way of organizing the cards (decks) when the data require more than one card, and to indicate how specific items are coded. Because a complete code for the specimen questionnaire would require more space than is available here, we have not shown the complete code for each card. In most cases, the full card would come to approximately 65 columns.

   A few aspects of the illustration merit particular attention. First, Card 01 records information about the sample selections and the household in general, largely drawn from the Cover Sheet. With the exception of the material on the head of the household (columns 27-31) and the composition of the household (columns 32-37), the data are provided by the interviewers rather than individual respondents. Second, at several points, the code illustrates the basic technique of "zeroing out" inapplicable responses. For example, in Card 03, column 10, some of the respondents state that their neighborhood is the same, rather than better or worse, as a place to live. For these respondents, columns 11 and 12, which ask about reasons for change, are inapplicable. The coders are instructed, therefore, to record "0" in columns 11 and 12 whenever the response in column 10 is "3". The same applies when the response in column 10 is "don't know" or "not ascertained" (9). "Zeroing out" inapplicable columns accelerates the coding process and removes potential points of ambiguity for the coders. Third, Card 04, column 10 shows the use of "zeroing out" in handling a complex filter-skip question on employment and unemployment (Question 13). In this and other cases, the coder is instructed to put zeroes in all the columns except those which apply to the respondent's situation (e.g., employed). This example also demonstrates how proper questionnaire design can facilitate coding. Many other items, by contrast, are perfectly straightforward and offer no particular problems in coding.

FURTHER READINGS

Muehl, D. (ed.), *Manual for Coders: Content Analysis at the Survey Research Center* (Ann Arbor, Mich.: Institute for Social Research, 1961).

NATIONAL URBAN SURVEY
CARD 01
SAMPLE DATA AND HOUSEHOLD CHARACTERISTICS

CODE ONE CARD 01 PER COVER SHEET

Column number	Item	CARD 01
1–3	Study Number (245)	
4–5	Card Number (01)	
6–9	Interview Number (CODE 4 digit number in red in upper right-hand margin) 0001 series. Interviews 8001 series. Nonsample cases 9001 series. Noninterview cases	

Column number	Item	CARD 01
10	Cover Sheet 1. Original 2. Additional	
11–12	(1) Name of Interviewer 01. John Smith 02. Maria Valdez etc.	
13–14	(2) Your Interview Number 01. 1 02. 2 etc.	
15	(3) Length of Interview 1. Less than 20 minutes 2. 20–29 minutes 3. 30–39 minutes 4. 40–49 minutes 5. 50–59 minutes 6. 60 minutes or more 9. N.A., D.K.	
16–19	Sample Information (CODE 4 digit number in "for office use only" box at top left) (Note: consult the sampling specialist for assistance in using this information in preparations for sampling error calculations.)	
20	(10, 10a) . . . more than one Dwelling Unit? How many . . . ? 1. One or none ("NO" to Q. 10) 2. Two 3. Three 4. Four 5. Five or more	
21	(12) Number of Calls 0. None 1. One 2. Two 3. Three 4. Four 5. Five or more	
22	(12) (Final) Result	

(12) (Final) Result

Code 0 in Col. 26	—1. Interview
Code 0 in Cols. 27–37	—5. Noninterview or nonsample

Item   **CARD 01**

23   (13) Type of Structure
         1. Detached single family house
         2. Apartment in partly commercial structure
         3. 2-4-family house or row house
         4. Apartment house of 5-9 units
         5. Apartment house of 10-19 units
         6. Apartment house of 20 or more units
         7. Other

24   (14) Type of Neighborhood
         1. Commercial, including stores, factories, warehouses,
            terminals, etc. (Exclude neighborhood stores and
            screened shopping centers.)
         2. Large apartment buildings; town houses of 3 or more
            floors; professional offices predominantly
         3. Small (2-4) multiple family units; row housing; town
            houses up to 2½ stories
         4. Built up area of single family houses with lot front-
            age on street of 200 feet or less; (define "built up"
            as half or more lots occupied)
         5. Sparsely settled; less than half the lots occupied;
            lots of over 200 foot frontage
         6. Isolated; no structures within 100 yards
         7. Other

25   (15) Access to Public Transportation
         1. Less than 5 minute walk
         2. 5 but less than 10 minute walk
         3. 10 but less than 20 minute walk
         4. 20 but less than 40 minute walk
         5. 40 or more minute walk; no public transportation

26   (16) Reason for Noninterview
            Nonsample:
         1. No such address; could not locate
         2. Address is without structure
         3. Nonresidential structure
         4. House or DU vacant
            Noninterview:
         5. NAH; no one answered
         6. RA; R unavailable
         7. Partial refusal
         8. Total refusal
         9. Inaccessible; administrative reason
         0. Inapplicable (interviewed) (Coded 1 in col. 22)

         (GO TO HOUSEHOLD LISTING)

Column number	Item	CARD 01

27   (Line 01) SEX of Head
    1. Male
    2. Female
    0. Inapplicable (Coded 5 in col. 22)

28   (Line 01) AGE of Head
    1. Less than 25
    2. 25–34
    3. 35–44
    4. 45–54
    5. 55–64
    6. 65 or over
    0. Inapplicable (Coded 5 in Col. 22)

29   (Line 01) MARITAL STATUS of Head

    ┌────────────────────┐
    │ Code 0 in Col. 3 │
    └────────────────────┘

    1. Single
    2. Married; common law union
    3. Widowed; divorced; separated
    0. Inapplicable (Coded 5 in col. 22)

30   (Line 01) EDUCATION of Head
    1. Yes, attending school now
    5. No, not attending school now
    9. N.A., D.K.
    0. Inapplicable (Coded 5 in col. 22)

31   (Line 01) HIGHEST YEAR Completed or Now Attending
    1. Less than 6 years
    2. 6–11 years
    3. 12 years
    4. Some college
    5. Completed college or more
    9. N.A., D.K.
    0. Inapplicable (Coded 5 in col. 22)

32–33   (Household Listing) Number of PERSONS, TOTAL, in Household
    01. 1
    02. 2
    03. 3
    etc.
    00. Inapplicable (Coded 5 in col. 22)

34   (Household Listing) Number of ADULTS in Household
    1. 1
    2. 2
    3. 3
    4. 4
    etc.
    9. 9 or more
    0. Inapplicable (Coded 5 in col. 22)

> NOTE: Code persons 18 years of age and over

<table>
<tr><td>

Column
number
35
</td><td>

Item   **CARD 01**
(Household Listing) Number of CHILDREN in House-
hold
</td></tr>
</table>

Column number	Item	CARD 01

35 (Household Listing) Number of CHILDREN in Household

1. 1
2. 2
3. 3
4. 4
etc.
9. 9 or more
0. None; inapplicable (Coded 5 in col. 22)

> NOTE: Code persons 17 years of age and under

36 (Household Listing) Age of YOUNGEST CHILD in Household

1. Youngest child 5 years of age or less
2. Youngest child 6–17 years of age
0. None; inapplicable (Coded 5 in col. 22)

37 (Household Listing) SPOUSE of HH Head Present or Temporarily Away?

1. Spouse present or temporarily away
5. No spouse
0. Inapplicable (Coded 1 or 3 in col. 29; 5 in col. 22)

NATIONAL URBAN SURVEY
CARD 02
CHARACTERISTICS OF RESIDENTS

CODE ONE CARD 02 FOR EACH HABITUAL RESIDENT IN HOUSEHOLD

Column number	Item	CARD 02

1–3 Study Number (245)

4–5 Card Number (02)

6–9 Interview Number

10–11 Line Number

12 Relationship to HH Head

1. Head
2. Spouse
3. Child of HH Head
4. Parent of HH Head
5. Brother or sister of HH Head
6. Other relative of HH Head
7. Unrelated to HH Head
9. N.A., D.K.

13 SEX of Individual

1. Male
2. Female
9. N.A., D.K.

Column number	Item	CARD 02

14–15   <u>AGE of Individual</u>
00. Less than one year
01. 1 year old
02. 2 years old
etc.
99. N.A., D.K.

> NOTE: Code age as of last birthday

16   <u>MARITAL STATUS of Individual</u>
1. Single
2. Married; common law union
3. Widowed; divorced; separated
9. N.A., D.K.

17   <u>ATTENDING SCHOOL NOW?</u>
1. Yes
5. No
9. N.A., D.K.

18   <u>HIGHEST YEAR COMPLETED?</u>
1. Nursery; kindergarten; other preschool
2. 1–3 years
3. 4–6 years
4. 7–9 years
5. 10–12 years
6. More than 12 years

19   <u>Highest Educational Cycle Completed?</u>
1. Graduate or professional degree
2. College degree
3. High school; 12 years completed
4. Junior high school; 9 years completed
5. Primary school; 6 years completed
9. N.A., D.K.
0. None completed; no schooling

20   <u>RESIDENCE STATUS</u>

> Error: Do not code card for temporary visitors

1. Permanent resident
2. Visitor
3. Temporarily away
9. N.A., D.K.

NATIONAL URBAN SURVEY
CARD 03
HOUSING, NEIGHBORHOOD, ECONOMIC SITUATION

CODE ONE CARD 03 PER INTERVIEW

Column number	Item	CARD 03
1-3	Study Number (245)	
4-5	Card Number (03)	
6-9	Interview Number	

10   (1) First of all, is this neighborhood a better or worse place to live than a year ago, or is it about the same?

1. Better

| Code 0 in Cols. 11, 12 |

— 3. Same

5. Worse

— 9. N.A., D.K.

11-12   (1a, 1b) How is that? Anything else?
(CODE TWO REASONS)

1. CRIME, VIOLENCE: more (less) robberies; can't (can) go out in the streets; better (worse) police protection
2. QUALITY OF HOUSING: people take better (worse) care of their homes; places are beginning to look run down (better); property values have gone up (down)
3. PHYSICAL ENVIRONMENT: air is more (less) clean; the city takes better (worse) care of the neighborhood, streets, garbage, parks, etc; more (less) pollution
4. RESIDENTS: a better (worse) type of person is moving in; people nicer (not as nice) to talk to; neighbors better (not as well) educated
5. TRANSPORTATION: harder (easier) to get around, get downtown; better bus, train service
6. EDUCATION FACILITIES: schools have improved (deteriorated); easier (harder) for children to get a good education
7. OTHER REASONS: a more (less) pleasant place, reason N.A.; reasons related to special family circumstances (family, friends moved in, out of neighborhood)
9. N.A., D.K.
0. Inapplicable (Coded 3 or 9 in col. 10); no second reason

Column number	Item	CARD 03

13

(2) Would you say that you and your family are better or worse off financially than you were a year ago, or about the same?
1. Better
3. Same
5. Worse
9. N.A., D.K.

Code 0 in Cols. 14-17

14-15, 16-17

(2a) How is that? REASONS FOR CHANGE: (CODE TWO REASONS)
(Reader: See double-column code in text, pp. 241-242)
99. N.A., D.K. (Use only in cols. 14-15)
00. Inapplicable (Coded 3 or 9 in col. 13); no second reason

18

(3) How about your house (apartment, room), is it better or worse than it was a year ago, or about the same?
1. Better
3. Same
5. Worse
9. N.A., D.K.

Code 0 in Cols. 19,20

19, 20

(3a) How is it different? REASONS FOR CHANGE: (CODE TWO REASONS)
(Single-column reasons code on the order of col. 11)
0. Inapplicable (Coded 3 or 9 in col. 18); no second reason

21

(4) Do you rent this house (apartment, room), are you paying for it on a mortgage or land contract, do you own it outright, or do you occupy it without having to pay anything?

Code 0 in Cols. 29-40 —— 1. Rents

Code 0 in Cols. 22-28 —— 2. Mortgage or Contract
—— 3. Owns outright

Code 0 in Cols. 22-40 —— 4. Occupies without paying
—— 9. N.A., D.K.

22-24

(5) How much do you pay a month for it?
(Code in DOLLARS)
000. Inapplicable (Coded 2-4 or 9 in col. 21)

25

(5) How much do you pay a month for it?
1. Under $25 per month
2. $25-$49 per month
3. $50-$99 per month
4. $100-$149 per month
5. $150-$199 per month
6. $200-$299 per month
7. $300-$399 per month
8. $400 per month or more
9. N.A., D.K.
0. Inapplicable (Coded 2-4 or 9 in col. 21)

Column number	Item	CARD 03

26   (5a) Does that include heat?
1. Yes
5. No
9. N.A., D.K.
0. Inapplicable (Coded 2–4 or 9 in col. 21)
etc.

NATIONAL URBAN SURVEY
CARD 04
EMPLOYMENT, INCOME

CODE ONE CARD 04 PER INTERVIEW

Column number	Item	CARD 04

1–3   Study Number (245)

4–5   Card Number (04)

6–9   Interview Number

10   (15) Now I'd like to ask you about your present job. Were you working last week, unemployed, laid off, retired, in school, or what?

Code 0 in Cols. 18–33 ———1. Working

Code 0 in Cols. 11–17, 26–33
——2. Unemployed
——3. Laid off
——4. Retired

Code 0 in Cols. 11–25
—5. Student
—6. Keeping house
—7. Other

Code 0 in Cols. 11–33 ———9. N.A., D.K.

11   (16) What kind of work were you doing; . . . ?
Code 1 digit number in red to left of Q. 16.

12   (17) What kind of business or organization . . . ?
Code 1 digit number in red to left of Q. 17.

13   (18) Did you have a second or other job last week?
1. Yes

Code 0 in Col. 15 ———5. No

9. N.A., D.K.

14   (19) How many hours did you work . . . PRINCIPAL JOB?
1. Less than 20 hours
2. 20–29 hours
3. 30–34 hours
4. 35–39 hours
5. 40 hours

Column number	Item	CARD 04

6. 41–49 hours
7. 50–59 hours
8. 60 hours or more
9. N.A., D.K.
0. Inapplicable (Coded 2–7 or 9 in col. 10)

15   (19) . . . hours worked . . . OTHER JOBS?
(Same code as col. 14)
0. Inapplicable (Coded 5 in col. 13; coded 2–7 or 9 in col. 10)

16   (19) . . . hours worked . . . TOTAL LAST WEEK?
(Same code as col. 14)

17   (20) Aside from paid leave or vacations, how many weeks, if any, were you out of work last year, from January through December?
0. None; inapplicable (Coded 2–7 or 9 in col. 10)
1. 1 week
2. 2 weeks
3. 3–4 weeks; 1 month
4. 5–8 weeks; 2 months
5. 9–13 weeks; 3 months; 1 quarter
6. 4–6 months
7. More than half a year
9. N.A., D.K.
etc.

NATIONAL URBAN SURVEY
CARD 05
POPULATION ATTITUDES

CODE ONE CARD 05 FOR EACH MARRIED RESPONDENT

Column number	Item
1–3	Study Number (245)
4–5	Card Number (05)
6–9	Interview Number

10   (31) Now we want to ask about your family. For example, in your own case do you want to have any (more) children?
1. Want (more) children
3. Uncertain
5. Does not want (more) children
9. N.A.
etc.

Chapter 10

Preparations for Analysis

This chapter focuses on the practical steps in moving from the coded data to the final tabulations. Before the coded data can be properly analyzed, they must be stored for easy retrieval and tabulation. Also, any errors should be identified and corrected; missing data will require special attention, and special variables often must be constructed. The following discussion covers these topics and provides a brief introduction to the data processing equipment commonly used for survey analysis.

INFORMATION STORAGE

The first step after coding is to record the information in such a way that it can be easily retrieved and tabulated. The most common option is to enter the data on cards by means of a keypunch machine. This specialized machine accomplishes the task by punching a series of holes designating the appropriate numbers or letters, as described in the last chapter. To check for errors in keypunching, the operation is repeated on a companion machine known as the verifier. Both machines are described later in the chapter. With this system, most of the data from a moderate sized survey can be stored on several thousand cards, with perhaps three to five cards per interview multiplied by the number of interviews.

In many centers the punch card system is being combined with or even replaced by the more sophisticated technology of magnetic tape and computers. The magnetic tape, or tape for short, can store the same information as the typical punch card, with one important difference: one tape can hold the contents of several thousand cards. It is still common for data first to be keypunched on cards and then later to be read onto tape for major processing, but the cards can be bypassed entirely.

A third and newer option is even more attractive. This is the disc, which has the advantages of greater storage capacity and more rapid access to the data on individual respondents. As with the tape, the coded information is typically punched on cards and then transferred to the disc.

Whether the storage system involves cards, tape, or discs, the stored data will usually contain errors from misunderstood responses from the questionnaire, errors of classification in coding, or even the transposition of digits or other mechanical errors. The procedures now to be described deal with identifying these errors and correcting them.

CONSISTENCY CHECKING

The aim of consistency checking is to seek out and correct impossible, inconsistent, or improbable punches, or their electromagnetic counterparts on a tape or disc. It is advisable to develop a plan of consistency checking before proceeding with statistical analysis. Errors discovered later may require the repetition of substantial parts of the tabulations, especially when they occur in small but important subgroups. Since the card, tape, and disc systems are conceptually similar, for simplicity the rest of this section deals only with data stored on cards.

Illegitimate Punches

A basic task in consistency checking is to detect illegitimate or "wild" punches on the cards. These are punches which do not correspond to any of the code categories for the column in question. Each column should be inspected for this type of error before further work is carried out. Wild punches may originate in coding or keypunching, machine error, or in improper corrections of previously discovered errors. For example, "7" may be punched in a column in which the only legitimate codes are 1, 2, 3, 4, 5, 8 and 9.

The procedure for detecting illegitimate punches is quite simple. Tabulations are carried out for every column or set of columns representing a field (variable). The resulting frequencies are then compared with the categories in the code for the corresponding variable. Cards with punches not shown in the code book are in error. Corrections can be made by obtaining the interview number of the offending card and checking it against the questionnaire or interview schedule. The original card would then be duplicated except for the column containing the error, and the correct punch entered in that column. Consistency checks can be performed on conventional punch card equipment, or they can be completed very quickly on electronic computers, provided the necessary programs are available.

Inconsistent Responses

Similar procedures are used to uncover logical inconsistencies between code categories in different columns. For example, the codes for the marital status of the household head and the ages of the head and spouse may be as follows:

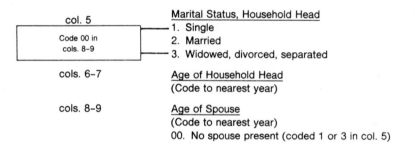

col. 5	Marital Status, Household Head
	1. Single
Code 00 in cols. 8–9	2. Married
	3. Widowed, divorced, separated
cols. 6–7	Age of Household Head (Code to nearest year)
cols. 8–9	Age of Spouse (Code to nearest year) 00. No spouse present (coded 1 or 3 in col. 5)

To be consistent between columns, those cards punched 1 or 3 in column 5 should also be punched 00 in columns 8-9, and those punched 2 in column 5 should be other than 00 in columns 8-9. This type of consistency check should be routinely completed for questions involving filters and skips (e.g., IF "YES," SKIP TO Q. 17) and for contingent codes such as the ones illustrated above. How far one goes beyond this point depends on the importance of a given variable to the overall analysis plan and the problems created by possible inconsistencies. In most surveys, it is not necessary to check out every conceivable inconsistency in the data.

Improbable Responses

The next set of checks concerns responses which are possible but improbable. It is very difficult to suggest a watertight policy for handling such responses because some may simply reflect an unusual situation. If most populations contain persons or families in improbable circumstances, we would expect some of them to appear in the sample. The challenge is to distinguish between truly unusual cases and improbable responses arising from errors in questionnaire administration, coding, or other facets of the survey.

The last example illustrates the difficulties with improbable responses. Should we try to determine whether there are any very young persons, say under fourteen years of age, among those coded 2 (married) in column 5? While in North America it is not impossible for a thirteen-year-old to be married and to head a household, it is most unlikely. If a consistency check uncovers one or more such individuals, it would be well worth returning to the questionnaire to see if other information would shed any further light on the situation.

Another example shows the variety of outcomes that could occur in investigating whether an improbable response is also incorrect. The national urban survey discussed in this book finds a laborer's family with a monthly income well beyond the average for his occupational category. This suggests that something may be in error, but the improbable response cannot automatically be regarded as incorrect. The original questionnaire should be consulted to determine whether an unusual situation exists or to provide the information by which to correct the card. The following outcomes are typical of what could be found in consulting the questionnaire on this example:

—the regular income received by the head of the family was typical for his group, but family income was inflated by the unusually high earnings of his son.

—the regular income of the family was typical, but a windfall income (lottery grand prize, inheritance, life insurance proceeds, etc.) was erroneously included in monthly income rather than in the separate category for windfall income.

—the occupation of the family head was misclassified. Closer checking revealed that he was a *labor negotiator,* and thus an executive, rather than a laborer.

—a mistake was made in coding the income figure. Because of illegible figures the decimal point separating dollars and cents was ignored so that all the digits were coded as dollars.

—an error was made in a previous correction of the card so that most of the punches on the card were shifted one column to the left.

It is impossible to set up ironclad rules about how much checking should be done for improbable responses. However, two broad guidelines can be suggested. First, it is particularly important to identify incorrect, improbable responses when the sample data are numerical, such as income and debt levels, and will be used to estimate averages or aggregate amounts for the population. An erroneous case of extremely high income or debt level will often increase the estimates of the average or aggregate figures by a noticeable amount for the total population, and by even more for population subgroups. These errors are less important when the data are to be reported only in frequency distributions. The presence of an erroneous case in the distribution will usually be less than the rounding error for percentages. Second, it is usually worth the effort to search for erroneous, improbable responses on *key* variables when: (a) the interpretations of the findings depend to a large extent on the accuracy of the information presented on these variables; and (b) the study is likely to be subject to criticism or attack, or viewed with skepticism if *any* errors are found in the critical tables. Even a single impossible or improbable case can provide the critic with ammunition to discredit the entire study.

MISSING DATA

Household surveys rarely obtain every item of information at every sample address. As a result, the analyst is often faced with the problem of missing data. Do these omissions damage the sample or invalidate the findings? Unfortunately, many researchers simply ignore this problem in processing, interpreting, and presenting their data. Noninterviews and nonresponse to specific items are often treated as if they had no particular implications for the findings ultimately presented to the reader. Here we argue for explicit attention to the various kinds of missing data, even when no adjustments are made in the final results.

Types

There are two principal types of missing data: noninterviews and response omissions. A *noninterview* occurs when the information required by the study is not obtained from a respondent selected into the sample. The most common reasons for noninterviews are the respondent's illness or absence from the sample address, his or her refusal to be interviewed, and administrative errors leading to lost interview schedules. Absence and refusals are by far the most common.

Response omissions consist of items for which information is not obtained in otherwise complete and satisfactory interviews. Such omissions may occur for any of several reasons: the respondent's inability or refusal to answer a specific question, illegible handwriting, the failure of the interviewer to follow proper procedures (as on SKIP questions), and so on.

A response of "don't know" may or may not be missing data. On some questions, such as those dealing with opinions and certain types of information, this response is as valid and useful as any other. If an item asks, in effect, "What is your opinion about the work of the United Nations?" and a respondent in rural India answers, "I have never heard of the United Nations," his reply provides important substantive data of direct relevance to understanding the United Nations. This situation can be distinguished from "don't know" answers where the respondent could reasonably be expected to have knowledge or an opinion about a matter. Such would normally be the case, for example, in questions about age, the number of people habitually residing in the household, marital status, and similar items which are very close to the respondent's own experience. "Don't know" responses to these questions are more properly classed as missing data. Even here, however, one must be careful. In the national sample survey of Peru carried out in 1970, the interviewers reported that substantial numbers of Quechua Indian women in the highland regions of the Andes did not know their age. For some purposes, their responses were missing data, but they could also be viewed as a significant index of their time horizon and information levels.

Another problem is that a reply of "I don't know" is often a disguised refusal to answer a delicate or offensive question. While for purposes of tabulation either a refusal to answer or a genuine "don't know" may be classed as missing data, a thorough understanding of the dynamics and validity of the interview may require close attention to the frequency of each type of response.

Problems

Two basic problems which may arise from missing data are bias and difficulties in working with other data. Bias would arise when the missing data are not a random sample of the data being sought. In theory, if the missing data constituted a random sample, their omission would be of little practical consequence, for the omission of a random part of a random sample leaves a smaller, but still random, sample. For example, if noninterviews, or response omissions, occurred randomly in the sample, their effect would be to reduce the sample size rather than to bias the results. Unfortunately, missing data cases are often not a random sample of the total sample. Those who refuse to be interviewed may differ from the total sample on age, income, education level, political attitudes, or other salient characteristics. The precise pattern of differences may vary from study to study and country to country; however, the important point is that those who participate in the research are likely to differ systematically from those who are selected into the sample, but who cannot be interviewed. Moreover, in many studies it is difficult, if not impossible, to measure the full degree of bias arising from noninterviews and other missing data. Even though the study directors do not and cannot know the full extent of such bias, they can usually find clues to it, as we shall show.

The second problem arises when the missing datum is needed as part of a larger set of data about the same respondent. This point can be illustrated with total family income. Many surveys attempt to determine family income levels by asking a series of questions about each family member and about different sources of income, such as wages and salaries, income from second jobs, freelancing, a business or profession, strike benefits, unemployment compensation, welfare, or dividends and interest. Either during the editing stage or during analysis, these separate figures are added together to form a measure or index of total family income. If one of the components is unknown for a given family, the resulting income measure may be understated or otherwise biased. Another example concerns the commonly used index of family life cycle. This is a composite measure using several factors, such as age, employment, and marital status of the head of the family, and the ages of the children. Again, if any one of the components is missing, such as the age of the youngest child, the family in question cannot be classified as to life-cycle stage and thus becomes a missing data case for all calculations involving that index.

Similar problems occur with multivariate analysis, that is, calculations making use of information on three or more variables. Suppose that we wish to tabulate family income by the age and occupation of the household head. If the information on either age, occupation or income is missing for a sample case, it will not be possible to classify that case in the cross tabulation. In this sense, the problems arising from missing data on individual items are cumulative. The index of family life cycle is invalidated for all cases lacking the needed information on age, marital status, children or any other component used. The total number of cases available for analysis involving indices, cross tabulations, and other multivariate operations may thus be substantially smaller than the total number of sample cases. The resulting decline in effective sample size has important ramifications for possible bias, and for sampling error, confidence intervals, and related problems of inference and estimation. For this reason, the reader of survey reports should look closely at the tables to determine the actual sample size in use rather than the total sample size for the study. Research reports often are less than candid in indicating the degree of sample mortality resulting from missing data on individual items.

Can anything be done to correct the problems introduced by missing data? Some adjustments are possible, but they should not obscure one important fact: missing data are missing and they cannot be fully replaced by statistical manipulation or guesswork. The most that can be accomplished is to reduce their negative effects on the total set of results.

Handling Noninterviews

Many household surveys lose 10 percent or more of the sample cases to noninterviews. The problems created by refusals, not-at-homes, and other kinds of noninterviews depend both on the total percentage of the sample cases involved and on their distribution in the sample. Other things being equal, it is obviously more damaging to have a 40 percent than a 10 percent noncompletion rate. Even with a low rate, however, serious problems can arise if most of the noninterviews are concentrated in one or two of the sample strata or population subgroups. Basically, three options exist for dealing with noninterviews: redefining the population, ignoring the missing interviews, and differential weighting.

Redefining the Population   One can imagine a situation, possibly in one of the less developed countries, where the interviews for an entire stratum, such as a self-representing city, are lost in transit. It might also happen that for political reasons, field work might be cancelled in a certain large section of a country. In such circumstances, it would be advisable to redefine the population to include only those areas covered by the study. If the study dealt with Peru, for example, and the missing area was the Amazon jungle region, the reader would be warned that the results of the study could not be generalized to that area.

Ignoring Missing Interviews   In practice, a common but highly unsatisfactory way of dealing with noninterviews is to ignore them. The final report of a study may simply point out a noncompletion rate of 10 or 25 percent with no further comment. Worse, some reports do not even mention the percentage of noninterviews. Clearly these forms of inattention are unacceptable.

Missing interviews should never be wholly ignored. One should include plans in a survey to collect whatever information is readily available about sample selections where interviews cannot be taken. Information about the missing interview addresses can then be compared with similar information from the addresses of the completed interviews to determine the extent to which the two groups differ. If the omissions appear to differ systematically from the rest, some adjustments such as differential weighting may be necessary before the main body of data is analyzed. If the evidence suggests that the missing cases are similar to the interviews, that is, approximating a random sample of the population, the analysis may proceed without special weighting. Only in this sense may the missing cases be ignored.

Even if the noninterviews are similar to the participants on such gross characteristics as housing and the socioeconomic level of the neighborhood, they may differ in important and nonrandom ways on attitudes and other less observable traits. For this reason it is very important that the final report disclose the *extent* of not-at-homes, refusals, and other missing interviews and the *evidence* as to how they are different from the rest of the sample. Such information will help the reader evaluate the representativeness of the interviewed sample. Failure to include this information may arise from ignorance, from carelessness or inertia on the part of the researcher, or even from lack of principle. These are all inexcusable for responsible researchers.

Differential Weighting With proper planning, it is usually possible to test for differences in the characteristics of sample cases and noninterviews. Some points of comparison are available from the sample addresses themselves: urban/rural location, city size, socioeconomic level of the neighborhood, and the predominant type of housing in that area (apartments, single-family dwellings, etc.). In addition, the interviewers can be asked to gather a variety of information about the home, neighborhood, and personal characteristics of the noninterview cases. They can, for example, provide subjective ratings of the socioeconomic status of the neighborhood, categorize the external features of the dwelling, and tactfully ask neighbors for information about the age, life-cycle stage, occupation, and a few other general characteristics of the household head. While extensive questioning of neighbors raises serious ethical problems and also leaves the study open to charges of invasions of privacy, a few questions about observable and publicly known characteristics would seem to be ethically permissible in most studies. In some countries the necessary information might also be found in city directories or other published sources. When this process is complete, respondents and noninterview cases can be compared on the dimensions observed or rated. If substantial differences appear, the same information can be used as the basis for postenumeration stratification and differential weighting of sample cases. This procedure is similar to the kinds of stratification discussed in Chapter 4, except that it takes place after, rather than before, the sample is selected.

Suppose that an equal probability survey yielded 400 sample selections each from rural and urban areas for a total sample size of 800. Let us further assume that the interviewers obtained a total of 280 interviews from the urban areas (70 percent response rate), and 350 from the rural areas ($87\frac{1}{2}$ percent response rate). (These response rates would not be unusual in the United States, where response rates for urban areas are often considerably lower than for rural areas.) The basic problem illustrated by the example is that if the urban interviews were simply added to the rural cases to make up the total sample, the rural portion would carry proportionately more weight than it deserves and the urban proportionately less.

To eliminate this imbalance, the researcher could introduce differential weighting of the sample cases. This first step would be to undertake a post-enumeration stratification by assigning each case to one of two categories: urban or rural. Then the weights (w) for each of the urban (u) and rural (r) interviews from the two strata could be determined so that the groups of urban and rural interviews would have equal weights. The equation below establishes that the number of urban interviews times the weight assigned to each urban interview will be the same as the number of rural interviews times the weight for each rural interview:

$$n_u w_u = n_r w_r$$
$$280 w_u = 350 w_r$$
$$w_u = \frac{350}{280} w_r = 1.25 w_r$$

That is, the weight per urban case (w_u) would have to be 1¼ times the weight of each rural case (w_r), such as by giving each urban interview a weight of 5 and each rural interview a weight of 4. The urban weights per interview would thus be one-fourth higher than those for the rural interviews. The effects of this process are identical to those seen with weighting in samples involving unequal probabilities of selection. These principles can be applied with more complex groupings to develop more sophisticated weights.

Handling Missing Responses

The methods discussed so far apply to noninterviews. We now turn to the options available for dealing with the occasional omissions appearing in otherwise satisfactory interviews. Here, too, three methods have been used: reporting the omissions in special categories, distributing the missing data, and assigning the missing data cases on the basis of related information.

Reporting Omissions in Special Categories The simplest procedure for dealing with missing answers to single questions is to include a special category in the tables for "not ascertained" responses. With this information, the reader can make his own judgment about the effects of item nonresponse on the interpretations drawn from a given table. Depending on the difficulty and sensitivity of the questions, the frequency of such nonresponse may be less than one-half of one percent to as high as 10 or 15 percent if "don't know" responses are included. When the percentages are fairly high, as is often the case in public opinion polls, it is especially important for this information to be reported. Some researchers are understandably reluctant to expose their flanks in this way, but it is a form of professional irresponsibility to cover up considerable missing data on single items.

Distributing the Missing Data Cases A second procedure, also commonly used, involves distributing the missing data cases proportionately among the categories of known data. Very often this procedure is applied unwittingly by simply excluding the missing data cases from the calculation of statistics such as percentages, means, and medians. Consider the case of an opinion question with five response options used with a sample of 1000. The results show that 900 (90 percent) of the respondents provide answers falling into one of these five categories, while the remaining 10 percent answer "don't know" or do not answer at all. Rather than reporting this last set of 100 responses in a separate category, the researcher may explicitly or implicitly ignore it by tabulating only the 900 cases available on this question. The difference between this and the first method of handling item nonresponse is illustrated in Table 10-1. These hypothetical findings show that when the missing data are reported separately (Method 1) 90 percent of the respondents fall into the five response categories and 10 percent into the category for missing data. In Method 2, the 100 missing data cases are left out of the calculation of percentages. The effect of using only the 900 "legitimate" cases is to distribute the 100 missing data cases proportionately across the five response categories in the calculation of percentages. As a result, the figures rise in each category with the second method.

The advantage of this distribution procedure is that it helps to simplify the analysis and presentation of the data. The problem, however, is that it assumes that the missing data cases are proportionately distributed across the regular response categories. Often this assumption is not justified. There

Table 10-1   Two Methods of Handling Item Nonresponse

Method 1: Reporting omissions		Method 2: Distributing missing data cases	
Response	Percent	Response	Percent
Strongly agree	18	Strongly agree	20
Agree	36	Agree	40
Neutral	9	Neutral	10
Disagree	18	Disagree	20
Strongly disagree	9	Strongly disagree	10
Don't know, N.A.	10		
Total	100		100
Number of cases	1000		900

is good reason to believe, for instance, that those who refuse to report their income have higher incomes on the average than those who do report it. Another drawback is that the distribution procedure may create a misleading picture of the attitudes, income, or other characteristics of the population. In public opinion polls dealing with a president's popularity, a large percentage of missing data cases may be politically, if not statistically, significant. Moreover, this method does nothing to solve the problems created by item nonresponse for the construction of indices and composite variables, or for multivariate analysis. On balance, it would seem preferable under most circumstances to report the missing data in a separate category or give some indication of the number of cases falling into the missing data category, rather than to obscure the extent of item nonresponse.

Assigning Missing Data Cases A third way of handling item nonresponses is to use other information to assign missing data cases to one or another of the regular response categories. This method is essentially one of betting. Since we know several characteristics about a respondent, we use some of this information to bet on his or her response to a given question. Suppose, for example, that a female respondent has failed to answer a question about whether or not she works outside her home. Rather than just listing this case as "not ascertained," or distributing it on the basis of average responses, we can use related information from her questionnaire to estimate whether she works outside the home or not, in this case "working" or "not working."

The first step is to decide which "known" characteristics will be used to make the bet, i.e., to calculate the probabilities of a "working" or "not working" answer to the question. Some good possibilities would be age, education, the number of children, the age of the youngest child, and the husband's income. For example, we might use an index of family life cycle involving age, marital status, and age of children. The second step is to use all the "known" cases to calculate the percentage of "working" and "not working" wives in each stage of the life cycle. The resulting information, which might appear in part as follows, can be interpreted as a contingency table expressing the probabilities that wives work:

Life cycle stage	Percent working	Percent not working	Total percent
1. Young married couple, no children	75	25	100
2. Married couple, children of preschool age	20	80	100
3. Married couple, school-age children Etc.	50	50	100

In our one case, we find that the female respondent is married, living with her husband and has one child of preschool age. The contingency table shows that 20 percent of the women in this category are working while 80 percent are not. To carry out the assignment process, we should choose a random number corresponding to the range of the percentages, that is, 00 to 99. If this number falls into the range of 00 to 19, the woman would be assigned to the "working" category; if it is from 20 to 99, she would be assigned as "not working." Her punch card could then be "corrected" with this datum recorded in place of the punch corresponding to "not ascertained." Thereafter in tabulations she would appear as a working (or not working) wife, according to the random number that turned up. Even more precise assignments could be made by using four, five, or more known characteristics to calculate the probabilities of working and not working.

This assignment procedure is obviously complex and time-consuming, although it can be greatly simplified through the use of a computer. But is it worth the effort? What is to be gained by clearing up a few dozen missing data cases? The answer depends on how critical the missing information is to the study's analysis plans. If these plans call for numerous indices to be built by combining single items, or if they rely heavily on multivariate analysis, there may be substantial benefits. If, on the other hand, the statistical analysis consists mainly of correlations, the advantages may be minimal. In any event, the assignment procedure does not *solve* the problem of missing data; it simply reduces the inconvenience which they create. Nevertheless, in many studies this procedure deserves consideration.

DATA PROCESSING EQUIPMENT

No aspect of survey research has changed more dramatically in recent years than data processing. The introduction of high-speed electronic computers has opened an enormous range of possibilities to the survey analyst, just as the advent of punch card equipment earlier had permitted tremendous advances over hand tabulations. Many of the complex statistical operations that were impractical or rarely carried out on conventional punch card equipment are now commonplace with computers, and many necessary but tedious operations once requiring hours of effort on conventional equipment can now be carried out in seconds on computers. Moreover, the availability of package programs for handling data from the social sciences allows even the beginner great flexibility in planning for analysis, provided that he has good guidance. A detailed discussion of the various types of computers and programs is beyond the scope of this book. We would only counsel the analyst to become thoroughly familiar with the facilities and options available very early in planning the survey.

If a computer is available, the researcher should consult data processing experts about its capacity for handling survey data, the programs available and their constraints (e.g., unacceptable punches), costs, time lags, etc., and the possibility of assistance with programming and using package programs. Both the overall capacity of the computer and the peculiarities of the statistical programs may argue for or against certain types of questionnaire items and codes. This is a further example of the "backward linkages" discussed in Chapter 2. Under no circumstances should plans for analysis be left until the end of the survey. A lack of foresight in the early stages may result in costly delays, recoding, and other inconveniences.

In some surveys, particularly in the less developed countries, computers may either be unavailable or beyond the budget of the small-scale researcher. The analyst need not despair. A great deal of work has been and still can be carried out on rather simple punch card equipment. With access to a few basic machines, the researcher with a limited budget can carry out many operations, including the building of scales and composite variables and tabulations for frequency distributions and means. Even the most advanced computer installations usually contain several pieces of conventional equipment, such as the counter-sorter, to handle relatively small and straightforward data processing operations. In fact, we would suggest that every new data analyst try his hand at these machines as a way of understanding the rudiments of data processing. This firsthand feeling is often most helpful in understanding the use of computers. We will now comment briefly on the main features of the standard punch card machines, and the basic operations carried out by computers.

The Keypunch and Verifier

The keypunch and the verifier make use of a keyboard similar to that of a typewriter and are operated by pressing a key corresponding to a given letter, figure, or other symbol. Both contain mechanisms for storing several hundred punch cards. These are fed automatically into the working portion of the machine. After each card is punched or verified, it is transferred to a storage space on the opposite side of the machine. The principal difference between the keypunch and the verifier lies in the operations carried out when the keys are depressed.

In the case of the *keypunch*, when one of the keys is pressed an electric mechanism causes the appropriate holes to be punched in the column located at the punch position. After punching one column (such as column 1 or column 45), the machine automatically moves the card to the next position (column 2 or column 46), and so on to the end of the card. The machine also has the equivalent of a typewriter's space bar so that columns may be skipped.

The *verifier* is similar to the keypunch, except that it makes no punches in the data columns. The cards coming from the keypunch are placed in the feed bin of this machine, and the process continues as though they were going to be punched again. The operator reads the coded information, as before, and strikes the keys corresponding to the symbols to be punched. The difference is that no holes are made by the working mechanism because the verifier is a machine for comparing rather than punching. When the key pressed corresponds to the punch found in that column, the card moves on automatically to the next column. If there is a difference between the key depressed and the hole(s) in the column, the machine stops, a light appears, and a notch is placed at the top of the column containing the error. When the entire card has been verified and no errors have been found, a notch is placed at the right edge of the card to indicate that the process is complete. If keypunching and verifying are carried out independently by different individuals, they should ensure that the information punched on the cards is identical to that on the code sheets or in a precoded questionnaire. The chances are minimal that both the keypunch and the verifier operators will make the same mistake. Any errors remaining, therefore, are almost always in the original data.

The Sorter

At various points in data processing, it is necessary to separate the cards or arrange them in order according to the punches found within a single column. The machine most commonly used for this operation is the *sorter*. It operates by quickly feeding the cards through a sensing device which reads the punch in the designated column on each card as it passes from its feed bin. The cards are then directed into the appropriate storage pockets corresponding to the 12 possible punches in a column, or into a thirteenth pocket for errors or special groups. By sorting sequentially on several columns, cards may be put in order on multiple-column fields. Depending on the model used, the sorter can read and sort anywhere from 650 to 2000 cards per minute. The *counter-sorter* is a sorter containing a special set of counters, one for each pocket, to count each card according to the pocket into which it falls. The numbers must be read from the counters and be recorded manually. This machine has no printing mechanism.

The sorter can also be used to obtain the frequencies of punches in a two-column code (field). If, for example, *age* is punched in columns 24-25, the frequencies for the various two-digit age codes can be obtained in the following manner. The first step is to sort the cards on column 24, that is, the first digit of the age code. Then each packet of cards obtained in this way is sorted separately on column 25. The reason for this two-step procedure is that the sorter can read only one column at a time. By recording the frequencies obtained through the second step, the analyst can prepare a table showing the number of cases falling into each of the possible two-digit age categories. The sorter and the counter-sorter also have other uses which make them very basic pieces of data processing equipment.

The Collator

The *collator* is used to find matching cards from two decks, to merge decks on some common basis, such as the interview number, and to determine whether cards are in ascending order on a given set of numbers. Like many of the conventional punch card machines, it is operated by providing instructions (column numbers, etc.) on a removable wired panel. While the collator is less commonly found than the sorter or the reproducer, it can be invaluable for the types of comparisons described here.

The Reproducer

The *reproducer* performs three basic operations relevant to survey analysis: the duplication of cards, transferring information from one card to another, and gangpunching. Because this machine has the capacity to read, punch, and compare all 80 columns of a card, the duplication of either a few cards or an entire deck is a simple matter. The machine can be wired to read each of the 80 columns on a card, punch the information onto a blank card, and compare the new and original punches for consistency before completing the same operation on the next card. Through essentially the same process, the machine can reproduce only a portion of a card onto blank stock, leaving the other columns empty.

The second basic operation is the transfer of information from one card to another for the same individual or sampling unit. Suppose, for example, that we wish to transfer information on age, columns 24-25 of Card 1, to columns 61-62 of Card 3. The only requirements for this operation are that the decks must be distinguishable in some way, such as a punch for the deck number, and that the separate decks contain the same interview numbers or some other identical information for matching. Both decks are then ordered separately by the identifying number, usually the interview number. One is then placed in the "Read" hopper, the other in the "Punch" hopper of the reproducer. The machine can be wired to read the appropriate columns on the first deck (columns 24-25) and punch the identical information into blank columns on the second deck (columns 61-62). The reproducer can also be instructed to compare the identifying information on both cards, such as four-digit interview numbers, to ensure that the transfers are being made between cards that are mates. The most common reason for carrying out this transfer process is to prepare the data for tabulations on conventional punch card equipment. With machines such as the counter-sorter, the variables to be analyzed together must be on the same card.

Gangpunching is an operation designed to punch a new variable by using the reproducer to recode or combine information from one or several existing columns. The same process may be carried out very quickly on a computer. The first step is to group the cards into logically distinct groups representing categories of a new variable or a reformulation of existing codes. For example, in the previous illustration age was recorded in a two-digit code in columns 24-25 of Card 1. We now decide that it would be more convenient to have age expressed in only six categories. The code book shows that column 70 of Card 1 is blank and could thus be used for gangpunching the new age code. The instructions for the gangpunching might read:

Sort on cols. 24–25 (age of family head) to form the following separate groups	Gangpunch col. 70 (now blank) as indicated below (age, in groups)
Under 25	1 (under 25 years)
25–34	2 (25–34 years)
35–44	3 (35–44 years)
45–54	4 (45–54 years)
55–64	5 (55–64 years)
65 and over	6 (65 years or more)

The same process could be followed in building a new variable based on information in several other columns of the same card. If the necessary information is not available on the card into which the new variable will be punched, it can be transferred there from another card, as indicated earlier.

The Tabulator

Prior to the introduction of computers, the more complex tabulations in survey analysis were often carried out on the large accounting machines or tabulators, such as the IBM Models 402 and 407. The tabulator is still a very useful machine for data processing centers which do not have easy access to a computer. It is controlled by a wiring panel which can be programmed for various types of operations. Typically, a research installation might have a half-dozen wiring boards, some prewired for standard operations, others free for special uses. A special feature of the tabulator is its capacity to be programmed to multiply. With proper wiring this capacity can be tapped to provide statistics such as weighted and unweighted distributions, weighted and unweighted means, and the sums of squares and cross products. The last two are important elements in calculating variances, standard deviations, and correlations. Another helpful feature is this machine's ability to print out the results of its calculations, or the content of an entire punch card with either numerical or alphabetic characters. In most large research centers, the tabulator has been displaced by computers performing the same operations in a much shorter time. Nevertheless, for a small-scale research operation, in areas of the world where computers are not common, or with a budget or experience too limited to permit access to computers, the tabulator can be a very valuable addition to a conventional punch card equipment pool.

Computers

High-speed electronic computers have revolutionized data processing in most fields, including sample survey analysis. From the standpoint of analysis, their most noteworthy characteristics are speed, reliability, flexibility, and capacity. While the conventional machines represented an enormous advance over hand tabulations, computers brought an even greater breakthrough. Today the variety of computer systems is so great, and the changes so frequent, that practical advice requires attention to specific situations and intended uses. In the following paragraphs, all we can hope to do is to provide an elementary introduction to the design and usage of computers for those with little prior knowledge of this field.

Figure 10-1 presents a diagram of a computer that might be programmed for survey analysis. The portion inside the dotted lines represents

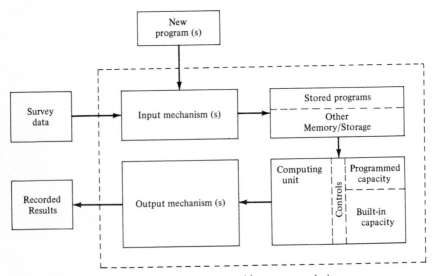

Figure 10-1 Diagram of a computer programmed for survey analysis.

the computer itself, while the boxes outside represent the information fed into or obtained from it.

A *computer* is an integrated set of equipment designed to perform high-speed mathematical or logical calculations or otherwise process information according to a predetermined program. A *program* is a set of instructions which organizes the computer to carry out the desired operations. Programs and the data to be analyzed are entered into the computer through an *input mechanism*. Programs are typically stored in a memory or *storage section* and become an integral part of the computer until erased or changed. The *computing unit* is that part of the computer which carries out the desired operations, while the *output mechanism* is the means by which the results are transmitted to the user. Depending on the particular program used, data may be entered directly into the computing unit for immediate processing, or read and stored for later use.

Input Mechanism Survey data may be entered into the computer in various ways. The early computers typically relied on card readers for the basic data input. This equipment operated by reading and transmitting the information contained on punch cards. More recently this method has given way to inputs from tape drives. These are similar in concept to the home tape recorder with two spools geared to run forward or backward, and magnetic heads between to record, read, or erase electronic impulses on the tape. While both the tape drive and the card reader perform similar tasks, the former carries out the operation much faster. It is still common, however, for data to be recorded on punch cards, and then transferred to tape via auxiliary equipment before entering the main computer.

Output Mechanism The first computers provided output to the user by punching cards. This system was replaced by printing machines tied into the computer and by tape drives which could store output information for later use on auxiliary equipment. In recent years, great strides have been made in eliminating the inefficiencies of earlier equipment. The new output technology has been particularly helpful in survey analysis, where input and output operations are relatively large compared to internal processing in the computer.

Storage The storage or memory capacity sets limits on the size and complexity of survey analysis. Computers require space to store programs, the data from each sample case, and the final results, as well as for intermediate operations. Because computers differ greatly in their storage capacity, it is important that this and other basic questions about data processing be raised with a specialist in the earliest stages of the survey.

An important advance in storage came with the development of the *disc* and random access to stored data. Consider the case of survey data stored on magnetic tape in the order of the interview numbers (respondent identification). This system does not provide equal access to each case on the tape. In order to reach a sample case with a low number, such as 0001, and another with a high number, such as 9999, it is necessary to run much of the tape from one reel to another so that the cases in question will pass the reading head. Present technology, which permits equal access to data in certain types of disc storage, greatly facilitates the data processing operations used in survey research.

Both the tape and the disc are portable, and may be moved with data from use to storage and back, and from one computer installation to another. Punch cards also remain a viable method of storing and transporting data, but they are seldom used with large computers.

The Computational Unit   The computational unit is the central nervous system of the various coordinated parts in a computer system. Through it, the desired operations are carried out. Computers are designed to perform such fundamental operations as reading, storing, adding, subtracting, dividing, comparing, erasing, transferring, and writing. The precise operations performed depend on the capacity of the equipment and on the instructions fed to it through the program. It is the responsibility of the programmer, who may also be one of the study directors, to provide instructions which will generate the information necessary for analysis.

Most computers used in survey analysis operate on a *binary system*. This is a numerical system capable of distinguishing between only two options: is or is not, or, in numerical terms, 1 or 0. As a result, most operations are carried out in binary arithmetic, with the original data translated from the decimal to the binary form, and the results translated back to the decimal system. Since the translation is carried out automatically, the survey analyst may never be confronted with the binary form in a user-oriented computer installation.

Conclusions   What attitude should the survey analyst adopt toward the use of high-speed computers in data processing? Any answer must take account of several factors: the availability of equipment and programming facilities, the budget available for analysis, the size and complexity of the projected operations, and the analyst's ability to work with this medium. Clearly, if the sample size is large and the analysis plans complex, if modern equipment and programming facilities are available, if there are ample funds for analysis, and if the analyst can translate his or her needs into appropriate requests as well as understand the output, there are overwhelming advantages to the use of the computer. If, on the other hand, there is no computer available, or if there is no budget for programming and computer time, or the study calls for relatively few simple tables, or the researchers do not have the time to learn how to relate to computer technology, the only or the best alternative may be to complete the analysis on conventional equipment. Each situation must be judged on its own merits. The best solution for a survey conducted in a well-equipped university in the United States may be totally inappropriate for another survey completed in India, Malawi, or Peru.

Several points should be remembered in comparing the relative advantages and disadvantages of computers versus conventional equipment. First, while computers are ultimately much faster and more versatile in processing data, they require a fair amount of time and expense for programming and other preliminary steps. These drawbacks will be reduced to the extent that the data requirements can be met by standard, "package" programs. Still, study directors have sometimes commented that the apparent advantages of computers were more than offset by programming costs and the delays involved in preparing the data, testing, and correcting programs, and in obtaining the necessary computer time.

Second, when the research staff has little or no experience in data processing, or when the plans for analysis are muddy or incomplete, it may be wise to begin with conventional equipment, such as a counter-sorter. For students conducting their first survey, there are real advantages to gaining firsthand familiarity with the data through work on the sorter, the reproducer, and similar equipment. Mistakes are sometimes more evident and easier to correct when at least some of the processing is carried out in this way. For more experienced researchers, on the other hand, there are few advantages to working with conventional equipment when computers are available.

Third, a study director should never overlook the possibility of carrying out certain tabulations by hand. When the total sample size is small and when the required tables call for very simple tabulations, this may be the simplest and most rapid means of analysis. In most large surveys, hand tabulations are rarely used, or are limited to a few preliminary tables. In short, decisions about whether to rely on computers or conventional equipment, or even hand tabulations, cannot be made in the abstract. They should take account of all of the factors noted, but especially the analysis requirements of the study and the availability of equipment and programming facilities.

DEVELOPING AND CHECKING VARIABLES

Another basic step in preparing for analysis is the construction and checking of variables. Sometimes the researchers will have had enough experience with the data in earlier studies to incorporate variables directly into the codes. Similar suggestions may arise from comparable studies published by others. Even with this coded information, it is usually worthwhile to carry out some checks on their general plausibility and their consistency with other information, whether in the same survey or from other sources. If, for example, the results from an income variable are inconsistent with findings from other surveys or governmental sources, it would be worthwhile to explore the sources of these differences, such as varying definitions of income or errors in data processing. Even if no errors or major discrepancies are found, the review provides a quality check on earlier stages of the work. Failure to take this step may lead to discarded tabulations or a time-consuming search for plausible explanations when the analysis is in its final stages.

Building New Variables

Not all variables to be used in the analysis can be completed during the original coding. Many composite variables, positional measures, and other indices are best handled in intermediate steps after the tabulations are available.

Composite Variables   Some variables are so complex or so poorly defined when the coding begins that it is advisable to leave the classification of sample cases on these measures until later. For example, the variable of life-cycle stage is a complex measure requiring information on age, marital status, and the presence and age of children. In most surveys, it would be better to code the required information on these single variables during the main coding stage, and to leave the construction of the life-cycle variable until the first stage of analysis. In this case, the punch cards representing individuals and families would be sorted on age, marital status, the age of the youngest child, and perhaps other variables to form the groups for each life-cycle stage. The resulting information, as noted earlier, could then be gang-punched in a blank column with a different punch for each category of the new variable. Even if such a composite measure had been handled during coding, it would still be advisable to carry out consistency checks involving the various components. Hence there would be no savings, and probably some loss in time, in constructing the more complex measure during coding.

Positional Measures   A positional measure is one which shows the relative standing of an individual or group on a quantitative variable. Some examples would be the quartile (top quarter, bottom quarter, etc.) position of individuals on the distribution of grade point averages, intelligence, or aptitude test scores, income, rent payments, or house value variables. With these measures, positional scores for individuals or groups cannot be assigned until the data for all the sample cases are available and tabulated.

Suppose, for example, that we wish to assign each sample case to quintile groups on family income. The resulting measure would show the top 20 percent, and so on, with regard to family income. To form the quintile groups, the punch cards would be sorted on the income variable so that they were arranged in ascending order. The resulting distribution would then be divided into five equal groups. The fifth of the cards with the highest income would be the top quintile group (top 20 percent), the next fifth the second quintile, and so on through the bottom fifth, which v/ould be the bottom quintile. Each of the five groups so formed would be gangpunched 1, 2, 3, 4, or 5 (or 9 for N.A.) in a blank column to form a new variable, income quintile position. A comparable operation could be carried out on a computer.

Other Indices   Finally, survey analysis will often require other indices or variables that can be defined operationally only after preliminary analysis and interpretation of the findings. Some of these may represent efforts to operationalize such abstract concepts as social class, occupational prestige, and ranking on traditionalism/modernity. The techniques for constructing appropriate indices may be straightforward, such as summing the number of *correct* answers on questions considered relevant to the concept. Often, however, the process is highly complex and time-consuming, requiring numerous theoretical assumptions as well as results showing the frequency distributions and the interrelationships among variables.

Scales   The process of building indices can be illustrated with a series of attitude and opinion items found in the specimen questionnaire (Chapter 6, Questions 36 to 43). The purpose of these items, as noted earlier, was to explore the possibility of developing combined indices of trust and modern experiences. The first step in determining the feasibility of combining two or more items into a single scale was to see whether they were closely related to each other. For this, the study directors obtained the coefficients of correlation (see Chapter 11) between all possible pairs of the eight items mentioned.[1] They assumed that if several items with a similar content showed reasonably high correlations with each other (such as .60 to .80), and higher correlations with each other than with other items, they would be justified in combining them into a single scale. On a priori grounds, they expected the following items to "hang together" in a single cluster representing *interpersonal trust:*

> **Q.36**   First, some say that if you aren't careful, people will take advantage of you. Others say that there is nothing to worry about because most people will not take advantage of you. How do you feel about this? (CODE NAME: TAKE ADVANTAGE)

---

[1] The procedure described here is a simplified version of a method of scaling discussed in most textbooks on psychometrics. More complex procedures, such as factor analysis, could also be applied.

Q.37 How do you feel about other people in this city—do you think that most people here can be trusted, that you can't trust most people here, or what? (CITY TRUST)

Q.38 Still thinking about the people you know in this city, do you feel that they will take the time to help others, or that they are too busy taking care of themselves? (HELP OTHERS)

Similarly, they expected to find reasonably high correlations among the following items dealing with *modern experiences:*

Q.39 What is the farthest that you have ever travelled from this city? (TRAVEL CITY)

Q.40 Suppose that you had enough time and money to go on a long trip, anywhere you wanted. Where would you like to go? (TRAVEL DESIRE)

Q.41 Have you ever flown in a plane? (AIR TRAVEL)

Q.42 Do you ever listen to the radio or watch television? (RADIO-TV)

Q.43 Do you ever read newspapers or magazines? (READING)

The results of the preliminary analysis showed that all three of the items dealing with interpersonal trust showed fairly high correlations with each other (average .70), and higher correlations with each other than with any of the remaining items. The fact that these correlations were not at very high levels, such as .95, suggested that the items were tapping somewhat different parts of the same general domain. On the basis of these findings, the analysts decided to combine the three trust items into a single scale. This was done in the following manner:

1 A preliminary check was made to determine the number of "don't know," "not ascertained," or "other" responses for the three items involved: TAKE ADVANTAGE, CITY TRUST, and HELP OTHERS. The results showed that no more than a total of 1 percent fell into these categories on any of the items.

2 The computer was instructed to remove all the cases coded "don't know," "not ascertained," or "other" on *any one* of the three items. The same step could easily be carried out by hand on a counter-sorter. As a result of this screening, a total of 2.7 percent of the respondents were excluded from the index, and coded as N.A., (0).

3 The computer was further instructed to reverse the codes for the responses on CITY TRUST and HELP OTHERS, as indicated in Table 10-2. The reason for this step was to arrange the data so that the responses on all three items would run in the same direction of trust. In the original form, a "1" on TAKE ADVANTAGE represented an indication of distrust, while the same code on the other two items indicated trust. The advantage of reversing the codes on CITY TRUST and HELP OTHERS rather than TAKE ADVANTAGE is that it will produce a scale in which higher scores

Table 10-2 Numerical Components of Interpersonal Trust Scale

Item	Response categories	Original code	Weight in index
TAKE	Take advantage	1	1
ADVANTAGE	Mixed	3	2
	Do not take advantage	5	3
CITY TRUST	Most can be trusted	1	3
	Half and half, etc.	3	2
	Most cannot be trusted	5	1
HELP OTHERS	Will take time to help	1	3
	Half and half, etc.	3	2
	Most cannot be trusted	5	1

indicate greater trust. In general, it is a good practice to have the values of scales increase in the direction of the variable described by the title, in this case *Interpersonal Trust.* As a result of this change, higher scores on the scale will mean greater trust. This is a minor point, but one which facilitates the interpretation of survey data, particularly when there are numerous variables.

4 The scale itself was constructed by having the computer sum the recoded values for each respondent on the three trust items and enter the total as a new variable. In this case, the score for any individual could range from a minimum of 3 (1 on all three items) to a maximum of 9 (3 on all three items).[2] If the analysis makes use of punch cards, the scores for the new variable could be punched in a blank column.

The preliminary results on the items thought to represent *Modern Experiences* were somewhat more complicated. Four of the items, TRAVEL CITY, AIR TRAVEL, RADIO-TV, and READING, showed reasonably high correlations among themselves, and higher correlations with each other than with the trust items. But the fifth item, TRAVEL DESIRE, presented two problems: (1) it correlated only about .20 with other items; and (2) it showed a 15 percent rate of nonresponse. Further checking revealed that many respondents with low levels of education were unable to adopt the hypothetical frame of reference required by this item. Because of these difficulties, and particularly because of the 15 percent item nonresponse, TRAVEL DESIRE was dropped from consideration.

The basic steps involved in constructing the *Modern Experiences* index were the same as those for *Interpersonal Trust.* Respondents who failed to answer any of the four component items (a total of 3.5 percent of the sam-

[2] Methodological purists might object to the practice of combining survey responses in this manner. Debates about the proper form of scaling items such as those under consideration have raged for years, and have produced no consensus, particularly among practitioners of survey research. While the general method outlined here might violate some of the canons of rigorous pychometrics, it has proven helpful as a data-reduction device in many surveys.

ple) were again separated and coded N.A. on the index. One important difference, however, arose from variations in the coding of the original items. TRAVEL CITY, for example, was coded with a scale of four points, while RADIO-TV had, in addition to the Yes-No filter, a six-point frequency scale. If the responses were summed in their original form, those with more scale points would de facto be given greater weight in the index. In order to meet the study directors' specification that each item carry equal weight, it was necessary to recode the four sets of responses. Recoding was also necessary to have the responses run in the same direction. Table 10-3 shows the details of the recoding, which produced an index with a maximum score of 12 and a minimum score of 0. Higher scores indicate greater exposure to modern experiences, such as the mass media and travel. These indices will be seen again in Chapter 11.

In sum, this chapter has reviewed the major tasks commonly carried out as preliminaries to statistical analysis, and the types of equipment used in

Table 10-3 Numerical Components of Modern Experiences Scale

Item	Response categories	Original code	Weight in index
TRAVEL CITY	Never left city	1	0
	Place in same province	2	1
	Place outside province	3	2
	Place outside country	4	3
AIR TRAVEL	Yes	1	(Coded below)
	No	5	0
	Once	1	1
	Twice	2	1
	Three times	3	2
	Four times	4	2
	Five times or more	5	3
RADIO-TV	Yes	1	(Coded below)
	No	5	0
	Every day	1	3
	2–3 days a week	2	3
	4–5 times a month	3	2
	Once or twice a month	4	2
	Once every 2–3 months	5	1
	Less often	6	1
READING	Yes	1	(Coded below)
	No	5	0
	Every day	1	3
	2–3 times a week	2	3
	4–5 times a month	3	2
	Once or twice a month	4	2
	Once every 2–3 months	5	1
	Less often	6	1

data processing. The preparatory steps include information storage; checks for impossible, improbable, or inconsistent responses; the types and problems of missing data as well as the more common methods for dealing with these problems; and the development and checking of variables to be used in subsequent analysis. The chapter concluded with an illustration showing how two composite indices are constructed from items in the specimen questionnaire.

FURTHER READINGS

Lansing, J. B., and A. T. Eapen, "Dealing with missing information in surveys," *Journal of Marketing*, October, 1959, pp. 21-27.

Moser, C. A., and G. Kalton, *Survey Methods in Social Investigation*, 2d ed. (London: Heinemann, 1971).

Parten, M. B., *Samples, Polls, and Surveys: Practical Procedures* (New York: Harper, 1950).

Chapter 11

Analysis and Reporting

One of the most challenging and rewarding tasks in survey research comes after the data have been collected and the coding, editing, and preliminary processing have been completed. This is the statistical analysis and interpretation of the data. The aim of this step is to provide a summary of the findings which satisfies the objectives of the research, and yet is as brief and understandable as possible. While a thorough discussion of statistical methodology and analytic techniques is beyond the scope of this book, we will review some of the more common tools and procedures for analysis, and suggest a simple strategy for their use.

BASIC METHODOLOGICAL ANALYSIS

Before moving ahead to the substantive findings, it is worthwhile to complete three basic and not very time-consuming methodological analyses of the data: the calculation of the sample yield and response rates, and a check for nonresponse bias. We suggest that these checks be made before the major tables are prepared or other calculations requested in order to detect unforeseen problems.

Sample Yield and Expansions

A useful starting point in the analysis of survey data is to calculate the sample yield and to expand the sample to obtain an estimate of the total number of dwellings, individuals, or other units in the population. The results can then be compared with previous estimates of the target population as a first analytic check.

Table 11-1 provides an example of these calculations for our illustrative survey of major cities in Pacifica. The second column, drawn from Table 5-1, shows the presurvey estimates of the total number of dwellings. The task now is to compare these figures with estimates based on the actual survey data. To do so, it is first necessary to tabulate the total number of occupied and vacant dwellings in the sample. The information required for these calculations is found on Card 01 (see Chapter 9). The total number of occupied dwelling units in the sample is obtained from column 26 of Card 01 by adding the total number of noninterviews (categories 5–9) and completed interviews (category 0). The number of vacant dwellings is simply the total number of cards punched 4 on the same column. Vacant addresses and nonresidential structures (categories 1–3 on column 26) can be disregarded in this check, for they are nonsample items which do not affect the sample of dwelling units (see Chapter 4). The number of occupied and vacant units, together with the total number of dwelling units, is shown in the third column of Table 11-1.

The procedure for obtaining population estimates on the basis of the sample data is straightforward. Since the sample called for the selection of one in every 265 dwellings in the target population, the estimate of the

Table 11-1 Sample Yield and Population Estimate

Domain	Estimated 1974 population in DUs*	Sample yield of DUs†			Population estimates‡	
		Total	Occupied	Vacant	Total	Occupied
1	324,000	1251	1200	51	331,500	318,000
2	98,000	375	360	15	99,400	95,400
3	107,000	406	390	16	107,600	103,400
4	120,000	448	430	5	118,700	113,900
Total	649,000	2480	2380	100	657,200	630,700

* Presurvey estimates from Table 5-1.
† Tabulated from Card 01, Col. 15 x Col. 26.
‡ Estimated from the survey; calculated by multiplying the sample yield in the previous column by the inverse (265) of the sampling fraction, $f = \frac{1}{265}$.

number of dwellings in the target areas is obtained by reversing this fraction and multiplying the sample yield by 265. The resulting estimates for the four domains of Pacifica are shown in the last column of Table 11-1. A comparison of these estimates with the expected yield in the second column reveals no major discrepancies. It should be remembered that the expected yields are themselves estimates based on projections from census counts which are several years old, and are therefore subject to error. More detailed analyses could be made (see Kish, 1965), but for present purposes we can be satisfied that no particular problems are presented by this expansion.

Response Rates

A second calculation, the response rate, should always be made before the final substantive tables are prepared. The response rate is the proportion of the eligible respondents in the sample who were successfully interviewed. In the case of Pacifica, the denominator is the total number of occupied dwellings, 2380. The numerator is the number of completed interviews which were coded and available for analysis (the number of cases coded 0 on column 26 of Card 01). With 2380 occupied dwellings and 2047 completed interviews, the response rate for the Pacifica survey is 86 percent, a good rate for a personal interview survey.

Nonresponse Bias

The question remaining is whether, even with an acceptable response rate of 86 percent, there is substantial bias arising from the noninterviews. Nonresponse bias can never be fully analyzed because, by definition, some of the necessary information is unavailable. Ideally, we would like to know if, on the crucial variables of the study, the nonresponse cases in the sample differed significantly from those successfully interviewed, but because we were not able to interview the former, we shall never know in a definitive way. Nevertheless, it is possible to use those bits of information available about the noninterviews to make appropriate comparisons between the two groups. In the Pacifica study, the interviews and noninterviews could be compared on two measures: *type of neighborhood* and *type of building structure*. If no significant differences appear in the comparisons made on these measures, the analysts can move ahead. If, on the other hand, most of the noninterviews are concentrated in a certain type of neighborhood, or live in different types of structures than the interview cases do, it might be necessary to apply the differential weighting system described on pages 273-274.

STATISTICAL ANALYSIS

Statistics provide the basic tools for summarizing survey data and for measuring the degree of association between variables and subgroups. The techniques commonly used in survey analysis can be described as simple, intermediate, and complex.

Simple Techniques

Survey analysts usually begin by describing the characteristics of the entire sample and the values obtained on the major analytic variables of the study. In the national urban survey, the following would be essentially descriptive of the sample: the age distribution of the heads of households, the composition of the families that they head, the proportion of adult men and women in the cities covered by the study, the proportion of the population comprised of children under six years of age, and the distribution of household heads across the various categories of occupations. Some key analytic variables would include the attitudes of household heads toward their neighborhood and toward their current financial situation, the employment status and unemployment rate of household heads, and family income over the past year.

The techniques most often used for describing the characteristics of the sample and the major study variables are frequency distributions and proportions, and measures of central tendency such as the mean and median. Measures of dispersion, such as the range or standard deviation, are also helpful as tools of descriptive analysis. It should be remembered that, while the data are based on the sample, they are useful only as they provide estimates about the population, and that they are subject to sampling errors. A discussion of sampling error for these and other estimates comes later in this chapter.

Frequency Distributions A simple way of reducing and summarizing data is through frequency distributions in either raw or percentagized forms. This technique involves showing the characteristics or categories of response for the variable under consideration (age, sex, occupation, attitudes, income), together with the number or percentage of sample cases falling into each category. If the sample cases carry differential weights, as would occur with unequal probabilities of selection, or from adjustments for differences in response rates, the distribution of weights rather than the number of sample cases should be used and percentages would be based on the weighted distribution. The simple distribution of sample cases is called the *marginals* for that item.

Table 11-2 shows the age distribution of heads of household in a weighted sample. The first column, called the stub, lists the categories of the variable being analyzed, age; the second and third show the number of sample cases and the distribution of weights for each category (the distribution of weights is seldom included in published tables). The last column is the percentage distribution based on the weights. If the data are unweighted, of course, the percentages would be calculated on the basis of the actual frequencies reported in the second column of the table.

Frequency distributions are particularly useful for survey data consisting of categories which cannot be treated as numbers, such as: race, marital status, reasons codes, attitudes, and occupations. But even with quantitative data such as age or income, meaningful subgroups can be designed, such as the various age or income categories.

Proportions Proportions offer another simple way of presenting survey data. Actually the proportion is a special case of the frequency distribution in which only a single characteristic or attribute is expressed as a fraction of the total, or 1.00 (e.g., .25, .60, .75). It may be converted to a percentage by multiplying that figure by 100 (e.g., 25%, 60%, 75%). Percentages are more commonly used and more easily understood than the decimal fraction. Some common uses of the proportion as a percentage are the unemployment rate and the percentage of households with preschool children. Table 11-3 illustrates this technique by showing the percentage of households in a national sample with various components of net worth. Once again, if the sample cases involve weights, the proportions or percentages should be based on the weighted data.

Measures of Central Tendency Measures of central tendency deal with averages or what is typical in a group. There are several measures of central

Table 11-2   Age Distribution of Heads of Households

Age of head of household	Number	Distribution of weights	Percentage distribution
Under 25	185	1,020	10
25–34	403	1,957	20
35–44	478	2,316	23
45–54	400	1,829	18
55–64	268	1,201	12
65 and over	313	1,677	17
All heads of household	2,047	10,000	100

Table 11-3 Percentage of Households with Various Net Worth Components

Component of net worth	Percentage of households with the component
Automobile	72
Owner occupied	
Nonfarm house	50
Farm	4
Other real estate	15
Business interest	7
Corporate stock	16
Liquid assets (cash; bank, savings, or credit union account; etc.)	73
Other financial assets	4
Debt of some type	68

Source: Data adapted from Katona, Lininger, and Kosobud (1962).

tendency, each of which rests on certain assumptions and has its own particular advantages and disadvantages. Those most commonly used in survey research are the arithmetic mean, the median, and the mode.

The *arithmetic mean* is the most widely used and easily understood average. It is obtained by dividing the sum of the scores or measurements on a given variable (age, income, installment debt) by the total number of sample cases for which this information is available. Thus, for example, the mean income in our urban sample would be obtained by adding the income of all the households and dividing by the total number of households for which income data are available. Missing data cases not completed by assignment procedures will normally be excluded from the calculations. The results would be distorted if the total sample size (n) were used to calculate means when some of the cases lacked income figures. With a weighted sample the mean would be computed by adding the product of income times the weight for each sample household and then dividing by the sum of the weights for the sample households covered.

The *median* is defined as the point above which and below which 50 percent of the cases lie. The median household income for the urban sample would be the income level which divides households into two equal size groups, half above that figure, half below. Calculations of the mean and the median based on the same set of data often yield different figures. The median income is usually lower than the mean income, because the mean is more affected by cases of extremely high income. Thus, it would make no difference to the median if a high-income household reported $50,000 or $500,000. Each would be above the median and counted as one case in that category either way. With the mean, however, the higher number could have a direct and significant impact on the result. For this reason, many survey researchers consider the median a more stable measure of central tendency for variables, such as income, which may have great variation in the sampling of extreme cases.

The *mode* is the score or measurement which occurs most frequently in a distribution. For example, this measure of central tendency is occasionally used in survey analysis to compare the ideal number of children reported by respondents in different social and cultural settings, to isolate the most typical patterns of personality characteristics or childrearing practices in studies of national character, and to show the most common responses occurring on quantitative attitude scales or to questions of candidate preference in election polls. Because the mode is the most crude of the three measures of central tendency reviewed here, it should be used sparingly. Its most direct application arises when the analysis focuses on the most common case rather than taking into account all cases in the distribution.

Dispersion Measures of dispersion are those which show the degree of variability or homogeneity in the data obtained on a certain characteristic. The importance of the concept of dispersion can be illustrated with information from the Pacifica survey. The survey data show that families in two of the cities had identical mean incomes of $7,000. At first sight it might appear that there is little difference in income between the two cities. But this is true only for this measure of central tendency. Closer analysis shows that in one city, family incomes generally fall between $5,000 and $10,000, while in the other, most are below $5,000 but with a few families reporting very high incomes ($50,000 to $100,000). Certainly any serious analysis of the socioeconomic situation would have to take account of this evident difference in dispersion.

At the simplest level, some notion of dispersion can be gleaned from an inspection of the frequency distribution of the variable in question, such as income. A graphic presentation of the results will show whether most of the figures cluster around the mean or median, or whether there is considerable spread in the data. A somewhat more precise, but still very limited, measure of dispersion is the *range*—the difference between the highest and the lowest figure in a distribution. In the Pacifica example, the range for family income would be the difference between the lowest and the highest income reported, such as $100 to $100,000 in one city and $600 to $25,000 in another. While the range is simple to calculate and interpret, it is only a rough measure of dispersion. The main limitation is that, because it relies on information from only two cases, it ignores the distribution between the highest and the lowest figures.

When the data involve quantitative measures such as dollars, feet, or pounds, often the most useful measure of dispersion is the *standard deviation,* described in Chapter 4. In the earlier example, the larger standard deviation for household income in one of the cities would quickly alert the analyst to an important difference in the data, even though the means were similar. Other measures of dispersion, such as the quartile deviation, are discussed in standard works on elementary statistics (cf. Blalock, 1960).

Intermediate Techniques

In most surveys, the goals of the study call for more than the kinds of descriptive analysis just reviewed. The next set of techniques is called into play when it is necessary to show the relationship between two variables, and perhaps express this relationship in mathematical form. As we move to this intermediate level of analysis, it is worth emphasizing the crucial role of some guiding theoretical framework, particularly in explanatory studies. Survey analysis requires much more than the routine application of statistical techniques. In fact, even to apply these techniques properly, the analyst must have some sense of where the analysis must go, including at least rudimentary hypotheses about casual influences in the data, and sufficient theoretical understanding to suggest alternative explanations of the data. As we shall see shortly, assumptions about the likely direction of influence are often involved even in designing a simple, two-variable table. Throughout the analysis, therefore, statistical and other analytic techniques must be part of a broader strategy of analysis.

Two-way Tables One of the most common and useful means of showing the relationship between two variables, such as education and age, is the *two-variable frequency table.* In its most typical form this table shows the frequency of one variable by categories of another. When the analysis makes assumptions about causal influence, the variable to be explained is known as the *dependent variable,* while the other is variously identified as the *independent, explanatory,* or *control variable.* Usually the table is set up so that the frequency distributions of the dependent variable are shown for various categories of the explanatory variable.

Table 11-4 shows the relationship between education, the dependent variable, and age, the explanatory or control variable. The assumption underlying this table is that a respondent's education can be explained, in part, by his or her age. In other words, we hypothesize that there is something unique in the experience of different age groups which has consequences for the educational attainment of the household heads, and not vice versa. The findings in the body of Table 11-4 are thus presented with education as the

Table 11-4 Education of Household Heads by Age Groups
(Distribution of household heads)

Education	Total	Age groups 18–34 years	35–54 years	55 years and over
No high school	30%	10%	26%	53%
High school*	46	55	51	31
College†	24	35	23	16
Total	100%	100%	100%	100%
Number of cases	2047	588	878	581

* Also includes people with less than nine years of schooling who have received other noncollege training.
† Includes all who attended college, whether or not a degree was received.
Source: Adapted from Katona, Lininger, and Mueller (1963).

dependent variable; that is, the percentages refer to levels of educational attainment rather than the different age groups. These findings show that educational attainment, measured by years of schooling, is progressively lower as the age of household heads increases. But the fact that there is wide variation on educational attainment within each age group suggests that other factors besides age must be at work.

Two-variable tables using the same data could also be set up in two other ways, depending on the hypothesized direction of causal influence. Suppose, for example, that a new theory emerged which said that education has an effect on longevity—that people with more education live longer. In testing this hypothesis, we would organize the table so that age was the dependent variable and education the control condition. The main differ-

Table 11-5 Distribution of Household Heads by Education and Age Groups

Education	All household heads	Age groups 18–34 years	35–54 years	55 years and over
No high school	30%	3%	11%	16%
High school*	46	16	21	9
College†	24	11	9	4
Total	100%	30%	41%	29%
Number of cases	2047	588	878	581

* Also includes people with less than nine years of schooling who have received other noncollege training.
† Includes all who attended college, whether or not a degree was received.
Source: Adapted from Katona, Lininger, and Mueller (1963).

Table 11-6   Shares of Income and Distribution of Households, by Income Groups

Income group	Shares of income	Distribution of households
Under $3,000	2%	14%
$ 3,000–4,999	4	12
$ 5,000–$7,499	10	16
$ 7,500–$9,999	13	16
$10,000–$14,999	29	24
$15,000 and over	42	18
Total	100%	100%

Source: Adapted from Katona, Mandell, and Schmiedeskamp, *(1970).*

ence would be that the rows and columns of Table 11-4 would be reversed so that the findings in the body of the table would refer to age rather than education. A two-variable table can also be set up with no assumptions about the direction of influence. In this case, the analyst would prepare the table with each cell percentagized to the total population ("percentagizing to the corner"), as in Table 11-5.

Shares   Tables of *shares* show the proportion or percentage of the total of a numerical variable held by certain categories of individuals, households, cities, and so forth. The concept of shares is different from that seen in earlier tables. While the latter showed the distribution of household heads, shares tables indicate the distribution of a numerical variable associated with each household head, city, etc.

Table 11-6 illustrates the difference in the two approaches. The first data column in the table presents the *shares* of total income received by households subdivided by income categories. The final column shows the distribution of households. A comparison of these data yields statements about the concentration of income. Twenty-six percent of all households receive less than $5,000 per year, and their combined incomes represent only 6 percent of the total income of all households. At the upper level of income, 18 percent of families receive $10,000 or more; this represents 42 percent of total household income.

Tables of shares can be used with income, age, education, occupation of household head, or any other variable on which it would be of interest to have figures on the concentration by subgroup. Travel market analysts, for example, might find it helpful to know the shares of airplane trips taken by businessmen, government officials, and other population subgroups.

Measures of Association   In some studies, the analyst may be satisfied to say that there is a relationship between age and education for household heads, and to state the direction of that relationship. But more often he or she will want to add something about the *strength* of the association. In this case, it is useful to calculate statistical measures of the degree and form of relationship.

Coefficient of Correlation   One frequently used measure of association is the *coefficient of correlation,* expressed in scores ranging from .00 to 1.00. This measures the extent to which variation on one variable, such as age, is matched by systematic variation on another, such as education. A coefficient of .00 would indicate no systematic relationship, 1.00 a perfect relationship, and .55 a moderately strong relationship. Details on the types and appropriate uses of this technique can be found in most works on introductory statistics. As a measure of association, it is neutral regarding the possible direction of influence or impact. Thus, in the above example it would not measure the impact of age on education, or vice versa, but merely the extent to which age and education were correlated.

Scattergram   The technique known as the *scattergram* allows for the visual observation of the form and degree of association between two variables. It can be prepared by hand on graph paper or mechanically by a computer. A basic requirement is that each of the variables to be compared, such as age and education, be expressed in numerical form. The *scatterplot* is prepared by taking each case in the sample and plotting the pair of values for the variables under study (e.g., age thirty-five, eight years of education on interview number 1445). The results may fall into many different patterns, four of which are shown in Figure 11-1. The dots in Figure 11-1(a) show a positive correlation because higher values of variable x are associated with higher values of variable y. The correlations in Figures 11-1(b) and 11-1(c), on the other hand, are negative, that is, a higher score on variable x is associated with a lower score on variable y. The scatterplot in Figure 11-1(d) represents a situation of no correlation between the two variables.

Another difference that can be seen in Figures 11-1(a) and 11-1(b) is in the compactness of the clustering. Because there is less scattering of cases in Figure (b) than Figure (a), there is a higher degree of correlation between the two variables. Other aspects of the shape of correlations will be taken up shortly.

Regression   A simple regression equation, or *regression* for short, is a cousin to the coefficient of correlation. The difference is that the regression calls for the *form* of the relationship to be specified, while the correlation coefficient deals only with the *degree* of association. Figure 11-1 shows two possible forms that may be readily visualized. The solid lines in Figures 11-1(a) and 11-1(b) depict *linear relationships,* or those which can best be summarized with a straight line. Findings which showed that, in general, as age increases, education decreases, would involve such a linear relationship. Although many of the cases in the 11-1(a) and 11-1(b) are not on the regression line, it can be seen that this straight line best fits the data at hand.

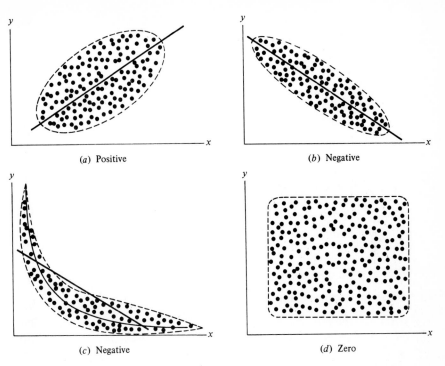

Figure 11-1 Four scattergrams showing different types and degrees of correlation.

In Figure 11-1(c), by contrast, the best-fitting straight line still leaves much to be desired. The scatter in this figure seems to show a *curvilinear relationship* between variables x and y. That is, the relationship between the two variables is much better summarized by a curved line than a straight line, as shown by the solid line in the figure. Such a pattern could occur, for example, with age and education. Curvilinearity would be seen if levels of education generally decreased with age, but at the same time there was a rapid drop in education levels up to age thirty-five, and only a gradual decline thereafter, as in Figure 11-1(c). Fortunately, statistical techniques are available to test for curvilinearity on a mathematical, rather than a visual basis.

Test Procedures Another set of intermediate techniques is found in the various methods for testing differences between groups on key variables. In analyzing survey data, it is often important to know if the differences observed between various sample groups arise from chance factors or from real differences in the population. In the Pacifica study, for instance, the analysts want to know if there are significant differences in the unemployment rates measured by the survey in San Pedro and Capital City, 7.5 percent and 5.2 percent respectively. Perhaps the most common approach to this problem is a procedure known as testing the *null hypothesis*. This begins with the hypothesis that no difference exists between the population groups being analyzed and that therefore any differences observed in the sample data arise from chance variations. To demonstrate that there are significant differences, it is necessary to *reject* this null hypothesis. The statistical techniques for carrying out such a test make use of the concepts of confidence levels and confidence intervals discussed in Chapter 4. If the differences observed in the data are greater than that which could be expected to occur by chance, the null hypothesis would be rejected and the differences would be assumed to be real, that is, statistically significant. In the Pacifica example, if rates of 7.5 percent and 5.2 percent could not be expected from sampling variability, this would mean that the differences in the unemployment rates observed in San Pedro and Capital City are likely to stem from genuine differences in the two cities rather than chance sampling factors. Through somewhat more complex methods, such as the analysis of variance, the same logic can be applied to testing for significant differences across three or more sample groups, such as occupational categories.

Complex Procedures

Only rarely in survey analysis will the search for relationships end with two-way tables or correlations between two variables. Several factors may push the analyst toward more complex analytic procedures. First, the hypothesized or observed relationship between the two variables may be difficult to interpret because of possible confounding by other variables. To demonstrate convincingly that age has a direct influence on levels of education, one would have to rule out other factors associated with age which might explain the drop in education. It may be, for example, that higher proportions of older people than younger people come from rural areas where either educational facilities were less available, or social norms favoring education were not as strong as in the cities. To rule out the influence of rural-urban background it would be necessary to take one or more of the steps to be suggested shortly. Second, the analyst may wish to show the joint or cumulative impact of two or more explanatory variables on the dependent variable. In explaining levels of income, the research team may want to find the four or five variables which, taken together, best account for income differences. Techniques such as multiple correlations or multiple regression analysis are well suited for this purpose. In this section we will review a few of the more common complex analytic techniques available to the survey researcher.

Three-or-more-way Tables   The most commonly used technique for determining the separate influence of several variables is the multivariate table.

This may take the form of a three- or four-variable or an even larger table. The major constraint in using such complicated tables, apart from the increased difficulties in interpretation, is the sample size. Three control variables such as age, sex, and education with five, two, and four categories each would require that the sample cases be divided into at least forty cells. Such division would be less of a problem if each cell received about one-fortieth of the cases, but this pattern almost never occurs in survey analysis. Before designing complex tables, therefore, it is always advisable to do some preliminary checks on the number of cases likely to be available for analysis in the resulting cells.

The uses of a multivariate table can be illustrated with data from the hypothetical survey of Pacifica. Because of the government's concern with population policies, the study directors are particularly interested in exploring the data on ideal family size. Questions 32 and 32a in the questionnaire were designed to obtain the ideal family size preferences of married respondents (see Chapter 6). The analysts now wish to test a key hypothesis about the relationship between modern experiences and ideal family size. Following current thought in the field, they predict that ideal family size will *decrease* with increasing exposure to modern experiences such as travel and the mass media. A two-way table relating the Index of Modern Experiences (see Chapter 10) to ideal family size provides striking confirmation of this hypothesis: the average number of children desired drops sharply as one moves from the lowest to the highest levels of modern experiences.

But the analysis cannot stop there. Anyone remotely familiar with the field of population could think of several alternative explanations for the same findings. The most obvious is that it is not modern experiences as such, but rather education, which accounts for the decline in preferred family size. A two-way table relating education to the dependent variable supports this hypothesis. Another possibility is that the real explanatory variable is age, rather than either modern experiences or education. According to this argument older people, having been raised in times when the society valued larger families, still carry this preference. At the same time, older people would have had fewer opportunities for modern experiences, and thus would score lower on the related index. Some might also say that because older people have already had their families, and because their *actual* family size is relatively large, they will rationalize their ideal family size preferences to bring them into line with this actual size. The number of alternative explanations could go on. In fact, part of the challenge and fun of survey analysis is precisely to concoct explanations which might demolish one's own favorite hypotheses, and to test them with real data.

Since the main challenges to the hypothesis about modern experiences and family size come from age and education, the best course is to test the hypothesis with controls for both variables. The sample size, in this case, is sufficiently large to permit the construction of a table with these controls.

Table 11-7 shows the mean family size preferences of married respondents cross-classified by modern experiences, age, and education. To avoid unnecessary cells, the Index of Modern Experiences was dichotomized by dividing the total distribution on this measure at the median. Anyone falling on or above the median was classified as "high," those below as "low." Age and education were divided three ways to permit a better test of the influence of these variables than would be possible if they were dichotomized. The result is a table with 18 cells. To simplify the presentation, we have assumed that there are no missing data cases on these variables, a rarity in analysis.

A careful inspection of the results in Table 11-7 points to several significant conclusions about the relationship between modern experiences and ideal family size:

1 Most importantly, there seems to be an interaction between modern experiences, education, and ideal family size. Modern experiences appear to exercise an influence on ideal family size mainly for those with only a primary school education, and to a lesser degree for those with high school education. No influence is seen with the college-educated groups. This pattern occurs with all three of the age groups as shown in the table. For example, among those between 18 and 34 who have had only a primary education, the average family size for respondents with "high" modern experiences is 3.3, while it is 4.3 for those in the "low" group. By contrast, among those with a college education in the 18-34 group, there is basically no difference in ideal size for the "high" and "low" groups on modern experi-

Table 11-7 Ideal Family Size Preferences (Mean Scores) of Married Respondents by Modern Experiences, Age, and Education*

Modern experiences	Primary	High school	College	$\bar{x}$	n
18–34 years					
High	3.3	3.2	3.1	3.2	375
Low	4.3	3.7	3.3	3.6	213
n	59	323	206		588
35–54 years					
High	3.8	3.7	3.6	3.7	450
Low	4.9	4.3	3.7	4.3	428
n	228	448	202		878
55 years and over					
High	4.5	4.3	4.2	4.4	198
Low	6.3	5.7	4.4	5.8	383
n	308	180	93		581

* Hypothetical data based on responses to Questions 32 and 32a in the specimen questionnaire, Chap. 6.

ences. The most plausible interpretation of this finding is that both college education and modern experiences have a similar, but not additive, effect on preferences for numbers of children. In other words, individuals who have either a college education or a fair amount of travel and exposure to the mass media will want, on the average, slightly over three children. A similar pattern, though with higher average figures, is seen in the other two age groups. The repetition of these patterns suggests that a real relationship exists, that the results are not just a chance outcome. Application of tests of significance to the observed patterns would add important evidence for the conclusions.

2   With controls for education, the negative association between modern experiences and ideal family size holds up very well across the three age groups. This relative independence of modern experiences and age can be seen by examining the differences between the high and low groups in each of the three education columns. Among those with primary education, the average family size of the high group on modern experience is consistently below that of the low group, even though there are average differences *across* the three age groups. The same pattern is seen among those with high school and college educations. Although some of the differences between the high and low groups may not be statistically significant because of the small numbers involved (e.g., only 59 cases of those 18–34 with primary education), the consistency of the differences across the nine possible comparisons strongly supports the notion that the effects of modern experiences are not explained away by the age of the respondents.

While this table may seem complicated, the pattern of the findings is rather clean by comparison with most survey data. In a typical large study, a few of the differences between the high and the low groups would be much less clear-cut, and there might even be the odd one which was reversed. But the example at least provides a general idea of the continuing process by which the search for plausible explanations is carried across the most difficult challenges until the analysts (or their critics) are satisfied that the interpretations are sound. Sometimes, of course, this very process of multivariate analysis leads to the discarding of the original hypothesis and the formulation of new interpretations. A working hypothesis is often nothing more than a key to the realm of interpretation. What happens after the analyst enters may lead to a complete recasting of the explanatory framework.

Residuals   An interesting, and sometimes neglected, technique which can be applied to numerical dependent variables is the calculation of *residuals*.[1] This is basically a two-step procedure for dealing with the effects of confounding variables. The first step consists of adjustments to correct for

---

[1] For applications of this technique see Lininger (1963) and Kreinin and Lininger (1963). The former also analyzes the data using multiple regressions and multiple classification analysis.

the influence of the confounding variable, such as education in the earlier example. The second step involves the tabulation of the adjusted dependent variable by the desired explanatory variable.

Let us take another example from the urban survey of Pacifica. In this case we wish to show the influence of two explanatory conditions, interpersonal trust and modern experiences, on unemployment. Once again there is good reason to believe that the relationship between these two explanatory conditions is confounded by other variables related both to them and unemployment. Preliminary analysis of the survey data and earlier studies suggest that age and occupation will be important controls for these tabulations. In Pacifica, there are higher rates of unemployment for both very young and older workers, and for certain categories of occupations, such as manual workers. While the effects of these variables could be explored through multivariate analysis along the lines discussed earlier, it is also possible to adjust the dependent variable (unemployment) to remove the effects of age and occupation. In essence, through the use of residuals, we would be developing a new dependent variable, that is, weeks of unemployment minus the influence of age and occupation.

The procedure for calculating residuals is not complicated. The first step is to classify the sample by age and occupation (e.g., manual workers aged 18-34, clerical workers aged 35-54, etc.). Next, the average unemployment for each age-occupation group is calculated. Then the *average* unemployment of the group to which it belongs is subtracted from the *actual* unemployment experience of each individual respondent, and the difference is entered as a new variable. Thus a household head whose weeks of unemployment are exactly equal to the average for his or her group would be scored a zero on the new variable. A person whose experience was one week above the average would have a positive residual of $+1$, while a person two weeks below the average would have a negative residual of -2, and so on. The new variable, which will average zero over the whole sample, can then be tabulated against the relevant explanatory conditions.

Table 11-8 shows the relationship between the adjusted dependent variable, unemployment, and the two explanatory variables mentioned earlier: interpersonal trust and modern experiences. The components and construction of the indices measuring these last two variables are discussed in Chapter 10. The results suggest that there is a systematic relationship between unemployment and modern experiences, but probably not between unemployment and interpersonal trust. With the adjustments for age and occupation, unemployment is relatively high for the lowest quartile group on modern experiences ($+1.2$ weeks), and relatively low for the highest quartile (-1.9 weeks), with the two middle quartiles between these figures. Tests not reported in the table show these differences to be statistically significant.

Table 11-8 Adjusted Unemployment by Modern Experiences and Interpersonal Trust

Index	Mean unemployment, adjusted, weeks*
Modern experiences	
Lowest quartile	+1.2
Next lowest quartile	+0.3
Next highest quartile	+0.5
Highest quartile	−1.9
Interpersonal trust	
Lowest quartile	+0.3
Next lowest quartile	−0.1
Next highest quartile	−0.4
Highest quartile	+0.2

* Unemployment residuals, adjusted by age-occupation unemployment means.
Source: Hypothetical data from Pacifica survey.

With the index of interpersonal trust, on the other hand, there is little variation in mean adjusted unemployment across the quartile groups, and no clear pattern of relationship. We would tentatively conclude that with controls for age and occupation, greater modern experiences are associated with lower unemployment, but that with the same controls, unemployment appears to be unrelated to trust. Given the importance of education to most socioeconomic variables, an essential next step would be to carry out similar controls for educational levels.

A major shortcoming of the residuals technique is that the first-stage variables, such as age and occupation in the example, are given priority in explaining variation in the dependent variable. The second stage variables, such as modern experiences and trust, are thus in a statistically disadvantaged position. Nevertheless, this method can be helpful as a conservative means of testing the persistence of relationships. For example, if the relationship between modern experiences and unemployment stands up with this kind of control, it is likely to emerge as strong, or stronger, with multivariable tables or other less conservative methods. Another advantage of residuals is that they can be calculated on conventional data processing equipment such as a cardsorter, keypunch, or reproducer and tabulator. For those lacking access to computers, this can be an important asset.

Multiple Regression *Multiple regression analysis* is the logical extension of the simple regression technique discussed earlier. The technique assumes that there is a numerical dependent variable which is to be explained, and that there are two or more variables whose purpose is either to explain variations in the dependent variable or to control for the influence of confounding conditions.[2] The basic aim of multiple regression is to develop an equation in which the predictor variables give the best possible accounting of the dependent variable. Suppose, for example, that we wish to use age, education, and modern experiences as predictors of unemployment. Through regression analysis we would seek a solution to the predictive equation from the study data. This would include regression coefficients with standard errors for each of the predictor variables, and a multiple correlation coefficient showing the effectiveness of all the predictor variables in accounting for the dependent variable. Through this method, which is too complex to review in detail here, it is possible to determine the relative power of each predictor in explaining the dependent variable, and the combined predictive power of all the explanatory variables. An explanatory variable with a coefficient of zero would have little or no effect on the dependent variable.

The main advantage of both multiple regression and multiple correlation in comparison with multivariable tables is that the former methods provide a direct means of assessing both the individual and combined impact of several predictors. Whereas tables such as the version with three control variables used in this chapter are cumbersome to use and complex to summarize, multiple regression equations are relatively simple and highly compact. Their main limitations are that they are difficult or impossible to use with data that are not well adapted to correlational analysis (e.g., occupational categories), and they do not provide the detail about subgroups available in multivariate tables.

To make the most effective use of multiple regression, it is advisable first to explore hypotheses in two-way or multivariate tables. The variables which finally enter into the regression equations should be well scouted and included for a particular purpose. Attempts to use this technique as a shortcut to analysis by including several dozen "potentially interesting" variables usually confirm the adage "garbage in, garbage out." Whatever the sophistication of modern computers—and they are particularly well adapted to regression analysis—there is no substitute for the human brain in the interpretation of survey data.

Multiple Correlations   Some study directors concerned about the presentation of survey findings to nontechnical readers may prefer to use multi-

---

[2] Neither the dependent variable nor the explanatory variables are required to have more than two scale points. Even categorical information, such as "yes" or "no" answers on a questionnaire, can be converted into "dummy variables" by assigning them scores of 1 and 0. Though there are some disadvantages to this procedure, it allows regression analysis, as well as correlational analysis, to be used with variables which might not seem at first sight to be amenable to these techniques.

ple correlation coefficients rather than multiple regression equations. From a mathematical standpoint, the two techniques are very similar; in fact, multiple correlations are calculated as part of the procedures for multiple regressions. The advantage of the correlations, however, is that they can be presented in a way which is more intelligible to the lay reader.

Their use can be illustrated with an example in which the index of modern experiences is the dependent variable. In this analysis, we would like to search out the main characteristics accounting for differences in travel and exposure to the mass media. Preliminary tabulations suggest that the best candidates for explanatory variables are age, education, income and interpersonal trust. Either by hand or with a computer, it is possible to begin with one variable—usually the one showing the highest correlation with the dependent variable—and then explore the independent contribution of the other variables. In the present case, education is clearly the best single predictor of modern experiences, with a correlation coefficient of .45. Beginning with education, we then attempt to determine the addition made by age, which correlates .36 with modern experiences. When these two variables are taken together as predictors, the multiple correlation with modern experiences increases to .52. This suggests that, while age has an independent contribution as a predictor, it also overlaps with education. When income is added to education and age, the correlation climbs to .55, and when interpersonal trust is added to the previous three predictors, the result is a multiple correlation of .57. A table showing the original (zero-order) correlations between the predictors and the dependent variable, and the increment added by each predictor to the multiple correlation, is often more readily understood than a presentation of regression coefficients.

Generalized Sampling Errors

Throughout this book it has been emphasized that survey findings are subject to errors arising from various sources. The three major sources in most surveys are (1) sampling variability; (2) missing data, whether from non-interviews or nonresponse to specific items; and (3) inaccuracies or distortions in the information supplied by respondents. Errors arising from missing data have already been discussed in Chapter 10, while those stemming from the questionnaire or the interview process were treated in Chapters 6 and 7. Unfortunately, it is nearly impossible to make numerical estimates of the errors attributable to missing data and inaccurate information. For this reason, the following discussion deals only with sampling errors.

Estimates of the sampling errors for key variables should form an integral part of the analysis and interpretation of survey data. As noted in Chapter 4, it is a relatively simple matter to calculate such errors for simple random samples. When the study involves a more complex sample, such as multistage area sample of the type described in Chapter 5, the determination of errors is considerably more complex. Both for this reason and for simplicity in the presentation of study findings to a large audience, survey analysts often publish tables of generalized sampling errors along with the findings. These are tables showing the approximate errors to be expected for most variables with groups or samples of different sizes.

Two generalized tables of sampling errors at the 95 percent confidence level are presented below. These were adapted from the Surveys of Consumer Finances of the Michigan Survey Research Center (cf. Survey Research Center, 1960; Katona, Lininger, and Mueller, 1961).

Table 11-9 presents sampling errors for *percentages* with subgroups of different sizes. The uses of this information can be illustrated with data on family size preferences. On the basis of the subgroup of about 400 cases, the directors of the Pacifica survey estimated that 70 percent of the household heads twenty-five to thirty-four years of age in that country wanted no more children. To place this finding in proper statistical context, it is important to know the sampling error of this estimate. Table 11-9 will give us an approximate figure for percentages and samples of this size. Interpolation from the columns for 300 and 500 interviews and the row for a reported percentage of 70 percent shows that the sampling error of this estimate is 7 percentage points. Thus there would be 95 chances out of 100 (the confidence level) that the population value of household heads twenty-five to thirty-four years of age wanting no more children falls between 70 percent plus or minus 7 percentage points, or 63 to 77 percent. Conversely, in 5 times out of 100 the percentage would be expected to lie outside this range.

Very often the analyst is more concerned with *differences* between survey estimates obtained with two or more groups than with the error attached to any single estimate. Here, too, it is possible to use a generalized table to

Table 11-9 Approximate Sampling Errors of Survey Findings
(The chances are 95 in 100 that the value being estimated lies within a range equal to the reported percentage plus or minus the number of percentage points shown below)

	Number of interviews					
Reported percentage	3000	1000	700	500	300	100
50	3	4	5	6	8	14
30 or 70	2	4	5	6	7	13
20 or 80	2	4	4	5	6	11
10 or 90	2	3	3	4	5	8
5 or 95	1	2	2	3	4	

obtain an approximate notion of which differences will be significant. Table 11-10 presents the percentage points needed to have significant differences at the 95 percent confidence level with different sample sizes. For example, in the sample subgroup of about 200 cases in Pacifica, 35 percent of household heads under twenty-five years of age reported that they wanted no addi-

Table 11-10   Sampling Errors of Differences*

[Differences required for significance (95 percent probability) in comparisons of percentages derived from successive surveys or from two different subgroups of the same survey]

Size of sample or group	Size of sample or group					
	3000	1000	700	500	300	200
	For percentages from about 35 percent to 65 percent					
3000	4	5	6	7	8	10
1000		6	7	8	9	11
700			8	8	10	11
500				9	10	12
300					11	13
200						14
	For percentages around 20 percent and 80 percent					
3000	3	4	5	5	7	8
1000		5	6	6	7	9
700			6	7	8	9
500				7	8	9
300					9	10
200						11
	For percentages around 10 percent and 90 percent					
3000	2	3	4	4	5	6
1000		4	4	5	6	6
700			4	5	6	7
500				5	6	7
300					7	8
	For percentages around 5 percent and 95 percent					
3000	2	2	3	3	4	4
1000		3	3	3	4	5
700			3	4	4	5
500				4	4	5
300					5	6

* The sampling error does not measure the actual error that is involved in specific survey measurements. It shows that (except for nonsampling errors, errors in reporting, in interpretation, etc.) differences larger than those found in the table will arise by chance in only 5 cases in 100.

tional children. The study directors now wish to know if the difference between this figure and the 70 percent seen with the previous group is statistically significant. Table 11-10 shows that with estimates based on groups of about 200 and 400 cases, it is necessary to have a difference of 12 to 13 percentage points to reach statistical significance. Since the difference between the estimates in this case is 35 percentage points, it is highly probable that it did not result from chance fluctuations in the sample.

A point which is sometimes forgotten in survey analysis is that the approximate sampling error for a finding depends on the actual number of interviews on which the percentages are based, rather than on the full sample. To facilitate proper interpretation of the findings, it is thus essential to report the number of cases involved in the major subgroups of the survey.

Finally, one complication in interpreting survey findings is that the errors seen with complex samples are typically higher than with simple random samples. For example, well-designed complex samples often produce estimates with sampling errors about one and one-half to two times those that would result from simple random samples of the same size (see Chapter 4 for a discussion of this point). Hence generalized tables developed for the interpretation of data based on complex samples should accurately reflect the errors for that kind of sample. This is another point that should be reviewed with a sampling expert before the final interpretations are made; ideally, it should have been discussed earlier when sampling plans were being considered.

A Strategy for Analysis

Thus far we have considered various simple, intermediate, and complex analytic techniques which can be used with survey data. For the newcomer, these techniques may be intelligible in themselves, but confusing in their relation to an overall analysis plan. At this point, we will draw a broader picture of the analysis effort, showing how such tools as the two-way and the three-way table can advance our understanding of a specific research problem. The following suggestions are offered more as an illustration than as a prescription for survey analysis. This time we will use an example drawn from a published survey report so that the interested reader can follow it further if he or she desires (cf. Mott et al., 1965). The basic strategy for analysis can be organized around seven questions:

1 *What is the problem?* Survey analysis should begin with a well-defined problem. To avoid wasted effort, the analysts should not request tabulations or other data without a clear sense of the questions to be answered by the information—what they want to know from the survey. This point may seem obvious as stated, but in many surveys enormous amounts of time are

lost on tabulations which prove to have little or nothing to do with the core analytic objectives of the study. Moreover, the first tables emerging from a survey are seductive; that is, they draw the analysts to them almost for their own sake. Once this process begins, study directors may go on for several weeks or more before they realize that they do not know where they are going with the analysis.

In the present example, the *general* question guiding the analysis is the effects of shift work on the health, attitudes, and social relationships of industrial workers and their families. Does shift work, as compared with regular day work, cause any harm to the worker and his family, and does it bring any distinctive benefits? The *specific* problem to be analyzed with survey data concerns the impact of shift work on two conditions: marital happiness and family integration.

2 *What is the fit between the problem and the data?* If the researchers were able to carry out their own survey, rather than working with existing data, and they began with a clear sense of the information needed to satisfy its objectives, the fit between the problem and the data should be reasonably good. If, on the other hand, the survey involves secondary analysis of data collected for some other purpose, the issue of the fit becomes quite salient. In this latter case, the analyst should make a realistic assessment of what can be accomplished through secondary analysis. If the data are only tangentially related to the guiding problem, detailed analysis involving dozens of tables may not be worthwhile. In the shift work study, the fit between the problem and the data was excellent, largely because of careful planning and extensive pretesting.

3 *What is to be explained (dependent variable)?* The distinction between what is to be explained and the explanatory conditions is provided by the theory or hypotheses of the study. Even in exploratory analyses, it is necessary at the beginning of analysis to decide which variables will be treated as *outcomes* and which will be regarded as *influences* on those outcomes. As noted earlier, methodologists often refer to the outcomes as *dependent variables* and the conditions affecting those outcomes as *independent variables.* An early decision about the core dependent variables in a study is essential to provide a concrete focus for the data collection as well as analysis.

In our example, the dependent variables or outcomes to be explained are two: marital happiness and family integration. They were defined and measured as follows:

> *Marital happiness:* the degree of satisfaction reported by the marriage partners concerning their relationship with each other. The index of marital happiness was constructed by combining responses to three questionnaire items dealing with this mutual relationship [Mott et al., 1965, p. 120].
> *Family integration:* the ability of the husband, wife, and other family members to work out an acceptable division of labor in the home and to coordinate their various activities. This concept refers primarily to the tasks carried out in the

home rather than to emotional relationships between marriage partners. The index of family integration was based on two items dealing with the division of labor in the family [ibid., p. 123].

4 *What are the main explanatory conditions (independent variables)?* The next task is to inventory all of the possible influences on the dependent variables that are of central concern to the study. For almost any dependent variable there will be a wide range of conditions which may exert some influence. The challenge for the analyst is to single out those explanatory conditions which comprise the main focus of the study, and then to determine the extent to which each exerts an independent influence and the way they operate collectively.

In our example, the central explanatory condition (independent variable) is the worker's shift or job schedule. Shift work was defined as any pattern of job hours regularly requiring work in the late afternoon or evening. The most common varieties are: (1) the fixed afternoon shift, a schedule in which the individual normally works from 4:00 P.M. until midnight; (2) the fixed night shift, a regular schedule involving work from about midnight to 8:00 A.M.; and (3) rotating shifts, or job schedules alternating between days, nights, and afternoons. Shift work is contrasted with day work, the most common or normal pattern of job hours.

The specific independent variable in the analysis of marital happiness and family integration was not the work schedule in itself, but its assumed departure from the usual rhythms of family activities in Western society. The study argued that the routines of the typical American family were closely bound to a schedule of day work. The timing of school for children, the organization of work in the household, the preparation of food and the serving of meals, social life, organizational activities, and many other critical aspects of family life follow the clock of day work. The researchers thus hypothesized that the greater the incompatibility between the demands of shift work and the prevailing expectations about the timing of family activities, the greater the problems for the marriage and for family work routine.

5 *What are the major intervening variables?* Before moving ahead to test hypotheses about the relationship between the independent and dependent variables, it is essential to consider certain "intervening" conditions. In our example, we would want to ask what other conditions *besides shift work* might make a difference in marital happiness and family integration among industrial laborers. It could be that, quite apart from job schedules, the marital happiness and family integration in this population will increase with age. If, as is usually the case, day workers are somewhat older than shift workers, the relationship between shift work and the dependent variables will be confounded. Similarly, some workers may choose to work on a night or afternoon shift to escape an unpleasant marital or family situation. In such cases shift work would be the effect rather than a cause of marital and family problems. Proper planning for analysis should include full consideration of alternative explanations for findings showing a relationship between the independent and dependent variables. Among the factors commonly considered as intervening variables or "controls" are education, length of service on the job, age, income, health, and racial or ethnic background.

6 *What specific hypotheses are to be tested?* An hypothesis is typically a statement proposing a relationship between the independent and the dependent variables. Such statements offer a concrete and helpful way of organizing the analysis of survey data. They also force the researchers to surface their own assumptions about what is likely to be found in the data and to express these ideas in testable form. Hypotheses need not be correct to be useful. The main requirement is that they utilize the central variables in the study to provide a clear focus for analysis.

In the shift work study, the following predictions were made about the relationship between work schedules and the indicators of marital happiness and family integration:

> **1** In general, day workers as a group will report greater marital happiness and family integration than will afternoon, night, and rotating shift workers.
>
> **1a** However, the first hypothesis must be modified as follows: some workers will remain on shift work because it provides a refuge from an unhappy marriage. When these workers are set aside, the differences in the marital happiness of shift versus day workers will be reduced or eliminated.
>
> **2** Among shift workers marital happiness and family integration will decrease as the amount of interference from work schedules increases. Men who report little or no interference will report about the same levels of happiness and integration as day workers of comparable age, education, etc.
>
> **3** Family integration will be more closely related to differences in shifts and in amounts of shift-related interference than will marital happiness [Mott et al., 1965, p. 116].

The assumption underlying the last hypothesis is that the happiness of a worker's marriage will be less susceptible than family integration to situational influences such as those produced by shift work. The reasoning was that marital happiness is primarily the result of emotional and other kinds of personal compatibility. Hence, while happiness may be indirectly affected by shift schedules, the impact should be less noticeable than in the case of family integration, a variable closely tied to the immediate environment.

7 *What is the relationship between the independent and dependent variables?* This is the stage in which the analyst actually requests tabulations and begins to interpret the data. An immediate question concerns the form of the tabulations. If tables are to be used, how should they be organized? Should all the categories of the independent, intervening, and dependent variables be used in preparing the tables, or should they be collapsed into a smaller number? How many tables will be needed? What practical steps can be taken to facilitate the interpretation of the results? In a brief discussion such as this, we can offer only a few general hints about such practical questions of analysis.

A useful first step is to obtain frequency distributions on all the variables to be used in the analysis. With samples of 1000 to 2000 and 7 to 10 categories per variable, it is almost never possible to carry out three-way tabulations (e.g., shift, age, and marital happiness) involving 7 or 8 categories per variable. For example, a table based on 4 categories of shift work, 7 of age, and 7 of marital happiness would result in 4 × 7 × 7, or 196 cells. Tables of this size not only create enormous difficulties of interpretation for the analyst and the reader, but result in many cells with one, two or no cases at all. Before requesting complicated cross tabulations, therefore, it is very wise and ultimately timesaving to reduce the total number of categories to manageable proportions. Instead of working with 8 or 9 subvarieties of shift work, the analysts in the study collapsed work schedules into four: days, afternoons, nights, and rotating shifts. Similarly, rather than using 6 or 7 categories of age, they broke age down into two groups: under forty and over forty. To simplify the analysis task even further, they reported scores on marital happiness and family integration in the form of means rather than percentages.[3] This step meant that marital happiness or family integration could be reported as a single score for a given group, such as day workers over forty. Some of this collapsing can, of course, be saved until the final tables are prepared for publication. Even so, it is rarely advisable to request tables with a total of more than 40 or 50 cells.

The analysis might then proceed to a series of two-way (two variable) tables. The aim now is to see if there is any relationship between the independent variable and the dependent variables, as well as between the main intervening (control) variables, such as age, and the dependent variables. This step is a helpful prelude to a more complex analysis of interactions among independent, intervening, and dependent variables. In our example, the following tables would be essential: work schedules (days, afternoon, nights, rotating shifts) against marital happiness and family integration; shift interference in family activities against marital happiness and family integration; and the worker's age against the same dependent variables. Table 11-11 shows a two-way table relating shift and marital happiness.

The results show that, as predicted, the marital happiness scores for day workers are higher than those for shift workers. However, other evidence suggested that 18 shift workers with low marital happiness scores remained on afternoon, night, or rotating shifts by choice, seemingly to escape an

---

[3] Some question could be raised about whether the type of data collected in the survey should be reported as means. However, this is a technical point that is not of direct concern in the present discussion.

Table 11-11   Mean Index of Marital Happiness by Shift

	Shift			
	Day	Afternoon	Night	Rotating
Index of marital happiness:				
Husband's report*	7.68	7.28	7.24	7.40
Number of cases	219	164	131	391

* Mean average scores. The higher the score, the greater the marital happiness.

Source: Mott, Mann, McLoughlin, and Warwick, *Shift Work: The Social, Psychological, and Physical Consequences,* 1965.

unpleasant marital relationship. When these cases were removed from the tabulations, the difference between the day workers and the shift workers was greatly reduced.

Before going on to more complex tabulations, the analyst should summarize the main trends emerging from the two-way tables. It is usually convenient to review these tables in logically organized sets or families. One way is to group them into sets dealing with the dependent variables, in this case marital happiness and family integration. The researchers can then study each table and note its interesting parts on the table itself or on separate index cards. Much of the interpretative effort may be wasted if the analysts fail to take notes while reviewing the tables. Even a day later they may find it difficult to reconstruct all the preliminary conclusions.

The next step is to work with three-way tables. At this stage the analysts should have a fairly clear idea of the relationships between the independent and the dependent variables, and between the intervening and dependent variables. They may wish to know: (1) if the relationships observed between the independent and the dependent variables remain essentially the same when controls are added for age, education, and similar conditions; and (2) if new and unsuspected relationships are seen when these controls are added. With regard to this second point, the results might show that, while there is no relationship between shift work and marital happiness for the sample as a whole, significant relationships do appear when controls are added for education. It could happen, though it did not in this study, that marital happiness is higher for day workers than for shift workers among men with less than high school education, and higher among shift workers than day workers for those with a high school education or more. This finding would argue for a more careful analysis of the relationship between education, shift work, and marital happiness.

Table 11-12 contains a three-way analysis drawn from the shift work study. This analysis sought to determine whether there was any relationship between (1) the amount of shift-related interference reported in husband-

Table 11-12 Mean Index of Family Integration,* by Difficulty in Husband-Wife Relations and Shift

Shift†	Difficulty in husband-wife relations			n
	Little	Medium	Great	
Afternoon	5.89	5.75	5.50	159
Night	6.21	5.86	5.18	126
Rotating	6.00	5.26	5.34	388
Total	6.01	5.49	5.35	673

* Mean average scores based on husband's report. The higher the score, the greater the family integration reported.

† The mean for the day shift was 6.02 (n, = 226). Because this shift was assumed to be the usual pattern, difficulty scores were not obtained for day workers.

Source: Mott, Mann, McLoughlin, and Warwick, Shift Work: The Social, Psychological, and Physical Consequences, 1965.

wife relations and (2) family integration.[4] Given the varying problems created by different shift patterns, the researchers considered it important to add a control for shift. The results show a clear relationship between interference in husband-wife relations and the index of marital integration. Workers who report the greatest amount of interference tend to report the lowest levels of integration. Those with little difficulty have an average integration score of 6.01, those with medium difficulty a score of 5.49, and those with little difficulty an average of 5.35. These differences are statistically significant. Table 11-12 also shows, however, that the effects of interference are most pronounced for night shift workers and least in evidence for afternoon shift workers.

The analysis would continue in this fashion until the researchers were satisfied that the data had been adequately interpreted. *When* this point arrives depends on the nature of the problem and the precise analytic objectives of the survey. A study aiming mainly at description may be able to stop sooner than a complicated explanatory study such as the one used in the example. Beyond the general guidelines suggested here, it is difficult to lay out specific procedures to be applied across the board. We would only add that the success of survey analysis ultimately requires a creative blend of theoretical imagination and statistical technique. One without the other usually proves to be sterile.

[4] The index of interference in husband-wife relations was based on the worker's report of shift-related difficulty in seven areas: companionship with his wife, assistance with housework, providing diversion and relaxation, protection of the wife from harm, mutual understanding, decision making, and sexual relations. The questions asked if the workers found it harder or easier on their shift, as compared with a steady day shift, to engage in these activities.

PRESENTATION OF THE RESULTS

The last stage of the survey consists of preparing a report summarizing the entire study. An immediate question about the report concerns its audience. Will it be addressed primarily to other professionals in the field, such as sociologists and political scientists, and others who can understand the technical details, or will it be aimed at a public interested only in the findings? A report prepared mainly for specialists will contain much more information about the methodology of the study than one designed for administrators. Whatever the audience, it is important to consider three basic questions in planning the report: the style in which it is to be written, the mechanics of presenting the material, and the organization of the topics.

Style

The basic aim of the research report is to communicate the findings to others as simply and directly as possible. Of course, what is simple and direct to one audience may be complex and confusing to another; hence the need for prior consideration of the audience. Above all else, the writing should be marked by clarity and accuracy. The writer should avoid long, involved sentences, unnecessary technical terms and disciplinary jargon, and anything else which stands in the way of clear writing. If elegance of expression comes naturally, it is welcome; if it is achieved at the expense of clarity, accuracy, and brevity, it should be avoided. Sometimes social scientists who study weighty and important topics feel that their reports should reflect the gravity of their subject matter. The writing which results from this attitude is probably the main source of complaints about social scientific publications, including many reports on surveys.

Closely related to the issue of style is that of interpretation. Should the findings be set up in orderly tables and then presented to the reader with little comment, or should there be fairly extensive interpretation? There is no answer to this question in the abstract; much will again depend on the audience and the purposes of the survey. Nevertheless, most reports should provide a reasonable amount of interpretation. As Moser points out,

> Most readers of a research report, fellow scientists or laymen, lack the time and perhaps the will-power to go through the tables and pick out the crucial results. But even if they had both it would be wrong to leave the interpretation entirely to them. There is after all more to a research than can be seen from the tables, and the researcher in interpreting his results is inevitably—and rightly—influenced by all that has gone before, by his acquaintance with the raw material behind the figures and by his own judgment [1958, p. 300].

Put another way, if the researcher does not interpret his own findings, there is a good chance that they will remain uninterpreted and even ignored. Moreover, on highly sensitive or controversial subjects, a failure to provide adequate context and interpretations may be an outright evasion of professional responsibility. If the findings are open to misuse and misinterpretation, the survey researcher is obligated to supply a reasonable interpretation.

Mechanics

Properly designed and used sparingly, devices such as graphs, charts, tables, and subheadings are often an effective aid to clear reporting. Before sending out the final copy, the writer should always doublecheck to ensure that a graph, chart, or table actually aids rather than encumbers the presentation. Sometimes, for example, tables are an important source of information for those who wish to check the fine points of the survey, but are too long or complicated to be included in the body of the report. These might well be included as an appendix.

Table Manners   A table, a graph, or a chart should be self-sufficient. The reader should not have to refer to the text or any other source to find out what it contains. While no set of suggestions can cover all of the situations likely to arise in reporting survey data, the following conventions are commonly used and should be adopted unless there is good reason to do otherwise.

Numbering   Tables should be numbered in Arabic figures (1, 2, 3) rather than Roman numerals (I, II, III). It is usually preferable to number them sequentially throughout the report, even when there are several chapters. However, practical considerations, such as separate typing of the different chapters, sometimes make it more convenient to have a double-numbered series within each chapter. The first number in this case would refer to the chapter, the second to the table number within the chapter (e.g., 10-2).

Titles   Each table should be completely described by a title which is as brief as possible. Occasionally, a subtitle is added in parentheses for explanatory material, such as the population group covered by the table. Both the title and the subtitle should refer to the entire table rather than to a single segment.

Rules   Rules are the lines drawn on a table. Horizontal rules of equal length should be drawn above and below the main body of the data, as in the tables used throughout this book. This convention ensures that the data are set off from the title and subtitle above and the footnotes, if any, below. Within the body of the table, another horizontal rule is commonly used to set off the column headings from the data. Shorter horizontal rules can also be used to separate the totals in a column from the rest of the information reported above it. Vertical rules are often used if there are more than two columns of data, but never with one column.

Column headings Each column should have a succinct heading. Very often column headings are written on two lines. The first describes the general topic used to organize the columns, such as "age of household head" or "level of family income." Then beneath this heading are the specific categories reported in the columns, such as 35-44 in the case of age. A special effort should be made to avoid a cluttered appearance in tables involving many columns. If there is any doubt about the comprehensibility of the headings, an explanatory footnote may be added.

Stub The first column on the left of a table is typically used to present the dependent variable. The heading for this column, the stub, is placed on the same line as other column headings, while the detailed categories are presented on successive lines below. These describe the findings reported as rows in the table, and may be indented. If the description of a category requires a second line, this should be indented about three spaces.

Capitalization The description of the table should be consistent with one of the following conventions: (1) capitalize all letters in the word TABLE and in the main title; or (2) capitalize only the first letters of the words therein, with the exception of articles and prepositions. If a subtitle follows in parentheses, only the first letter of each word should be capitalized. In the case of column headings or stub categories, only the first letter of the first word should appear in capitals.

Footnotes In general, footnotes should be used as little as possible in tables and similar material. When included they should be single-spaced and located directly beneath the rule ending the body of the table. They should run the full width of the rule except for an initial indentation.

Signs Dollar signs, plus and minus signs, and percent signs should appear with the first datum in each column and again for the total. They should not be repeated for each figure in the column, except where there is some possibility of confusion. If the table contains several series of data, such as income figures for each city in our urban sample, the appropriate signs should be repeated in each series. Percent signs should appear to the right of the column of data, while dollar, plus, and minus signs should be to the left.

Data source The source of the data reported in the table, unless apparent from the context, normally appears as the last footnote. It is introduced by an approximately indented label, such as *Source*: . . . , or *The question was:* . . . . Some of these "table manners" are illustrated in the tables included in this book.

Headings and Subheadings Another useful technique for improving the clarity of the report lies in the use of descriptive headings and subheadings throughout the text. In fairly complex reports, as many as three or four

levels of headings may be needed to cover the various topics. These should
be clearly distinguished by some uniformly applied convention, beginning
with the early drafts of the report. The following example, used in preparing
the manuscript for this work, illustrates how four levels might be differenti-
ated:

MAJOR HEADING, CENTERED ON PAGE, ALL CAPS

Secondary Heading, Centered, First Letters in Caps

Next Subheading, at Left Margin, Underlined

    With this subheading, the text begins as an indented paragraph on the
next line. Note that the first letter of each word in this subheading is capitalized.

    Fourth level heading. This subheading is indented as a regular para-
graph, and is run into the text of the paragraph. It is underlined and punc-
tuated with a period, but only its first letter is capitalized.

Organization of the Report

As with style, the organization of the report depends on the problem to be
discussed and the intended audience. The arrangement suggested here is
probably the simplest and most concise when the survey is fairly straightforward in content, and the audience consists of a combination of professionals
and the general public. In certain cases, it may be worthwhile to issue the
report in two volumes, one focusing on the results with only cursory attention to methods, and the other a detailed commentary on sampling, questionnaire design, and other technical details of interest to the specialist.

    In our recommended model of presentation, the report is organized
around four main headings: the research problem, methods, findings, and
discussion and implications. The report would then end with a summary,
followed in some cases by one or more appendices.

    The Problem The report should open with a clear statement of the
purposes of the study and the specific problem or problems being examined.
The reader should not have to wade through a lengthy review of the literature before finding out the central topic of the study. Once the problem is
laid out, a summary of the background of the study, including findings from
related studies and guiding theoretical concerns, helps to provide a sharper
focus for the reader.

    Methods The report next reviews the methods used to gather the data.
The level of detail should be sufficient to answer the questions likely to arise
in the minds of the readers, but not so much as to be overwhelming or to

detract attention from the later sections. Normally, it is enough to provide a general statement at this point, and then to review the details in one or more appendices. A minute examination of sampling fractions, for example, would usually be out of place in the body of the report, as would a full reproduction of the questionnaire or interview schedule. It may be helpful, on the other hand, to state the population covered by the sample, the general procedures followed in arriving at the selected households, and the type of question used. The following are some specific items that might be covered in this section:

—the design of the research: the main features of the research plan and how they relate to each other.

—sampling procedures: the population covered and how sampling units and respondents were selected.

—data collection methods: self-administered questionnaires, personal interviews, a combination of interviews and observation, etc.

—estimation procedures: a brief discussion of the procedures and assumptions by which the sample information is converted to estimates about the population, including some comment on sampling error and confidence intervals.

—time periods: the major stages of the study and the period of reference for the data obtained.

Research Findings The discussion of the findings will often extend over several chapters and include tables, graphs, and charts. It is always worthwhile to spend a fair amount of time planning for a logical and coherent presentation of each topic. Before launching into a detailed discussion of any single topic, it is important to have a general sense of where the report is going. Otherwise, both the writer and the reader may become mired in the details of a minor issue or table. At this point it is helpful to develop a skeleton outline of topics, logically organized in the form of major points and subpoints. This outline will quickly reveal whether or not the author has thought out the flow of the discussion. A good outline can also easily be converted into a coherent set of headings and subheadings for inclusion in the text.

One point which cannot be overemphasized is the necessity of presenting a balanced and honest summary of the data. Regardless of his own point of view, the writer is obligated to present all pertinent evidence related to the research question. This is admittedly a tall order when the topic is controversial or one on which the researchers themselves have very strong feelings. The task becomes even more complicated when some of the findings which are less than pleasing to the authors are also those which could be most susceptible to misuse by unscrupulous campaigners or critics. Nevertheless, the integrity of the entire research enterprise rests on the honesty with which the data are collected, analyzed, *and* reported. Suppressed tables, selective discussion of relevant findings, soft-pedaling of negative data, and similar tactics have no place in responsible survey reporting. As noted in Chapter 2, they may occur despite the best intentions of the researcher, but they should not occur knowingly. The chance of unwitting bias or lapses in reporting can also be greatly reduced by circulating the draft for critical comments among individuals known to be unbiased or even hostile to the viewpoint of the writers.

Discussion and Implications A good research report will usually contain a discussion of the principal conclusions to be drawn from the findings, as well as their theoretical or practical implications. While this discussion is sometimes included as part of the findings, it is normally preferable to save it for a separate section.

Some questions which might be considered for this section include the following: Does the study add any totally new information to existing knowledge on the question? Does it challenge earlier interpretations or the conventional wisdom in the field? Does it suggest certain directions for action or new implications for theory? Does it point to new areas for research? Before concluding, the authors should indicate any major limitations of the research, and any restrictions on the generalizability of the findings to the population or other likely areas of concern. Here and elsewhere, they should be frank and honest in reporting any weaknesses in the research proceedings. Too often, social researchers feel that they will lose face if they expose their flaws. Ironically, in the long run, honest reporting may produce fewer problems of face-saving than cosmetically improved research reports.

Summary Some writers prefer to end the body of the report when they reach the end of the section on discussion and implications, while others lean toward a separate summary. This is really a matter of taste. To be useful, the summary should provide a recapitulation of the main points emerging from the previous sections, and should be brief. It should not discuss any of the points in detail.

Appendices Most research reports contain one or more appendices reproducing the questionnaire and describing the technical details of the research procedures. Some of the material commonly reserved for appendices are sampling procedures and sampling errors, complex statistical procedures and tables which are too long to fit conveniently into the main body of the report, a detailed discussion of field work procedures and problems, and an analysis of the costs of the survey.

The completion of the research report brings the researcher to the end of his journey and us to the end of ours. We will end with a note of encouragement for those setting out on this road for the first time. The map we have provided may make the journey seem interminable and the landmarks ill-defined. This feeling is inevitable but should not be permanent. As in any other field, training in survey research requires close attention not only to the guiding theory but also to the working details of research. In the first survey, these details may seem to be limitless and to loom very large on the horizon. With practice they recede into the background, and eventually become second nature to the experienced researcher.

FURTHER READINGS

Blalock, H. M., Jr., *Causal Inferences in Nonexperimental Research* (Chapel Hill: University of North Carolina Press, 1967).

Davis, J. A., *Elementary Survey Analysis* (Englewood Cliffs, N.J.: Prentice-Hall, 1971).

Rosenberg, M., *The Logic of Survey Analysis* (New York: Basic Books, 1968).

BIBLIOGRAPHY

Almond, G., and S. Verba, *The Civic Culture* (Princeton, N.J.: Princeton University Press, 1961).

Anderson, R. B. W., "On the comparability of meaningful stimuli in cross-cultural research," *Sociometry,* **30,** 1967, 124-136.

Arnold, R. G., "The interview in jeopardy: A problem in public relations," *Public Opinion Quarterly,* **28,** 1964, 119-123.

Athey, K. R., J. E. Coleman, A. P. Reitman, and J. Tang, "Two experiments showing the effect of the interviewer's racial background on responses to questionnaires concerning racial issues," *Journal of Applied Psychology,* 1960, 244-246.

Back, K. W., and K. J. Gergen, "Idea orientation and ingratiation in the interview: A dynamic model of response bias," *Proceedings of the Social Statistics Section of the American Statistical Association,* 1963, 284-288.

Back, K. W., and J. M. Stycos, "The survey under unusual conditions: Methodological facets of the Jamaica human fertility investigation" (Ithaca, N.Y.: Society for Applied Anthropology, Monograph No. 1, 1959).

Bauer, R. (ed.), *Social Indicators* (Cambridge, Mass.: M.I.T. Press, 1966).

Baum, Samuel, K. Dopkowski, W. G. Duncan, and P. Gardiner, *The World Fertility Survey Inventory: Major Fertility and Related Surveys Conducted in Asia 1960-1973* (London: Occasional Paper of the World Fertility Survey, No. 3, April 1974a).

Baum, Samuel, K. Dopkowski, W. G. Duncan, and P. Gardiner, *The World Fertility Survey Inventory: Major Fertility and Related Surveys Conducted in Africa 1960-1973* (London: Occasional Paper of the World Fertility Survey, No. 4, April 1974b).

Baum, Samuel, K. Dopkowski, W. G. Duncan, and P. Gardiner, *The World Fertility Survey Inventory: Major Fertility and Related Surveys Conducted in Latin America 1960-1973* (London: Occasional Paper of the World Fertility Survey, No. 5, April 1974c).

Baum, Samuel, K. Dopkowski, W. G. Duncan, and P. Gardiner, *The World Fertility Survey Inventory: Major Fertility and Related Surveys Conducted in Europe, North America and Australia 1960-1973* (London: Occasional Paper of the World Fertility Survey, No. 6, April 1974d).

Becker, H. S., and B. Geer, "Participant observation: The analysis of qualitative field data," in R. N. Adams and J. J. Preiss (eds.), *Human Organization Research* (Homewood, Ill.: Dorsey Press, 1960), 267-289.

Bell, D., "Twelve modes of prediction—A preliminary sorting of approaches in the social sciences," *Daedalus,* Summer, 1964.

Belson, W. A., "Tape recording: Its effect on accuracy of response in survey interviews," *Journal of Marketing Research,* **4,** 1967, 253-260.

Bendix, R., and S. M. Lipset, "Political sociology," *Current Sociology,* **6,** 1957, 79-99.

Berreman, G., "Behind many masks: Ethnography and impression management in a Himalayan village," in D. Warwick and S. Osherson (eds.), *Comparative Research Methods* (Englewood Cliffs, N.J.: Prentice-Hall, 1973), 268-312.

Blalock, H. M., Jr., *Causal Inferences in Nonexperimental Research* (Chapel Hill: University of North Carolina Press, 1967).

Blalock, H. M., Jr. (ed.), *Causal Models in the Social Sciences* (Chicago: Aldine, 1971).

Blalock, H. M., Jr., *Social Statistics* (New York: McGraw-Hill, 1960).

Blankenship, A. B., *How to Conduct Consumer and Opinion Research* (New York: Harper, 1946).

Bogart, L., *Silent Politics: Polls and the Awareness of Public Opinion* (New York: Wiley-Interscience, 1972).

Brunner, E. deS., "Social research dollars and cents," *Public Opinion Quarterly,* **26,** 1962, 97-102.

Bucher, R., C. E. Fritz, and E. L. Quarantelli, "Tape recorded interviews in social research," *American Sociological Review,* **21,** 1956, 359-364.

Cain, L. D., Jr., "The AMA and the gerontologists: uses and abuses of 'A Profile of the Aging: USA'," in G. Sjoberg (ed.), *Ethics, Politics, and Social Research* (Cambridge, Mass.: Schenkman, 1967), 78-114.

Campbell, A. A., and G. Katona, "The sample survey: A technique for social science research," in L. Festinger and D. Katz, *Research Methods in the Behavioral Sciences* (New York: Holt, 1953), 15-55.

Campbell, D. T., and D. W. Fiske, "Convergent and discriminant validation by the multitrait-multimethod matrix," *Psychological Bulletin,* **56,** 1959, 81-105.

Campbell, D. T., and J. C. Stanley, *Experimental and Quasi-Experimental Designs for Research* (Chicago: Rand McNally, 1963).

Cannell, C. F., F. J. Fowler, Jr., and K. H. Marquis, *The Influence of Interviewer and Respondent Psychological and Behavioral Variables on the Reporting in Household Surveys* (Washington, D.C.: U.S. Department of Health, Education and Welfare, National Center for Health Statistics, Series 2, No. 26, 1968).

Carlson, R. O., "The issue of privacy in public opinion research," *Public Opinion Quarterly,* **31,** 1967, 1-8.

Carter, L. F., "Survey results and public policy decisions," *Public Opinion Quarterly,* **27,** 1963, 549-557.

Centro de Investigaciones Sociales por Muestro, *Población Economicamente Activa en Lima Metropolitana* (Lima: Servicio del Empleo y Recursos Humanos, 1967).

Coleman, J. S., et al., *Equality of Educational Opportunity* (Washington, D.C.: U.S. Government Printing Office, 1966).

Coombs, L., and R. Freedman, "Use of telephone interviews in a longitudinal fertility study," *Public Opinion Quarterly,* **28,** 1964, 112-117.

Cronbach, L. J., N. Rajaratnam, and G. C. Gleser, "Theory of generalizability: A liberalization of reliability theory," *The British Journal of Statistical Psychology,* **16,** 1963, 137-163.

Davis, K., "Problems and solutions in international comparisons for social science purposes," Population Reprint Series No. 273, University of California at Berkeley, International Population and Urban Research, Institute of International Studies, undated.

Deutscher, I., "Asking questions cross-culturally: Some problems of linguistic comparability," in D. Warwick and S. Osherson (eds.), *Comparative Research Methods* (Englewood Cliffs, N.J.: Prentice-Hall, 1973), 163-186.

Dexter, Lewis Anthony, "The use and function of social science consultants," *The American Behavioral Scientist,* **9** (6), 1966.

Eckler, A. Ross, *The Bureau of the Census* (New York: Praeger, 1972).

Economic Behavior Program, *Economic Survey Data* (Ann Arbor: University of Michigan, Survey Research Center, 1960).

Economic Behavior Program, *1960 Survey of Consumer Finances* (Ann Arbor, Mich.: Institute for Social Research, 1960).

Frey, F. W., "Surveying peasant attitudes in Turkey," *Public Opinion Quarterly,* **27,** 1963, 335-355.

Friedrichs, R. W., *A Sociology of Sociology* (New York: Free Press, 1970).

Goffman, E., *Strategic Interaction* (Philadelphia: University of Pennsylvania Press, 1969).

Gorden, R. L., *Interviewing: Strategy, Techniques, and Tactics* (Homewood, Ill.: Dorsey, 1969).

Gullahorn, J. E., and J. T. Gullahorn, "An investigation of the effects of three factors on response to mail questionnaires," *Public Opinion Quarterly,* **27,** 1963, 294-296.

Hansen, M. H., W. N. Hurwitz, and W. G. Madow, *Sample Survey Methods and Theory* (New York: Wiley, 1953, 2 vols.).

Hauck, M., and S. Steinkamp, *Survey Reliability and Interviewer Competence* (Urbana, Ill.: University of Illinois, Bureau of Economic and Business Research, 1964).

Henriot, P. J., "Political questions about social indicators," *Western Political Quarterly,* **23,** 1970, 235-255.

Hildum, D., and R. W. Brown, "Verbal reinforcement and interviewer bias," *Journal of Abnormal and Social Psychology,* **53,** 1956, 108-111.

Hirschman, A. O., *The Strategy of Economic Development* (New Haven: Yale University Press, 1958).

Horowitz, I. L. (ed.), *The Rise and Fall of Project Camelot: Studies in the Relationship between Social Science and Practical Politics* (Cambridge, Mass.: M.I.T. Press, 1967).

Horowitz, I. L. (ed.), *The Use and Abuse of Social Science* (New Brunswick, N.J.: Transaction Books, 1971).

Hunt, W. H., W. W. Crane, and J. C. Wahlke, "Interviewing political elites in cross-cultural comparative research," *American Journal of Sociology,* **70,** 1964, 59–68.

Hyman, H. H., *Interviewing in Social Research* (Chicago: University of Chicago Press, 1954).

Hyman, H. H., *Survey Design and Analysis: Principles, Cases, and Procedures* (Glenco, Ill.: Free Press, 1955).

Inkeles, A., "Fieldwork problems in comparative research on modernization," in A. R. Desai (ed.), *Essays on Modernization of Underdeveloped Societies* (Bombay, India: Thacher & Co., Ltd., 1971), Vol. 2, 20–75.

Jones, E. L., "The courtesy bias in South-East Asian surveys," *International Social Science Journal,* **15,** 1963, 70–76.

Kahn, R. L., and C. F. Cannell, *The Dynamics of Interviewing* (New York: Wiley, 1967).

Katona, George, C. A. Lininger, J. N. Morgan, and Eva Mueller, *1961 Survey of Consumer Finances* (Ann Arbor, Mich.: Institute for Social Research, 1961).

Katona, George, C. A. Lininger, and R. F. Kosobud, *1962 Survey of Consumer Finances* (Ann Arbor, Mich.: Institute for Social Research, 1962).

Katona, George, C. A. Lininger, and Eva Mueller, *1963 Survey of Consumer Finances* (Ann Arbor, Mich.: Institute for Social Research, 1963).

Katona, George, Lewis Mandell, and Jay Schmiedeskamp, *1970 Survey of Consumer Finances* (Ann Arbor, Mich.: Institute for Social Research, 1970).

Katz, D., "Do interviewers bias poll results?" *Public Opinion Quarterly,* **6,** 1942, 248–268.

Keesing, F. M., and M. M. Keesing, *Elite Communications in Samoa: A Study of Leadership* (Stanford, Calif.: Stanford University Press, 1956).

Kelman, H. C., *A Time to Speak* (San Francisco: Jossey-Bass, 1968).

Kinsey, A. C., W. B. Pomeroy, and C. E. Martin, *Sexual Behavior in the Human Male* (Philadelphia: W. B. Saunders, 1948).

Kish, L., "A two-stage sample of a city," *American Sociological Review,* **17** (6), 1952, 761–769.

Kish, L., *The Survey Sample* (New York: Wiley, 1965).

Kreinin, M. E., and C. A. Lininger, "Ownership and purchases of new cars in the United States," *International Economic Review,* **4** (3), 1963, 310–324.

Lambert, W. E., "Measurement of the linguistic dominance of bilinguals," *Journal of Abnormal and Social Psychology,* **50,** 1955, 197–200.

Landsberger, H. A., and A. Saavedra, "Response set in developing countries," *Public Opinion Quarterly,* **31,** 1967, 214–229.

Lazersfeld, P. F., "Repeated interviews as a tool for studying changes in opinions and their causes," *American Statistical Association Bulletin,* **2,** 1941, 3–7.

Lenski, G. E., and J. C. Leggett, "Caste, class and deference in the research interview," *American Journal of Sociology*, **65**, 1960, 463–467.

Lerner, D., "Interviewing Frenchmen," *American Journal of Sociology*, **62**, 1956, 187–194.

Lerner, D., *The Passing of Traditional Society* (New York: Free Press, 1958).

Lesser, G. S., "Advantages and problems of cross-cultural research," (Cambridge: Harvard University, Mimeo, 1967).

Lininger, Charles A., *International Research and Intellectual Imperialism* (New York: Mimeo, 1973).

Lininger, Charles A., *Unemployment Benefits and Duration* (Ann Arbor, Mich.: Institute for Social Research, 1963).

Lininger, Charles A., and Donald P. Warwick, *Introduction to Survey Research* (Lima: Servicio del Empleo y Recuros Humanos, 1965).

Mauldin, W. Parker, "Fertility studies: Knowledge, attitude, and practice," *Studies in Family Planning* (New York: The Population Council, **1** (7), 1965).

Miller, A. R., *The Assault on Privacy* (Ann Arbor: The University of Michigan Press, 1971).

Mitchell, R. E., "Survey materials collected in the developing countries: Sampling, measurement, and interviewing obstacles to intra- and inter-national comparisons," *International Social Science Journal*, **17**, 1965, 665–685.

Morgan, J. N., "Contributions of survey research to economics," in C. Y. Glock (ed.), *Survey Research in the Social Sciences* (New York: Russell Sage, 1967), 217–268.

Moser, C. A., *Survey Methods in Social Investigation* (London: Heinemann, 1958).

Mott, P. E., F. C. Mann, Q. McLoughlin, and D. P. Warwick, *Shift Work: The Social, Psychological, and Physical Consequences* (Ann Arbor: The University of Michigan Press, 1965).

Muehl, D. (ed.), *A Manual for Coders* (Ann Arbor: The University of Michigan, Survey Research Center, 1961).

Myrdal, G., *Asian Drama: An Inquiry into the Poverty of Nations* (New York: Pantheon, 1968, 3 vols.).

Newcomb, T. M., K. E. Koenig, R. Flacks, and D. P. Warwick, *Persistence and Change: Bennington College and Its Students after Twenty-five Years* (New York: Wiley, 1967).

Newcomb, T. M., *Personality and Social Change* (New York: Dryden, 1943).

Oppenheim, A. N., *Questionnaire Design and Attitude Measurement* (New York: Basic Books, 1966).

Orlans, H., "Ethical problems in the relations of research sponsors and investigators," in G. Sjoberg (ed.), *Ethics, Politics, and Social Research* (Cambridge, Mass.: Schenkman, 1967), 3–24.

Osgood, C. E., G. J. Suci, and P. H. Tannenbaum, *The Measurement of Meaning* (Urbana: University of Illinois Press, 1957).

Parten, M., *Samples, Polls, and Surveys: Practical Procedures* (New York: Harper, 1950).

Patchen, M., in collaboration with D. C. Pelz and C. W. Allen, *Some Questionnaire Measures of Employee Motivation and Morale: A Report on Their Reliability and Validity* (Ann Arbor, Mich.: Survey Research Center, Institute for Social Research, The University of Michigan, Monograph No. 41, 1965).

Payne, S. L., *The Art of Asking Questions* (Princeton, N.J.: Princeton University Press, 1951).

Phillips, D. L., *Knowledge from What? Theories and Methods in Social Research* (Chicago: Rand McNally, 1971).

Przeworski, A., and H. Teune, "Equivalence in cross-national research," *Public Opinion Quarterly*, **30**, 1966-1967, 551-568.

Rainwater, L., and W. L. Yancey, *The Moynihan Report and the Politics of Controversy* (Cambridge, Mass.: M.I.T. Press, 1967).

Richardson, S. A., B. S. Dohrenwend, and D. Klein, *Interviewing: Its Forms and Functions* (New York: Basic Books, 1965).

Roeher, G. A., "Effective techniques in increasing response to mailed questionnaires," *Public Opinion Quarterly*, **27**, 1963, 299-302.

Rorer, L. G., "The great response-style myth," *Psychological Bulletin*, **63**, 1965, 129-158.

Rosenthal, R., *Experimenter Effects in Behavioral Research* (New York: Appleton-Century-Crofts, 1966).

Rossi, P. H., M. Boeckmann, and R. A. Berk, "Some ethical implications of the New Jersey-Pennsylvania Income Maintenance Experiment." Paper prepared for the Conference on the Ethics of Social Intervention, Battelle Seattle Research Center, May 1973.

Rossi, P. H., and W. Williams (eds.), *Evaluating Social Programs: Theory, Practice, and Politics* (New York: Seminar Press, 1972).

Rugg, D., and H. Cantril, "The wording of questions in public opinion polls," *Journal of Abnormal and Social Psychology*, **4**, 1942, 469-495.

Schuman, H., "The random probe: A technique for evaluating the validity of closed questions," *American Sociological Review*, **31**, 1966, 218-222.

Selltiz, C., M. Jahoda, M. Deutsch, and S. W. Cook, *Research Methods in Social Relations*, rev. ed. (New York: Holt, 1959).

Sheldon, E. B., and H. E. Freeman, "Notes on social indicators: Promises and potential," in C. H. Weiss (ed.), *Evaluating Action Programs: Readings in Social Action and Education* (Boston: Allyn and Bacon, 1972), 166-173.

Shils, E., "Social inquiry and the autonomy of the individual," in D. Lerner (ed.), *The Human Meaning of the Social Sciences* (New York: Meridian Books, 1959), 114-157.

Sjoberg, G. (ed.), *Ethics, Politics and Social Research* (Cambridge, Mass.: Schenkman, 1967).

Sjoberg, G., and R. Nett, *A Methodology for Social Research* (New York: Harper & Row, 1968).

Smith, D. H., and A. Inkeles, "The OM scale: A comparative sociopsychological measure of individual modernity," *Sociometry*, **29**, 1966, 353-377.

Sobol, Marion Gross, "Panel mortality and panel bias," *Journal of the American Statistical Association*, **54**, 1959, 52-68.

Star, S. A., and H. M. Hughes, "Report on an educational campaign: The Cincinnati plan for the United Nations," *American Journal of Sociology*, **55**, 1950, 355-361.

Stephan, F. F., "Public reactions and research interviewing," *Public Opinion Quarterly*, **28**, 1964, 118.

Suchman, E. A., "Action for what? A critique of evaluative research," in C. H. Weiss (ed.), *Evaluating Action Programs: Readings in Social Action and Education* (Boston: Allyn and Bacon, 1972), 52–84.

Sudman, S., "Reducing the costs of surveys," *Public Opinion Quarterly*, **27**, 1963, 633–634.

The Population Council, *A Manual for Surveys of Fertility and Family Planning: Knowledge, Attitudes, and Practice* (New York: The Population Council, 1970).

U.S. Department of Health, Education and Welfare, *Measurement of Personal Health Expenditures* (Washington, D.C.: National Center for Health Statistics, Series 2, No. 2, 1963).

Walton, R. E., and D. P. Warwick, "The ethics of organization development," *The Journal of Applied Behavioral Science*, **9**, 1973, 681–698.

Warwick, D. P., "*Tearoom Trade*: Means and ends in social research," *Hastings Center Studies*, **1**, 1973, 27–38.

Warwick, D. P., and H. C. Kelman, "Ethical issues in social intervention," in G. Zaltman (ed.), *Processes and Phenomena of Social Change* (New York: Wiley, 1973), 377–417.

Warwick, D. P., and S. Osherson (eds.), *Comparative Research Methods* (Englewood Cliffs, N.J.: Prentice-Hall, 1973).

Webb, E. J., D. T. Campbell, R. D. Schwartz, and L. Sechrest, *Unobtrusive Measures: Nonreactive Research in the Social Sciences* (Chicago: Rand McNally, 1966).

Weiss, C. H. (ed.), *Evaluating Action Programs: Readings in Social Action and Education* (Boston: Allyn and Bacon, 1972).

Whyte, W. F., *Street Corner Society* (Chicago: University of Chicago Press, 1955).

Williams, J. A., Jr., "Interviewer-respondent interaction: A study of bias in the information interview," *Sociometry*, **27**, 1966, 338–352.

Williams, J. A., Jr., "Interviewer role performance: A further note on bias in the information interview," *Public Opinion Quarterly*, **32**, 1968, 287–294.

Williams, W., *Social Policy Research and Analysis: The Experience in the Federal Social Agencies* (New York: American Elsevier, 1971).

Wilson, E., "Problems of survey research in modernizing areas," *Public Opinion Quarterly*, **22**, 1959, 230–234.

Wyatt, D. F., and D. T. Campbell, "A study of interviewer bias as related to interviewer's expectations and own opinions," *International Journal of Opinion and Attitude Research*, **4**, 1950, 77–83.

Yntema, Dwight B., "Survey of unemployment compensation in Michigan, 1955" (Holland, Michigan: Hope College, Department of Economics and Business Administration, Mimeo, 1957).

Young, P., *Scientific Social Surveys and Research* (Englewood Cliffs, N.J.: Prentice-Hall, 1944).

Zurcher, L. A., and C. M. Bonjean, *Planned Social Intervention: An Interdisciplinary Anthology* (Scranton, Pa.: Chandler, 1970).

Index

Interviewer:
  work assignments for, 229-230
  work groups for, 229
Interviewer-respondent interaction, 127-131, 135,
    183-189, 192-199
Interviewer effect, 190, 197
  in successive samples, 61
Interviewer probe, 47, 213-215
  clarification as, 214
  coding and, 253-254
  elaboration as, 213-214
  encouragement as, 213
  errors in, 200
  leading, 215
  neutrality of, 215
  random, 139-140
  recording of, 217
  repetition as, 214-215
  respondent accuracy and, 159
  silent, 213
Interviewer training, 224-229
  background materials for, 225
  coder check in, 228, 229
  continuing, 228-229
  group sessions for, 225-226
  initial program for, 224-228
  interviewer's kit for, 225
  questionnaire review in, 228
  trial interviews in, 227-228
  written examination in, 227
Interviewer's Manual, 203
  on choice of respondent, 207
  on elaboration, 214
  on gaining entry, 206
  in interviewer training, 225-227
  on thoroughness in questioning, 212
Interviewing, 3
  abbreviations for, 216
  asking the questions, 210-213
    as written, 211
  blank questions in, 213
  call-backs in, 208
  with cluster sampling, 99-100
  concluding procedures for, 218-219
  costs, 35
  explaining the survey in, 207-208
  gaining entry in, 205-206
  guidelines for, 203-219
  handling objections in, 208-209
  information on noninterviews in, 273
  introducing the study in, 205
  introductory letter for, 204
  knowing purpose of questions in, 211
  opening remarks in, 205-210
  postponements in, 208
  preliminary preparations for, 203-204
  preparing the community for, 203-204
  preparing the respondent for, 204
  privacy in, 191, 209-210
  probe recording in, 217
  probing in (*see* Interviewer probe)
  providing transitions in, 212-214
  questionnaire sequence and, 211-212
  refusals in, 209
  response adequacy in, 213-215
  response recording in, 200-201, 215-217
  in sample survey versus census, 13
  seasonal factor in, 36-37
  suggesting answers in, 212
  with tape recorder, 217-218
  "third parties" and, 192, 209-210
  thoroughness of questioning in, 212
  time factor in, 35
  timing of, 205
  use of questionnaire in, 211
  weather factor in, 36
Interviewing situation, 183, 191-192
  age factor in, 195

Interviewing situation:
  cultural factors in, 194-196
  external indicators in, 194
  interpersonal exchanges in, 196-199
  negative forces in, 198-199
  objectivity in, 187, 198
  positive forces in, 197-198
  race factor in, 195
  rapport in, 196, 198, 210
  sex factor in, 195, 210
  social class and, 195-196
  sociocultural factors in, 194-196
    study topic and, 194-195

Jamaica Fertility Investigation, 189, 199, 209
Japanese, 202
Jones, E. L., 202
Judgment sampling, 72

Kahn, R. L., 137, 149, 186, 200, 201
KAP (knowledge, attitude, practice) survey, 5
Katona, George, 41, 159, 312
Katz, D., 195
Keesing, F. M., 202
Keesing, M. M., 202
Kelman, H. C., 6, 54
Keypunch machine, 265, 278
Keypunching, 265-266
  consistency checking in, 266-269
  illegitimate punches in, 266-267
  improbable response checking in, 267-269
  inconsistent response checking in, 267
  "wild" punches in, 266
Kinsey, Alfred, 142, 186
Kish, L., 64, 67, 80, 101, 294
Klein, D., 142
Koenig, K. E., 136
Kosobud, R. F., 41
Kreinin, M. E., 307n.

Labor force, concept of, 27
Lambert, W. E., 166
Landsberger, H. A., 167
Lazersfeld, Paul, 3, 64
Le Play, Frédéric, 2
Leadership research, 202
Leading question, 142
Leggett, J. C., 196
Lenski, G. E., 195
Lerner, Daniel, 16, 186
Less-developed countries (*see* Developing countries)
Lesser, G. S., 164, 166
Letter of introduction, 204
Life and Labor of the People of London (Booth), 2
Likert, Rensis, 3
Likert scale, 154
Linear relationships, 302
Linguistic equivalence, 166-167
Lininger, Charles A., 6, 41, 43, 159, 307n., 312
Lipset, S. M., 16
List, 75-76
  cluster, 76
  element (*see* Element list)
Literature review, 26
Living expenses, 34
Loquacity, differential, 167

McLoughlin, Q., 47
Madow, W. G., 76
Mail questionnaire, 131-132
  response rate, 129, 131-132
Malaysia, 167
Managers survey, 135-136, 154